Carbohydrates as Organic Raw Materials

Edited by
Frieder W. Lichtenthaler

Distribution:

VCH, P.O. Box 10 11 61, D-6940 Weinheim (Federal Republic of Germany)

Switzerland: VCH, P.O. Box, CH-4020 Basel (Switzerland)

United Kingdom and Ireland: VCH (UK) Ltd., 8 Wellington Court, Cambridge CB 1 1HZ (England)

USA and Canada: VCH, Suite 909, 220 East 23rd Street, New York, NY 10010-4606 (USA)

ISBN 3-527-28280-7   ISBN 1-56081-131-5

Carbohydrates as Organic Raw Materials

Edited by

Frieder W. Lichtenthaler

Developed from a Workshop Sponsored
by Südzucker AG, Mannheim/Ochsenfurt
at the Technische Hochschule Darmstadt,
April 11-12, 1990

Weinheim · New York · Basel · Cambridge

Editor:
Prof. Dr. Frieder W. Lichtenthaler
Institut für Organische Chemie
Technische Hochschule Darmstadt
D-6100 Darmstadt, Germany

The *cover illustration* shows the hydrophobicity potential profile of sucrose projected onto its contact surface in an sixteen color code ranging from violet (most hydrophobic) to red (most hydrophilic region). For generation of the solid state molecular geometry of sucrose the neutron diffraction data of G. M. Brown and H. A. Levy, *Acta Crystallogr., Sect. B* **29** (1973) 790–797 were used. The picture was generated on a Silicon Graphics workstation using the molecular modelling program MOLCAD, developed by Prof. J. Brickmann, of the Technische Hochschule Darmstadt. For further comments see p. 14 ff. of this monograph.

Published jointly by
VCH Verlagsgesellschaft, Weinheim (Federal Republic of Germany)
VCH Publishers, New York, NY (USA)

Editorial Director: Dr. Hans-Joachim Kraus
Production Manager: Max Denk

Library of Congress Card No. applied for.

British Library Cataloguing-In-Publication Data
Conference on Progress and Prospects in the Use of Carbohydrates as Organic Raw Materials (1990 Darmstadt, Germany)
    Carbohydrates as organic raw materials.
    I. Lichtenthaler, Dr. Frieder W.
    547.1

    ISBN 3-527-28280-7

CIP-Titelaufnahme der Deutschen Bibliothek

Carbohydrates as organic raw materials: developed from a workshop at the Technische Hochschule Darmstadt, April 11 – 12, 1990 / sponsored by Südzucker AG, Mannheim/Ochsenfurt.
Ed. by Frieder M. Lichtenthaler. – Weinheim; New York; Basel; Cambridge: VCH, 1991
    ISBN 3-527-28280-7 (Weinheim ...)
    ISBN 1-56081-131-5 (New York)
NE: Lichtenthaler, Frieder W. [Hrsg.]; Technische Hochschule ‹Darmstadt›

Composition and printing: Colordruck, Kurt Weber GmbH, D-6906 Leimen.
Bookbinding: J. Schäffer GmbH & Co. KG, D-6718 Grünstadt
Printed in the Federal Republic of Germany

Foreword

The workshop conference "Progress and Prospects in the Use of Carbohydrates as Organic Raw Materials" was held on April 11–12, 1990 at Darmstadt jointly by the Technische Hochschule Darmstadt and the scientific committee of CITS (Commission International Technique de Sucrerie). The organizers of this event were Prof. Frieder W. Lichtenthaler of the TH Darmstadt and Dr. Huber Schiweck, head of our central laboratories at Obrigheim/Pfalz. The present book aims at making the lectures held at this conference available to a broader audience.

Considering the fact that there is no lack of conferences on "Carbohydrate Chemistry" and "Renewable Resources", and that the last-mentioned topic is finding ample room in press, what were the reasons that prompted Südzucker to sponsor this conference? We believe that most of the scientific congresses have too broad and too academic a base as to attract a larger audience from industry. This causes these meetings to miss one of their aims, i. e., to bring about an intensive exchange of ideas between representatives from academia, chemical industry, and producers of renewable commodities such as plant fibres, vegetable oils, starch, and sugar.

On the other hand, what we need are new ideas, an intensive research and development phase, and close cooperation between the three aforementioned groups to achieve our long-term goal of making renewable resources an integral component in the production of chemicals. That would not only alleviate the pressure on fossil raw materials, it would also be an appropriate contribution towards the utilization of the agricultural surplus production.

Currently, in Germany, the utilization of renewable resources is characterized by the fact that they comprise only about 10 % of the raw materials used by chemical industry and that only a fraction of this figure stems from German agriculture. This fraction corresponds to an agricultural area of about 34,000 hectares for rapeseed oil, 140,000 hectares for starch containing products (potatoes, cereals), and 5,000 hectares for sugar beet. With such figures the problem of agricultural surplus production cannot be solved.

The broad interest this workshop has found equally by scientists and in the industry justifies the hope that similar conferences may help us achieve the aim outlined here.

October 1990
Mannheim/Ochsenfurt

Dr. Klaus Korn
Member of the Board
SÜDZUCKER AG

Preface

Upon being approached to host and organize a meeting on Carbohydrates as Organic Raw Materials, the reflex response was positive despite of discouraging thoughts on the organizing efforts required. Such a meeting, I felt, was an important forum for transmitting the newest thoughts in an area to which chemical industry has still a rather cautious, stand-by type relationship with little indication for a rapid change. There are plausible reasons for this, of course, the major ones being that comparatively few basic organic chemicals are derivable from our regrowing resources on an economic, industrial scale, and that more elaborate, high value-added products – as of now – cannot be generated on a competitive economic basis. The challenge emerging therefrom is obvious, and has led to multifold efforts towards broad-scale, practicality-oriented basic research at academic institutions, mainly making it increasingly difficult for the industrial chemist in particular, to keep abreast of the new developments in the field.

To steer against this situation the attempt was made to cover the various endeavours towards industrial utilization of carbohydrates in a 2-day workshop by presentations of leading authorities in the field whereby emphasis was placed on the chemistry of low-molecular-weight carbohydrates, i.e. the large-scale accessible mono- and disaccharides. Despite of thus inevitably excluding starch and cellulose chemistry, the resonance to the workshop was exceedingly positive: with a turnout of 200, half of them from European countries other than Germany, and about two third from industrial institutions, the participation surpassed the expectations.

The workshop eventually resulted in the production of this monograph, containing all of the lectures presented, save one. The scope of this monograph, however, goes far beyond the material covered in the oral presentations, since all of the authors have substantially extended their written accounts to include new results as well as novel lines of research. In addition, one topic not presented at the workshop has been included because it was considered relevant to the focus of the book: the computer modeling of sucrose and related substances and their structural representation of various properties on their surfaces, which – in my evaluation – adds a new dimension in the visual perception of sugars and in understanding their chemistry and biology.

The workshop – and hopefully this monograph also – was the first of what promises to be a continuing series, the second one being held at the University of Lyon in July 1992 with G. Descotes as the chairman. I am pleased to have been able to foster the incipiency of an "industrial information turntable" which was made a success by the combined efforts of Südzucker AG, most notably Dr. H. Schiweck of their central laboratories, and of myself with participation of essentially all of my research group, where Mrs. Gerda Schwinn and Dipl.-Ing. Uwe Kreis bore the major organizational burden. For their diligent and resourceful assistance with the many tasks that had to be done for the workshop and for this monograph, I would like to express my sincere appreciation.

It is hoped that this monograph will foster, unify, and consolidate the research and developments in this new and challenging area of utilization of "Carbohydrates as Organic Raw Materials". Where will it lead? One thinks of the next decade with optimism. A careful read through this book should give at least a glimpse of what to expect.

October 1990   F. W. Lichtenthaler
Darmstadt

Contents

1

Old Roots – New Branches :
Evolution of the Structural Representation of Sucrose

Frieder W. Lichtenthaler, Stefan Immel, and Uwe Kreis

Institut für Organische Chemie, Technische Hochschule Darmstadt
D-6100 Darmstadt, Germany

*A picture may instantly present what a book
could set forth only in a hundred pages.*

Ivan Sergeyevich Turgenev

Summary. Following a historical account on the establishment of the constitutional formula of sucrose and of its conformational features, the present possibilities for interactive graphics display of its molecular geometry, based on X-ray structural data, are given. In addition, the MOLCAD program-computed contact surface is presented as well as the electrostatic potential and, most relevant for structure-sweetness relationship considerations, its hydrophobicity potential profile on the contact surface. Finally, an attempt is made towards a preliminary assessment of the computer-generated distribution of hydrophilic and hydrophobic regions over the surface of the sucrose molecule (cf. cover photo) in terms of the "sweetness triangle" AH-B-X concept.

1. Introduction

Knowledge of the structure of a molecule is of fundamental importance, inasmuch as its *composition* (molecular or sum formula), *constitution* (type of bonding between elements), *conformation* (three dimensional arrangement of elements) and its *dynamic stereochemistry* (time and solvent dependent aspects of conformation) not only determine the physical, chemical, and biological properties, but – ultimately – the industrial applications as well. This most pertinently applies to the fairly complex molecule of sucrose, which with an annual production of over a 100 million tons is certainly the most readily available, lowest cost, crystalline organic compound.

Sucrose being a leading world commodity for centuries, it is not surprising that the roots of its "structural tree" – to adopt the metaphoric picture used in the title – reach back to the very beginnings of organic chemistry as it gradually unfolded with the advent of reasonably exact elemental analyses. The first, albeit not very accurate combustion analyses by Prout in 1827[1] and their interpretations by Liebig,[2] Peligot,[3] Berzelius,[4] and Dubrunfaut[5] eventually led to the correct molecular formula $C_{12}H_{22}O_{11}$. Its expansion into a satisfactory constitutional formula mystified a number of great organic chemists and proceeded in a number of distinct phases, each reflecting the progress in experimental methodology: methylation and periodation studies leading to the correct ring structures,[6] investigations on the chemical and enzymatic hydrolysis of sucrose establishing the anomeric configurations about the intersaccharidic junction,[6] X-ray- and neutron diffraction-based structural data unravelling ring conformations and non-bonded interactions between the saccharide portions[7] and NMR studies providing first insights into the dynamic stereochemistry of sucrose in solution.[7] The newest phase of structural visualization of the sucrose molecule appears to have been launched by the availability of reasonably advanced computer-based graphic and computational tools that allow molecular modeling with the potential of eventually understanding structure-activity (sweetness) relationships.

In the sequel, an account is given on the gradual evolvement of the structure of sucrose from its molecular formula $C_{12}H_{22}O_{11}$ to the color-coded visualization of its hydrophobicity potential profile (cf. cover) with particular emphasis to the current possibilities in computer-aided molecular modeling.

2. The Constitutional Formula of Sucrose

As early as 1883, Tollens[8] advanced the glucoseptanosyl fructofuranoside formula **1** for sucrose, followed, 10 years later, by Emil Fischer's proposition[9] that sucrose rather is a glucofuranosyl fructofuranoside (**2**).

Neither formulation contained configurational implications, nor were their ring structures adequately supported by experimental evidence of which Fischer was well aware: "Ich halte die schon von Tollens vor längerer Zeit aufgestellte, allerdings nur ungenügend begründete Structurformel mit einer kleinen von mir vorgenommenen Änderung im Wesentlichen für richtig."[9]

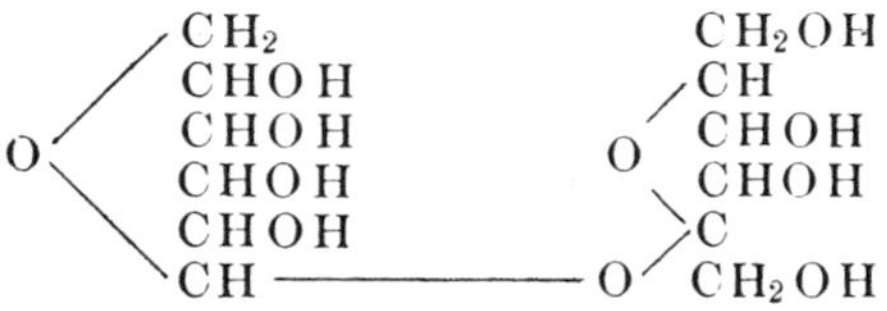

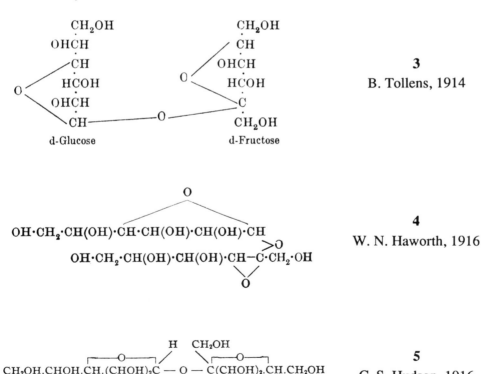

1

B. Tollens, 1883

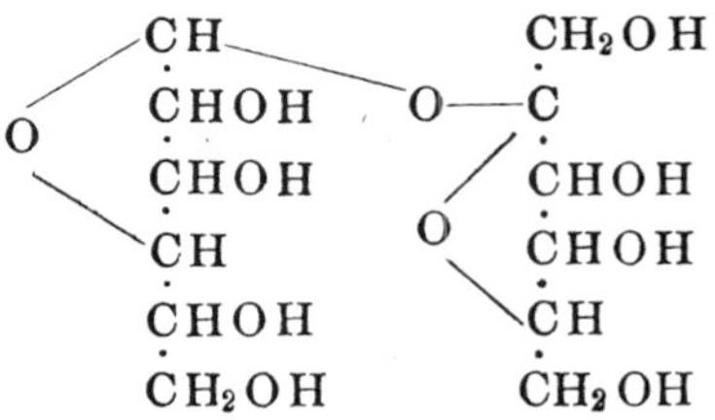

2

E. Fischer, 1893

Around 1915 structural repesentations of the type **3**,[10] **4**,[11] and **5**[12] appeared to be the state of the art. Thereby it is worthy to note that the Tollens formulation **3** contains the configurational features for the glucose and fructose portions as established by E. Fischer in 1891,[13] whereas Haworth[11] and Hudson,[12] obviously, were not prepared yet to accept Fischer's projection formulae representation advanced 25 years earlier.

d-Glucose   d-Fructose

3

B. Tollens, 1914

$$O$$

$$OH \cdot CH_2 \cdot CH(OH) \cdot CH \cdot CH(OH) \cdot CH(OH) \cdot CH$$
$$OH \cdot CH_2 \cdot CH(OH) \cdot CH(OH) \cdot CH - C \cdot CH_2 \cdot OH$$
$$O$$

4

W. N. Haworth, 1916

H   CH₂OH

$CH_2OH.CHOH.CH.(CHOH)_2C - O - C(CHOH)_2.CH.CH_2OH$

Glucose chain
(G).

Lactonyl
carbon (B').

Fructose residue
(F).

5

C. S. Hudson, 1916

It was not until 1926 – a hundred years after Prouts combustion analyses[1] – that Haworth and his coworkers,[14] on the basis of their classical methylation studies, deduced the correct ring structures for sucrose as depicted in the representations 6[14] (without configurational assignments), 7[15] (implementing configurational relationships), and 8.[15] The latter, in fact, appears to be a first, cautious formulation of what was to become the "Haworth perspective formula", that in the ensuing years found general acceptance in the form propagated by Pigman,[16] Purves,[17] and others. The particularly lucid, didactically auspicious formulation 10 used by Morrison and Boyd[18] in the first edition of their textbook on Organic Chemistry (1959) appears to be the first instance where conformational concepts were incorporated.

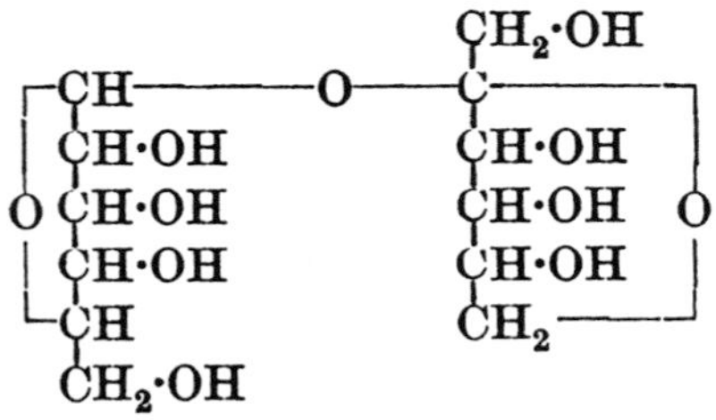

6

W. N. Haworth, 1926

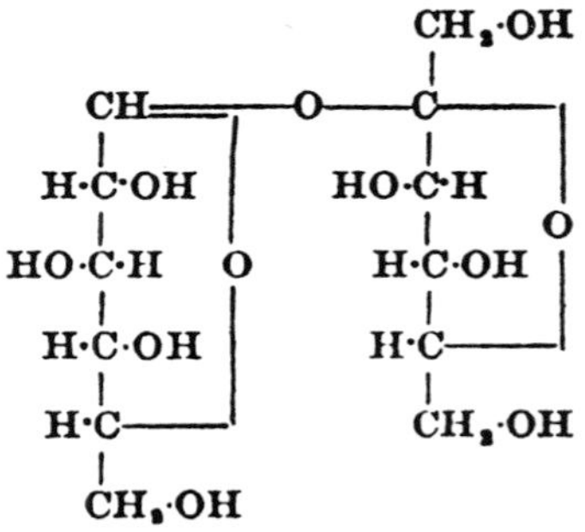

7

W. N. Haworth, 1929

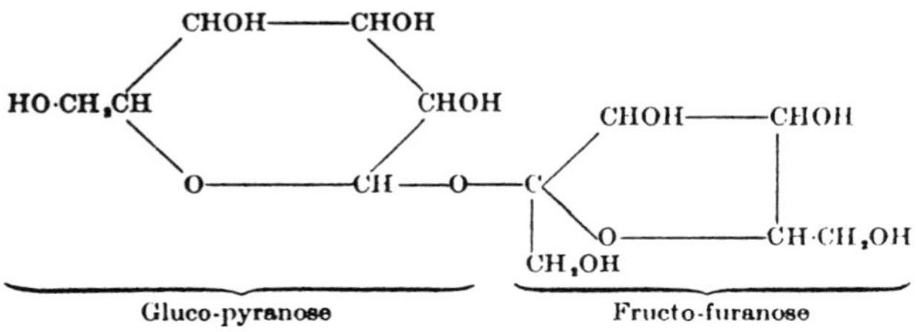

8

W. N. Haworth, 1929

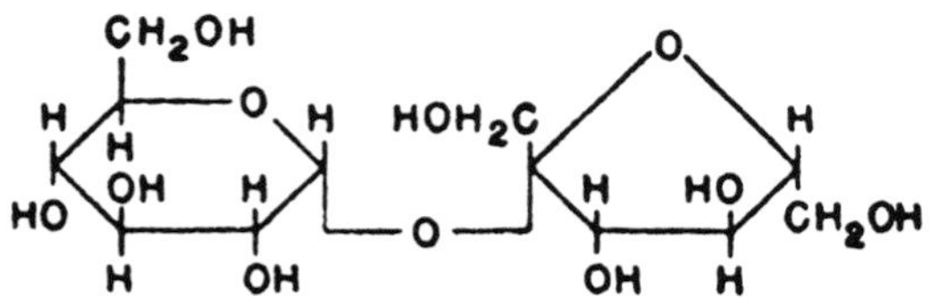

Gluco-pyranose   Fructo-furanose

9

W. W. Pigman, 1948

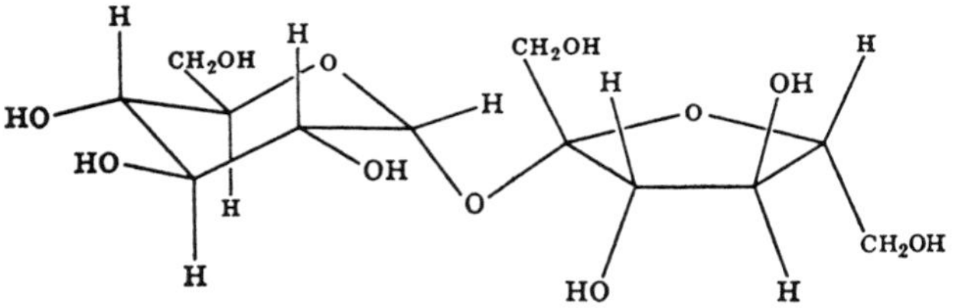

10

R. T. Morrison,

R. N. Boyd, 1959

(+)-Sucrose
α-D-Glucopyranosyl β-D-fructofuranoside
β-D-Fructofuranosyl α-D-glucopyranoside
(no anomers; *non-mutarotating*)

Except for writing the respective C-hydrogens, these formula representations are still used today, as, for example, the formulation **11** favoured by Hough,[7] which is sufficiently clear and simple to straightforwardly develop the chemistry of sucrose (see this monograph, pp. 33 ff.) and which, as evidenced by **12**,[19] may even be subjected to unusual substitution to account for carbohydrate symposium meeting places.

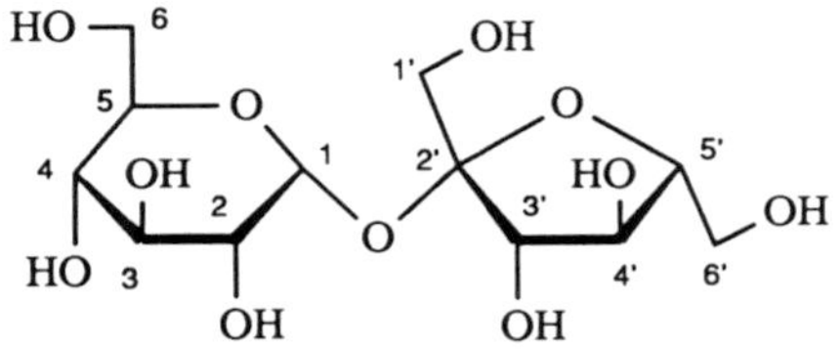

11

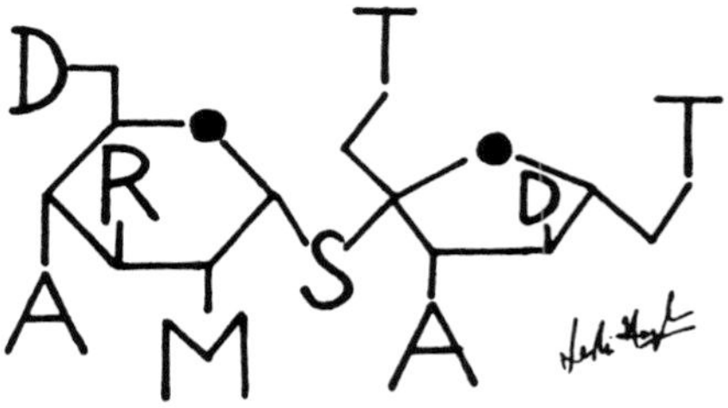

12

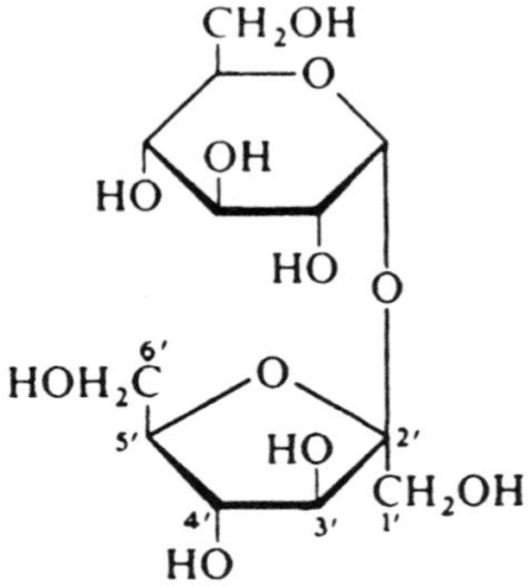

13

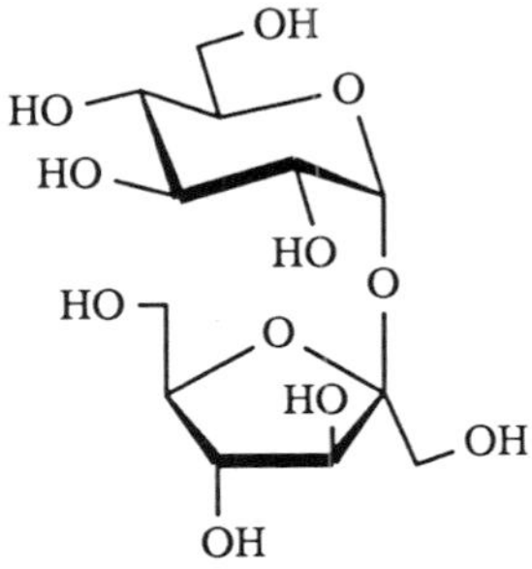

14

Another representation that is reasonably clear and economical in space are formulae **13**[20] and the ChemText[21] graphics program-generated **14**. Both do not give, of course, the correct disposition of the glucose and fructose portions in the solid state or in solution, but have the didactic advantage that the configurational relationships of the two monosaccharide units, i.e. identical arrangements at C-3 to C-5 of pyranose and furanose rings, become particularly obvious.

3. The Conformation of Sucrose

A new possibility to further probe into the structural subtleties of sucrose opened up by crystal structure determinations. The first measurements accurate enough to locate the positions of the hydrogen atoms, were based on neutron diffraction data by Brown and Levy in 1963,[22] which upon later refinement[23,24] resulted in structure **15**: the glucose portion adopts a $^4C_1$ conformation whereas that of the fructofuranose residue is a $^4T_3$ twist. The two rings are approximately at right angles, yet the overall structure, most notably, is determined by two strong intramolecular hydrogen bonds: one between the terminal 6'-OH of fructose to the pyranoid ring oxygen of 1.89 Å lenght, the other, even shorter, reaching from the primary 1'-OH of fructose to O-2 of glucose (1.85 Å), as depicted in formulae **16**[23], **17**[25], and **18**[26]. These hydrogen bonds serve to fix the molecule in a well-ordered, rigid conformation.

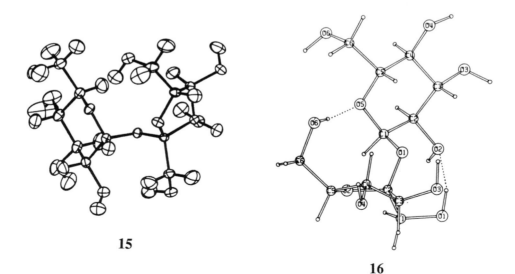

15

16

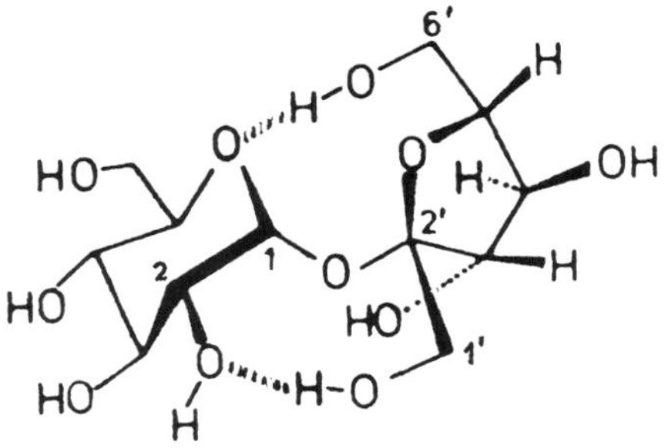

17 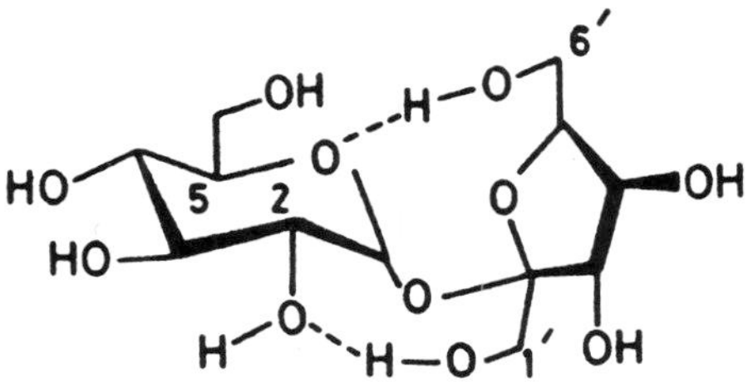   18

As inter- and intramolecular hydrogen bonding in the solid state is influenced by packing effects, it cannot be expected that the fixed overall shape as illustrated by formulations **15 - 18** is retained in solution, since the hydroxyl groups could satisfy their hydrogen bonding requirements by bonding with the solvent. Accordingly, it is surprising that through probing with several NMR criteria in conjunction with hard sphere *exo*-anomeric (HSEA) calculations, Bock and Lemieux[27] provided strong evidence that the overal conformation of sucrose is approximately the same in dimethyl sulfoxide and water solutions, and that this conformation is similar to that observed in the solid state. Only the longer, and, hence, weaker 5-O^g ···· HO-1^f hydrogen-bond is disintegrated by solvation.[27] These conclusions are supported by detailed analysis of ^{13}C-NMR spin-lattice relaxation experiments of aqueous sucrose solutions, showing its conformation to be independent of temperature and concentration in the 0.1-1 molar range,[28,29] and by recent SIMPLE ^{1}H-NMR isotope-shift measurements.[30,31] The latter revealed the presence of two intramolecular hydrogen-bonded conformations **19** and **20** for DMSO as well as for aqueous solutions of sucrose, in which the 2-oxygen of the glucose portion (i.e. O-2^g) acts as the acceptor for either the 1'-OH or 3'-OH of the fructose moiety:

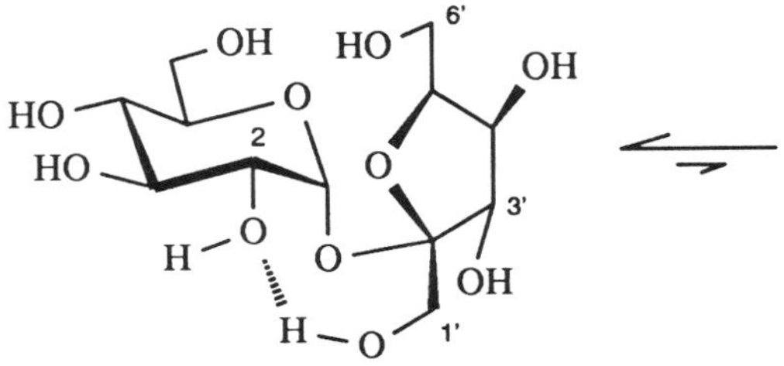

19 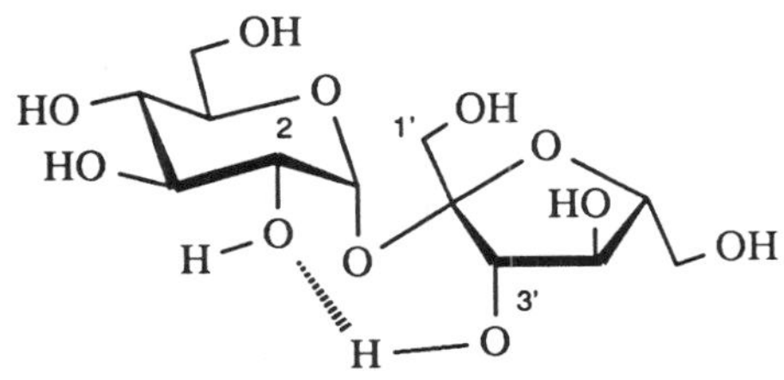   **20**

These two forms, which are graphic representations adapted from ref. 7, exist in a competitive equilibrium in which **19** is favoured over **20** in an approximate 2 : 1 ratio.[30,31] The ready interconversion of these two conformations becomes sterically understandable when using a more realistic sucrose representation, in which the clockwise arrangement of the hydrogen bond network[30] has been incorporated into the pyranoid ring:

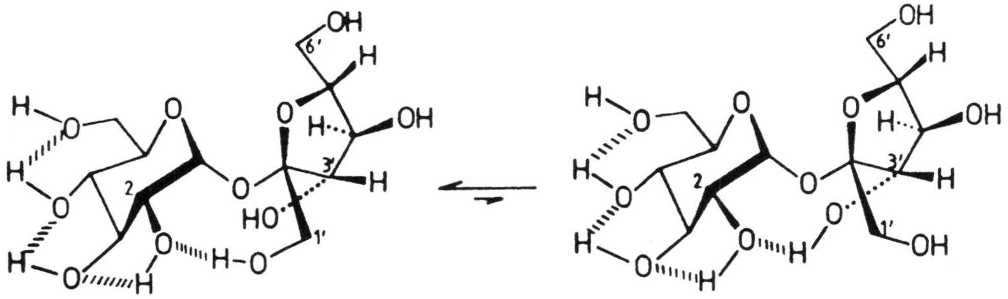

19   **20**

The proximity of 1'-OH and 3'-OH is clearly evident, the switch from one form to the other requiring a minor rotation within the intersaccharide linkage only, thus largely retaining the overall molecular geometry. Consistent with this picture of sucrose as a relatively rigid molecule in the crystal and in solution are recent molecular dynamics simulations in vacuum,[32] where the intramolecular hydrogen bond is retained in all but one of the five local minimum energy conformations.

4. Possibilities for Interactive Graphics Display of Sucrose

Whilst crystallographers pioneered techniques to visualize, scrutinize, and manipulate three-dimensional molecular models – the ORTEP program[33] used in the sucrose representation **15** being a notable, early example – only their combination with computational chemistry led to molecular modeling at its present impressive level.[34-37] It offers the chemist an expanding arsenal of user-friendly tools with which molecules can be displayed in unexcelled, three-dimensional representations, relieving from the "drawing artistry" required to jot down on paper formulae of type **19** and **20**, for example. The additional possibility to impose vital chemical properties into the three-dimensional molecular representations, such as the electrostatic potential or – biologically more relevant – the hydrophobicity distribution over its surface, opens up an entirely new dimension of visualizing not only the structure of sucrose, but essentials of its chemical and physiological properties.

The various possibilities for structural display of the sucrose molecule are elaborated in the sequel utilizing the MOLCAD program* developed by Brickmann and coworkers[38] and the neutron diffraction-based structural data of Brown and Levy.[23] It is particularly simple to computer-generate pictures on a screen that correspond to the classical Dreiding model presentation (Fig. 1a). Therein, special attention is given to the intramolecular hydrogen bonds prevalent in crystalline sucrose, each being signified by dotted lines and the respective distances: 5-O^g ···· HO-6^f (1,89 Å) and 2-O^g ···· HO-1^f (1.85 Å). Undoubtedly, this representation is more exact and more lucid than the formulations **16 - 18**, mentioned above.

Fig. 1b displays the space-filling Corey-Pauling-Koltun (CPK)-type model of sucrose, that is as easily generated either in the orientation used in Fig. 1a, or in any other, producable by rotation on the screen. These CPK models, however – whether by computer-generation or by actual built using a model kit – display only the van der Waals surface of sucrose, which as such is not fully accessible to the solvent.

The Contact Surface of Sucrose

A more "realistic" molecular surface is the contact- or Connolly-surface, that can readily be created by computers; it is defined[40-42] as the smooth molecule surface, which is accessible to a probe sphere with a radius of 1.4 Å, representing a solvent molecule like water in size. Such a contact surface – to be distinguished from the solvent-accessible surface[43] defined in a slightly different way – is displayed in Fig. 2 in two versions, the first (Fig. 2a) giving the overall shape in dotted form, with the steric information of Fig. 1a inserted, the second (Fig. 2b) depicting the space-filling CPK type contact-surface, indicating the different atoms contributing to the outer sphere in the usual atomic color-code.

---

* The program MOLCAD (for MOLecular Computer Aided Design) is an interactive, fast computer program for building and manipulating molecules and molecular systems. It is particularly suited to analyze and represent different physical molecular properties such as the electrostatic or hydrophobic potential on three-dimensional solid molecule surfaces, even of large molecules like proteins, zeolithes, and polymers. MOLCAD runs on Silicon-Graphics 4D-workstations and can be licenzed from the authors.[39]

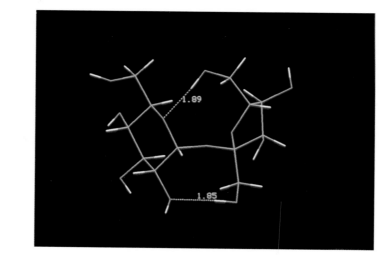

a)

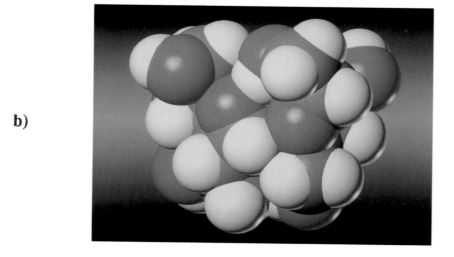

b)

Fig. 1. MOLCAD generated models of sucrose in the solid state, based on the neutron diffraction data of Brown and Levy.[23] Pictures are photographed from the computer screen. (a) Dreiding model representation including the two intramolecular hydrogen bonds (dotted lines) and their interatomic distances (in Å). (b) Space-filling CPK type form in the same orientation as above (white: hydrogen; grey: carbon; red: oxygen).

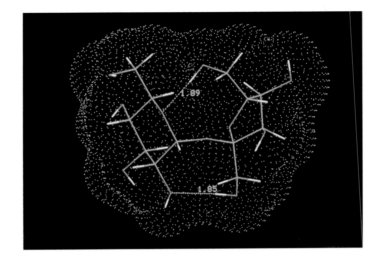

a)

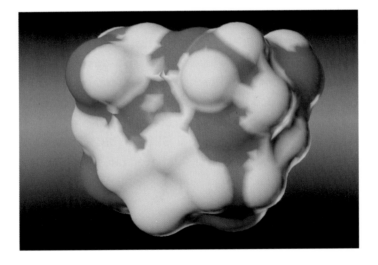

b)

Fig. 2. Contact surface (roughly equivalent to solvent-accessible surface) of crystalline sucrose in dotted form with Dreiding model insert (a), and in the space-filling CPK model representation (b), the individual atoms contributing to the outer sphere being indicated by the usual atomic colors (orientation of sucrose corresponds to that in Fig. 1).

Electrostatic Potential of Sucrose

Aside purely sterical factors, the distribution of electrophilic and/or nucleophilic sites over the surface of a molecule is a decisive element in assessing its reactivity and the regioselectivities attainable in reactions. In this context, knowledge of the charge density or electrostatic potential on the molecular surface[44] is of vital importance. Calculation of this potential from the atomic charges obtained by semiempirical MNDO-methods using the MOPAC[37,45] program can indeed be accomplished, and, most conveniently, translated into an sixteen color-code, reaching from red (most negative) to violet (most positive). In the case of sucrose this is borne out in the pictures displayed in Fig. 3.

The correlation of the electrostatic potential patterns thus unveiled in sucrose with the known "sucrochemistry" is tempting, particularly ferreting out reactivity and selectivity profiles. As of now, however, caution is to be exercised, since the pattern of Fig. 3 is valid for the crystalline state and the overall electrostatic distribution is apt to vary with the solvent. In water, the rigid solid-state structure of sucrose is largely preserved due to retention of one of the intramolecular hydrogen bonds (vide supra, formulae **19** ⇌ **20**, thus the overall solvation pattern by water appears to be similar to that in the crystal lattice between sucrose molecules. In pyridine, however, i.e. a solvent in which many derivatizations are being performed, the solvation may be distinctly different. Such differences in solvation between water, dimethyl sulfoxide, and pyridine have recently been established for the five α-linked glucosyl-fructoses isomeric with sucrose[46] and are likely to be found in sucrose as well. In addition, the electrostatic distribution over the surface of sucrose may vary with the reagent utilized for ensuing reactions. Nevertheless, electrostatic potential surfaces show promise of having substantial predictive value in further exploiting the chemistry of sucrose.

The Hydrophobicity Potential of Sucrose

The interaction of a biologically active molecule with its receptor – as, for example, of sucrose with the taste bud initiating the sweet sensation – is not only determined by the spatial complementarity of their three-dimensional molecular geometry and the electrical conformity of their charges on the surface, but also by the hydrophobic interactions between their contact surfaces.[47,48] Since there is now sufficient evidence to assume that the sweet-taste receptor is proteinaceous and that the main driving-force for the initial binding is hydrophobic in nature,[49-51] the elicitation of the sweet-taste responce is likely to involve an interaction between a hydrophobic cleft in the surface of the protein and a hydrophobic portion of sucrose.[27,51] Consequently, for a new insight on the sweetness of sucrose, the van der Waal surface and the electrical potential of the effector molecule is not adequate but must be complemented by an appropriate estimation and representation of the hydrophobicity around the molecule.

The hydrophobic effect[52] of a molecule describes its tendency to orient in non-homogenous ways, like water to exclude hydrophobic substances and vice versa. Usually, the hydrophobicity of a molecule is only seen as an over-all quality and expressed as the partition coefficient in the water/*n*-octanol system. First attempts to calculate this "one-dimensional" property led to an incremental system, assigning a hydrophobic potential to molecular fragments or atoms.[53] Taking into account that the hydrophobicity potential of each fragment or atom decreases exponentially with increasing distance, it is possible to calculate it at every point of the molecule's surface.[47,48] It must be stressed though, that the hydrophobic potential is empirically defined and calculated; it is not proportional to the electrostatic potential, since, for example, $-NH_3^+$ and $-COO^-$ have opposite electric charges, but a similar contribution to hydrophilicity.

Application of this computational methodology[47,48] to sucrose and expressing it in the sixteen color-code on its surface, results in the hydrophobicity (or molecular lipophilicity) potential represented in Fig. 4, the respective legend explains the meaning. Especially informative is the half-opened model with the stick and ball model insert, which allows unequivocal correlation of the most hydrophobic (i.e. violet) section of the molecule to the fructose portion, involving the region characterized by the CH-protons $H-1^f$, $H-1'^f$ ···· $H-3^f$ ···· $H-5^f$ ···· $H-6^f$, and $H-6'^f$. The hydrophilic (red) section is distinctly separated therefrom, quasi at the "other side" of the molecule (Fig. 4), with its major intensity centered around the 3-oxygen of glucose.

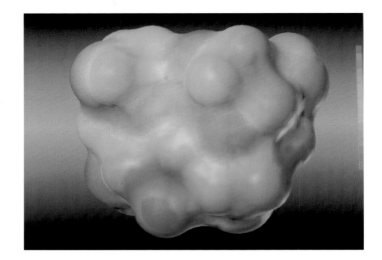

a)

b)

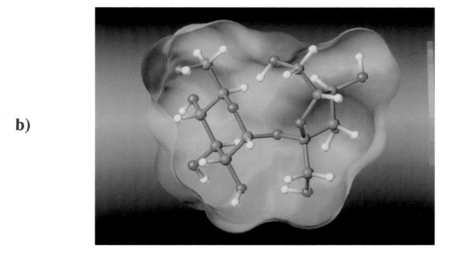

Fig. 3. (a) Electrostatic potential molecular surface of sucrose, as obtained by MNDO-calculations, in color-coded form, red respresenting the negative, violet the positive maximum in charge density in relative terms. (b) For clarity and unequivocal orientation, a ball-stick model is inserted into the molecule in its half-opened form. As is clearly apparent the violet, i.e. most electropositive area is centered around the glucosyl-2-OH.

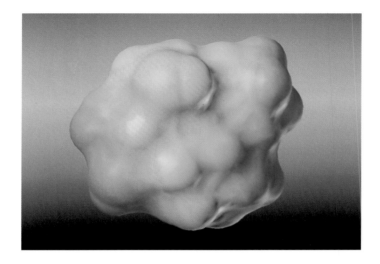

a)

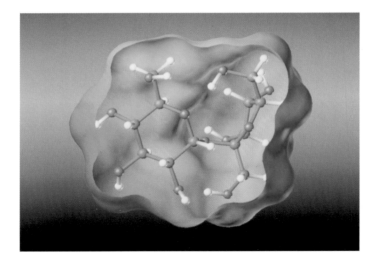

b)

Fig. 4. MOLCAD-generated hydrophobicity potential of sucrose in the crystalline state, red representing most hydrophilic, blue the most hydrophobic portion of the molecule (in relative values). For identification of the atoms contributing to hydrophobicity the ball-stick model is inserted into an opened form. The orientation of sucrose used here is different from that in Figs. 1-3; it was chosen to illustrate the opposite hydrophilic and hydrophobic sites of the molecule.

5. Preliminary Assessment of the Hydrophobicity Potential Profile of Sucrose

The topography of the region in sucrose amenable to engagement in hydrophobic bonding has been assessed[27] on the basis of CPK models to include parts of the surface described by H-3[f], O-5[f], O-5[g], and the surface involving H-1[f], H-1'[f], O-1[f] ···· O-2[g], H-1[g], and H-2[g]. Comparison with the computational data presented in Fig. 4a and b, however, reveals them to be distinctly different, particularly with respect to the non-contribution of the glucose portion to the hydrophobicity of the molecule. In view of the fact that – at least in the hands of the authors of this article – it is essentially impossible to delineate with any certainty hydrophobic or hydrophilic regions from a CPK model of sucrose, the computationally generated hydrophobicity profile of Fig. 4 is deemed the more realistic one. By consequence, the hydrophobic cleft of the taste protein may conceivably be portrayed to correspond to the hydrophobic region in the fructose part of sucrose (cf. Fig. 5):

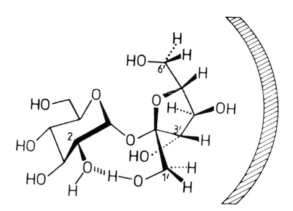

Fig. 5. Conceivable hydrophobic cleft of the taste bud protein corresponding to the hydrophobic fructose portion of sucrose.

In keeping with this notion, it would have to be surmised that an increase of hydrophobicity in the fructose portion – i.e. along the violet region in Fig. 4 – favours binding to the taste bud receptor protein and, hence, enhances sweetness. The data collected in Table 1 appear to support this rationalization, since replacement of the fructose hydroxyl groups at the 1'-, 4'-, and/or 6'-position by chlorine uniformly leads to compounds sweeter than sucrose.[54] Thereby, it is probably not without significance that, with a factor of 3500, the sweetness enhancement is largest in the 4',6'-dichloro-sucrose, i.e. the compound in which hydrophobicity is substantially increased with maintenance of the fructose-1'-OH, involved in hydrogen bonding to the glucose-2-OH in the crystal and in aqueous solution (cf. above).

Table 1. Relative sweetness of sucrose in which the fructose hydroxyl groups at 1', 4', and 6' are replaced by halogen[54]

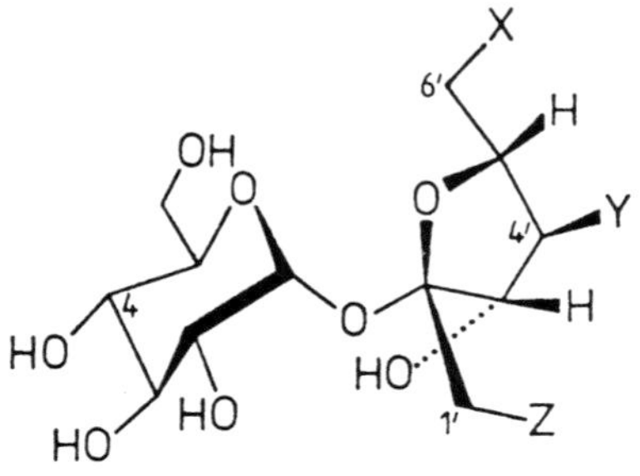

X	Y	Z	Relative Sweetness
OH	OH	OH	1 (sucrose)
Cl	OH	OH	20
OH	OH	Cl	20
Cl	OH	Cl	76
Cl	Cl	OH	3500
Cl	Cl	Cl	100
Br	Br	Br	30

In this context, the glucose-4-OH group, close to the hydrophilic region of sucrose and contributing to it (cf. Fig. 4) appears to play a weighty role, because its replacement (with inversion) by chlorine substantially extends the fructose-located hydrophobic region into the glucose portion.

Table 2. Relative sweetness of 4-chloro-4-deoxy-*galacto*-sucroses in relation to the halogen substitution pattern in the fructose portion[54]

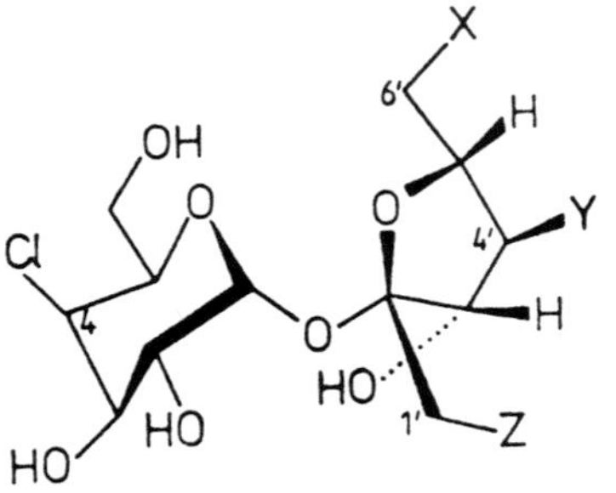

X	Y	Z	Relative Sweetness (sucrose = 1)
OH	OH	OH	5
Cl	OH	OH	120
Cl	Cl	OH	220
Cl	OH	Cl	650 (sucralose)
Cl	F	Cl	1000
Cl	Cl	Cl	2000
Cl	Br	Cl	3000
Cl	I	Cl	7500

The sweetness enhancements obtained thereby are amazing (cf. Table 2), most remarkable being the steady increase of sweetness with the degree of substitution, particularly evident when going from 4,1',6'-trichloro-*galacto*-sucrose ("sucralose®", 650 x sweeter than sucrose) to its 4'-fluoro- (1000 x), 4'-chloro- (2000 x), 4'-bromo- (3000 x) and 4'-iodo- (7500 x) derivatives.

The taste properties of deoxy-sucroses and of a variety of its methyl ethers[51] allow similar rationalizations as to their relationships between sweetness and substitution (i.e. hydrophobicity) pattern. An assessment is not given since sweetness effects are substantially smaller and require considerably more detailed modelings to arrive at meaningful conclusions.

In this context, it is tempting to scrutinize Shallenberger's AH-B concept[55] or the more advanced AH-B-X theory of Kier[56] in terms of the electrostatic and the hydrophobicity potential profile emerging from the calculations (Figures 3 and 4), particularly in view of X being a hydrophobic factor which acts in harmony with the AH-B unit to guide or lock the sweet compound into the proteinaceous receptor site. What clearly evolves from the distinctly hydrophobic (i.e. violet) region of sucrose in Fig. 4, is that the X-part of the AH-B-X triangle is to be placed into the fructose portion, conceivably within the 1^f-CH_2-3^f-H area; this X-site, obviously, can vary within a certain range and reach over, via the 6^f-CH_2, to the glucose portion (the axial side of C-4^g, in particular), to account for the dramatic increase of sweetness on increase of hydrophobicity in the 4^f-, 6^f-, and 4^g-portions, e.g. by chlorine substitution (cf. Tables 1 and 2).

Correspondingly, the AH-B unit of the glucophore – AH being the proton donor, B the proton acceptor in simultaneous hydrogen-bonding to a complementary AH-B system of the receptor protein – is most likely to be located at the diol grouping made up by the 2-OH and 3-OH of the glucose portion which turns out to be the most hydrophilic (red) region of the molecule (cf. Fig. 4). The pronounced tendency of the equatorial glucose 2-OH to engage in intramolecular hydrogen bonding is not only prevalent in sucrose (cf. Fig. 1a), but in other disaccharides containing α-glucosyl residues (cf. Table 3) as well. Apparently, any sterically available OH-group in the second sugar moiety can participate in this hydrogen bonding interaction. It is interesting to note that the glucosyl-2-OH (2-O^g in Table 3) can take part as the proton acceptor, as in sucrose, turanose, and α-maltose, or as the proton donor (isomaltulose, β-maltose, sucralose) with comparable intensity, whereas there exists some indication that the 2-OH group of glucose reverts to the acceptor in solution (cf. sucralose).[63,64]

Table 3.   Intramolecular hydrogen bonding of the glucosyl-2-OH (2-O^g) in the X-ray structures of α-glucosyl-disaccharides

compound	intramolecular hydrogen bond	distance (Å)	ref. (X-ray data)
sucrose	2-O^g ···· HO-1^f	1.85	23
isomaltulose	2-O^{g}H ···· O-2^f	2.28	57
turanose	2-O^g ···· HO-4^f	2.11	58
β-maltose	2-O^{g}H ···· O-2$^{g'}$	1.84	59
methyl β-maltoside	2-O^{g}H ···· O-3$^{g'}$	1.86	60
α-maltose	2-O^g ···· HO-3$^{g'}$	2.06	61
phenyl α-maltoside	2-O^g ···· HO-3$^{g'}$	1.74, 1.90[a]	62
sucralose[b]	2-O^{g}H ···· O-3^f	1.86	63

[a]   Corresponding to two different molecular conformations realized in the crystal lattice.

[b]   The direction of the intramolecular hydrogen bond is reversed to 2-O^g ···· HO-3^f on dissolution in dimethyl sulfoxide.[64]

The glucosyl-2-OH thus being capable of functioning – in the solid state – as the acceptor or donor part in intramolecular hydrogen bonding, both, the AH or the B part may, conceivably, be assigned to it. Support in favour of the glucosyl-2-OH being the AH portion of the AH-B-X tripartite system can be derived from the distribution of electrostatic potential over the surface of sucrose as displayed in Fig. 3. There, unequivocally, the intensely violet glucosyl-2-OH area is the center of the highest positive charge density, and, accordingly, should exhibit a pronounced tendency to participate in hydrogen bonding as the donor, i.e. AH part. These rationalizations, by consequence, entail the B portion of the AH-B-X tripartite system to be allotted to the glucosyl-3-OH. This leads – taking into account the two forms of sucrose prevalent in aqueous solution[30,31] – to the assignments made in Fig. 6.

In these assignments, the notion is fostered that the intramolecular hydrogen bond, as drawn in (a) and (b), plays an integral role in eliciting the sweetness response such that it strongly enhances the H-donor (AH) capacity of the glucosyl-2-OH in its hydrogen-bonding interaction with a complementary acceptor group within the taste bud receptor. Indeed, all intensely sweet halo-sucroses listed in Tables 1 and 2 are capable of establishing the 2-O^g ···· HO-1^f (a) or the 2-O^g ···· HO-3^f (b) arrangement, which for some of these compounds (inclusively sucralose) has even been proved to prevail in dimethyl sulfoxide solution.[64]

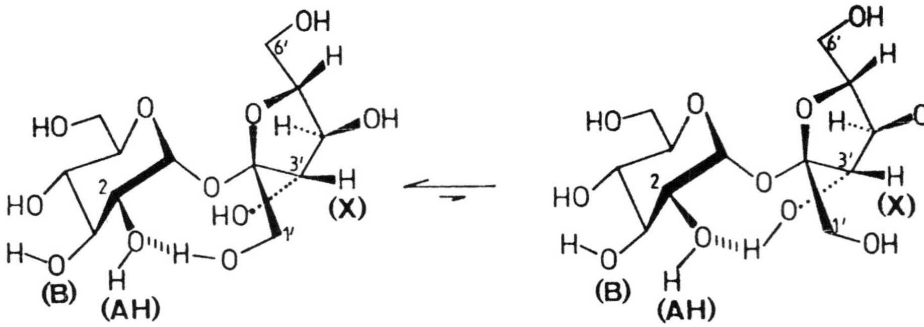

Fig. 6. Location of the AH-B-X glucophore ("sweetness triangle") in sucrose emerging from the electrostatic potential distribution (Fig. 3) and from the hydrophobicity potential profile (Fig. 4)

Additional support may be found, for example, in the observation that 3-ketosucrose is similar in sweetness to sucrose (retention of the hydrogen bond acceptor B-site),[65] whilst on steric intervention at the glucosyl-3-OH (e.g. by inversion to *allo*-sucrose) sweetness is lost altogether.[65] Continuing this line of thought, it would have to be surmised, that 3-O^g-methyl sucrose should be sweet, whilst the 2-O^g-methyl derivative, like the 2-deoxy- and the 2-keto-sucroses should be not – predictions that remain to be proved.

Whilst these rationalizations appear to be reasonable, especially the extension of the tripartite AH-B-X theory to comprise an intramolecular hydrogen bonding for enhancement of the H-donor capacity of the AH part – at least for sucrose derivatives – it is to be noted that the computation-based assignments of Fig. 6 differ substantially from those made on the basis of a huge number of structure-sweetness considerations[7,51,54] (see also pp. 47 ff. of this monograph). At the present state of affairs, there can be no doubt, that further investigations in general, and along this computer modelling vein in particular, are needed – not only to probe into the validity of assignments in Fig. 6 but to prove the validity of the AH-B-X concept altogether, its major shortcoming being the absence of any predictive value. Nevertheless, the approach outlined here towards unravelling structure-sweetness relationships has high potential for providing new insights into the subtleties of sweetness and, hopefully, for leading to concepts of higher predictive value than those presently available.

6. 3D-Representation and Hydrophobicity Potentials of Isomaltulose, Leucrose, Glucosyl-α(1→1)-mannitol, and Glucosyl-α(1→6)-sorbitol

In view of the far-reaching implications inherent in the hydrophobicity potential profile of sucrose it was deemed appropriate to probe with this approach into some sucrose-related sugar substitutes, such as isomaltulose (21), leucrose (22), and the terminally α-glucosylated mannitol (GPM, 23) and sorbitol (GPS, 24):

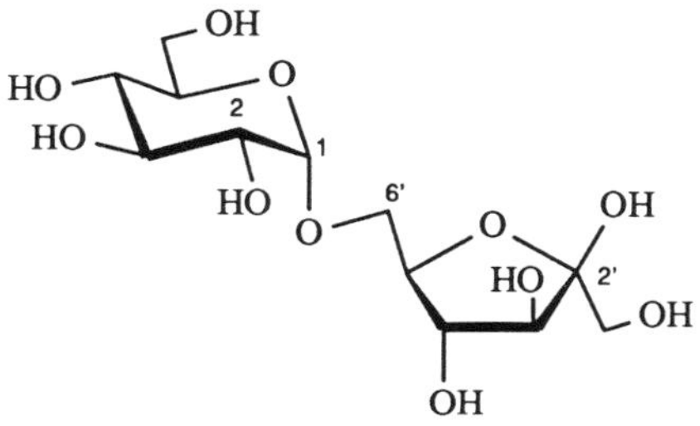

Isomaltulose (21)

D-Glucopyranosyl -
α (1→6) -
D-fructofuranose

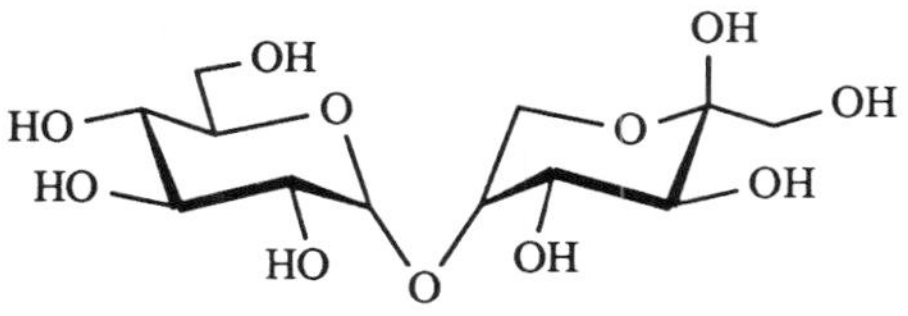

Leucrose (22)

D-Glucopyranosyl -
α (1→5) -
D-fructopyranose

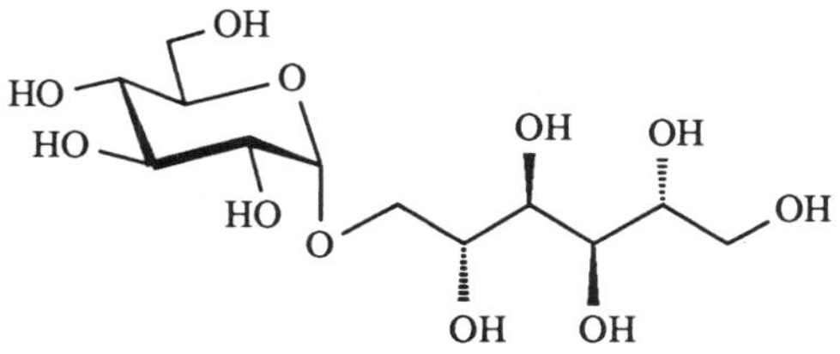

GPM (23)

D-Glucopyranosyl –
α(1→1) -
D-mannitol

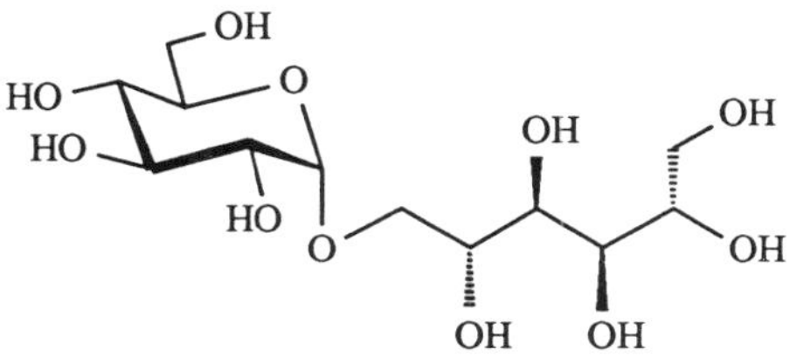

GPS (24)

D-Glucopyranosyl –
α(1→6) -
D-sorbitol

Due to their sweetening properties, all four compounds are accessible on an industrial scale,[66,67] the latter two already being approved for use as a low-caloric, low-cariogenic sweetener. Since determinations of the crystal structures are available for each of the compounds,[57,68-70] their 3D-geometrical characteristics in the representations used for surcrose, as well as their hydrophobicity profiles can be computer-generated with the X-ray coordinates as input.

In the case of isomaltulose (**21**), which crystallizes as a monohydrate, the X-ray structure[57] reveals (Fig. 7a) the glucose moiety in the $^4C_1$ conformation and the fructofuranose portion in a $^4T_3$ twist form – as is the case of sucrose. The glucosyl-2-OH engages in a comparatively weak (2.28 Å, cf.Table 3) intramolecular hydrogen bonding to the fructosyl-2-O, the glucosyl-2-OH functioning as the donor.

The molecular hydrophobicity potential of isomaltulose is presented in Figures 7b and 7c, wherein the water of crystallization has been removed to better account for the overall shape of the molecule before dissolution, i.e. before solvation with water molecules. Another reason for omiting of the crystal water in the hydrophobicity potential surface representations is to get a more feasible visualization of the hydrophobic interactions with a protein, this not only being pertinent for isomaltulose, but for GPM as well. The separation of hydrophobic and hydrophilic regions on the contact surface is clearly evident, yet their correlation with the sweet response in the taste bud receptor – isomaltulose has about 40 % of the sweetness of sucrose – must clearly await detailed investigations of the molecular shape in aqueous solution, i.e. whether the intramolecular hydrogen bond (cf. Fig. 7a) that gives the molecule a rigid overall conformation (cf. Fig. 7) is retained or not.

The $\alpha(1{\rightarrow}5)$-intersaccharidic linkage in leucrose (**22**) entails pyranoid forms for both monosaccharide units as evidenced in its X-ray structure[68] by the $^4C_1$ conformation of glucose and a $^2C_5$ chair for the fructose portion (Fig. 8a). Unlike isomaltulose, leucrose, in its crystalline β-anomeric form, does not develop an intramolecular hydrogen bond. In the hydrophobicity potential profile (Fig. 8b and c) a distinct distribution of hydrophobic and hydrophilic regions is observed, whose significance again remains to be interpreted.

The molecular geometry of GPM dihydrate (**23** · 2 H_2O) as revealed by X-ray structural analysis,[69] is depicted in Fig. 9a. This representation nicely reveals the mannitol chain in a nearly planar zigzag conformation and the two molecules of crystal water to be fixed via hydrogen bonds establishing a rather unusual "double water bridge" between two oxygen atoms that are five bonds apart. The corresponding 2-epimeric sorbitol analog, GPS (**24**), crystallizes anhydrous and exhibits in its X-ray structure[70] (Fig. 10a) a linear extended-chain backbone running from C-2 of glucose to the penultimate O-2, whereby the terminal sorbitol-CH_2OH is bent off to avoid unfavourable 1,3-interactions between O-2 and O-4 that would be operative in the extended-chain rotameric form, realized in **23**.

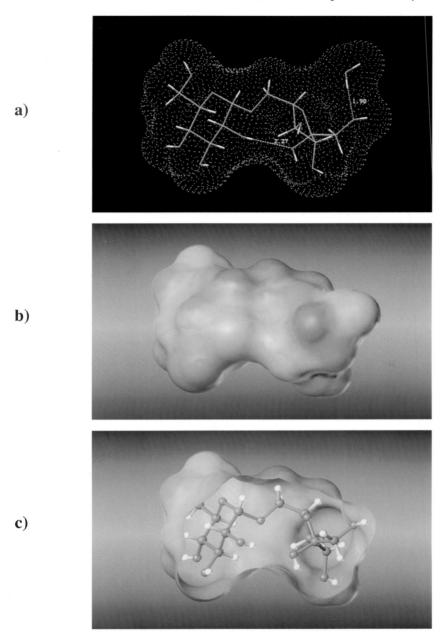

Fig. 7. (a) Dotted contact surface of β-D-isomaltulose monohydrate (**21** · H$_2$O) with Dreiding type model as insert, based on X-ray structural data.[57] (b) Color-coded molecular lipophilicity potential profile (red for hydrophilic, blue for hydrophobic regions). (c) Hydrophobicity profile representation in open form with a ball-stick model inserted. In (b) and (c), the water of crystallization has been left off to simulate the molecule's shape before solvation.

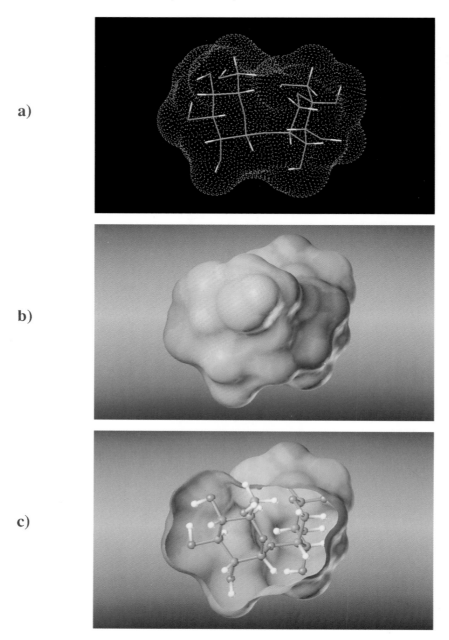

Fig. 8. Leucrose (**22**): X-ray data-derived[68] solid state contact surface with stick model insert (a), and hydrophobicity potential profile (b, c).

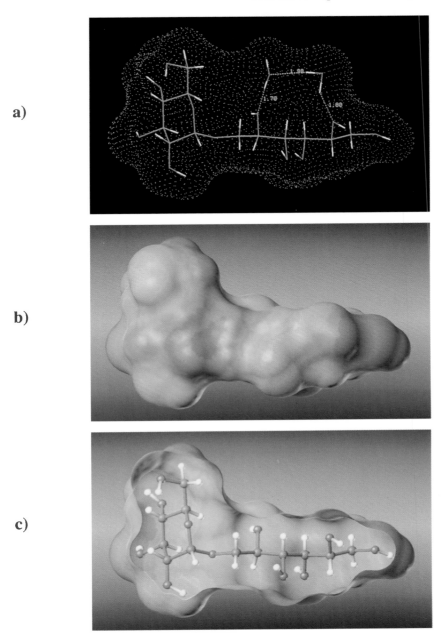

Fig. 9. Glucopyranosyl-α(1→1)-mannitol (GPM) dihydrate (**23** · 2 H$_2$O): (a) dotted contact surface for solid state with stick model insert as developed from X-ray structural data;[69] the double water bridge spanning O-2 and O-5 of the mannitol portion is accentuated. (b) and (c) molecular lipophilicity potential profile in color-code (violet for most hydrophobic region), leaving off the water of crystallization to extricate the basic molecular geometry.

a)

b)

c)

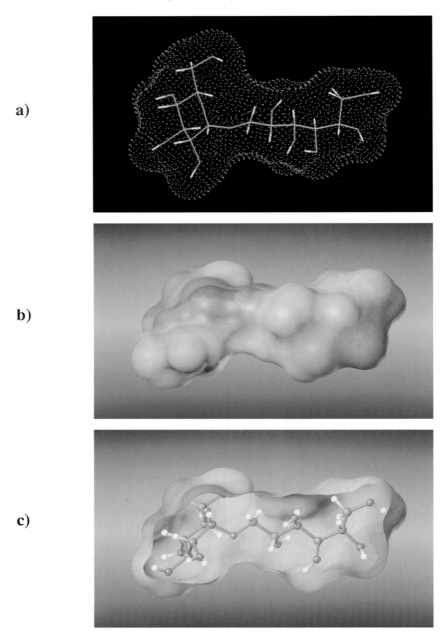

Fig. 10. Glucopyranosyl-α(1$\rightarrow$6)-sorbitol (GPS, **24**): (a) X-ray structure[70] derived contact surface and molecular geometry for the solid state, the stick model insert clearly showing the kink of the terminal hydroxymethyl group in the sorbitol chain. (b) and (c) color-coded hydrophobicity profile in an orientation different from that in (a) to account for full visualization of hydrophobic (violet) and hydrophilic (red) regions.

The hydrophobicity profiles for GPM and GPS in Figures 9 and 10 appear to be surprisingly similar if in the former the crystal water is left off for comparability, both then representing the molecular shape of the molecules before solvation (i.e. before dissolution in water). That, in fact, the side chain conformation in aqueous solution is very similar to that found in the solid state, is evidenced by the close correlation of the coupling constants along the alditol chains[71] with the respective X-ray-derived[69,70] dihedral angles. Nevertheless, an assessment of the lipophilicity potentials in terms of their biological significance has to await further studies.

7. Epilog

In the preceeding pages of this account, we have attempted to demonstrate that fairly advanced modeling modes for the molecular geometry of sucrose, for its contact surface, its electrostatic potential, and, most notably, for its hydrophobicity potential profile, provide unusually powerful tools for *building, visualizing,* and *analyzing* models of carbohydrates that are particularly suited to bring structure-activity (sweetness) relationships to an entirely new level. Thereby, the individual 3D-representations given are not limited to expose new insights only, but to stimulate new lines of thought *and* new experiments to prove them. Irrespective of whether these ideas and experiments are particularly ingenious, of aesthetic attraction, and / or practicable (or not), they are apt to be a stimulation for the progress in the chemistry and biochemistry of sucrose, not the least in its use as an organic raw material.

Acknowledgments. This work was supported by the *Bundesministerium für Forschung und Technik.* We are particularly grateful to Mrs. Gerda SCHWINN for applying her incomparable typing skills, to Dipl.-Ing. Ev LICHTENTHALER for expert drawings of stereoformulae, to Dipl.-Ing. M. WALDHERR-TESCHNER and Dipl.-Inform. M. KNOBLAUCH for their kind assistance in securing the graphic displays, to Prof. Dr. J. BRICKMANN for using his MOLCAD program and the Silicon Graphics workstation, and to Prof. H. J. LINDNER for lucid discussions on the subject.

References

1. W. Prout: On the Ultimate Composition of Simple Alimentary Substances; with some Preliminary Remarks on the Analyses of Organized Bodies in General. *Phil. Trans.* **1** (1827) 355-388.

2. J. Liebig: Ueber die Constitution des Aethers und seiner Verbindungen. *Poggendorffs Ann. Phys. Chem.* **31** (1834) 321-360.

3. E. Peligot: Untersuchungen über die Natur und die chemischen Eigenschaften der Zuckerarten. *J. Prakt. Chem.* **15** (1838) 65-113.

4. J. Berzelius: Über einige Fragen des Tages in der Organischen Chemie. *Poggendorfs Ann. Phys. Chem.* **47** (1839) 289-322.

5. A.-P. Dubrunfaut: Mémoire sur les Sucres. *Compt. Rend. Acad. Sci.* **29** (1849) 51-55; **42** (1856) 901-905.

6. I. Levi, C. B. Purves: The Structure and Configuration of Sucrose. *Adv. Carbohydr. Chem.* **4** (1949) 1-35.

7. C. E. James, L. Hough, R. Khan: Sucrose and its Derivatives. *Prog. Chem. Org. Nat. Prod.* **55** (1989) 117-184.

8. B. Tollens: Über das Verhalten der Dextrose zu ammoniakalischer Silberlösung. *Ber. Dtsch. Chem. Ges.* **16** (1883) 921-924; formula: p. 923.

9. E. Fischer: Über die Glucoside der Alkohole. *Ber. Dtsch. Chem. Ges.* **26** (1893) 2400-2412; formula: p. 2405.

10. B. Tollens: *Kurzes Handbuch der Kohlenhydrate.* Verlag J. A. Barth, Leipzig, 1914, p. 363.

11. W. N. Haworth, J. Law: The Constitution of Disaccharides, I. The Structure of Sucrose. *J. Chem. Soc.* **109** (1916) 1314-1325; formula p. 1319.

12. C. S. Hudson: Some Numerical Relations among the Rotatory Powers of the Compound Sugars. *J. Am. Chem. Soc.* **38** (1916) 1566-1575; formula p. 1567.

13. E. Fischer: Über die Configuration des Traubenzuckers und seiner Isomeren. *Ber. Dtsch. Chem. Ges.* **24** (1891) 2683-2687.

14. a) W. Charlton, W. N. Haworth, S. Peat: A Revision of the Structural Formula of Glucose. *J. Chem. Soc.* **1926**, 89-101; formula p. 99.

    b) W. N. Haworth, E. L. Hirst: The Structure of Fructose, γ-Fructose, and Sucrose. *J. Chem. Soc.* **1926**, 1858-1868; formula p. 1864.

15. W. N. Haworth: *The Constitution of Sugars.* Arnold and Co., London, 1929, pp. 70-71.

16. W. W. Pigman: *Chemistry of the Carbohydrates.* Acad. Press, New York, 1948, p. 446.

17. Ref. 6, p. 25.

18. R. T. Morrison, R. N. Boyd: *Organic Chemistry.* Allyn and Bacon, Inc., Boston 1959, p. 789.

19. L. Hough: Presentation at the 4th European Carbohydr. Symp., Darmstadt 1987; cf. *Nachr. Chem. Tech. Lab.* **38** (1990) 860.

20. P. M. Collins: *Carbohydrates.* Chapman and Hall, London 1987, frontispiece and formula p. 466.

21. Chemist's Personal Software Series: ChemText 1.3. Molecular Design Ltd., San Leandro, Cal., USA; *Labo* (Hoppenstedt Verlag, Darmstadt) **12** (1988) 32-36.

22. G. M. Brown, H. A. Levy: Sucrose: Precise Determination of Crystal and Molecular Structure by Neutron Diffraction. *Science* **141** (1963) 921-923.

23. G. M. Brown, H. A. Levy: Refinement of the Structure of Sucrose Based on Neutron-Diffraction Data. *Acta Crystallogr., Sect. B* **29** (1973) 790-797.

24. J. C. Hanson, L. C. Sieker, L. H. Jensen: Sucrose: X-Ray Refinement and Comparison with Neutron Refinement. *Acta Crystallogr., Sect. B* **29** (1973) 797-808.

25. F. W. Lichtenthaler: Disaccharidalkohole – Molekulare Geometrie und mögliche Beziehungen zwischen Struktur und Süßkraft. *Dtsch. Zahnärztl. Zeitschr.* **37** (1982) S46-S49; formula p. S48.

26. L. Hough: The Sweeter Side of Chemistry. *Chem. Soc. Rev.* **14** (1985) 357-374.

27. K. Bock, R. U. Lemieux: The Conformational Properties of Sucrose in Aqueous Solution: Intramolecular Hydrogen Bonding. *Carbohydr. Res.* **100** (1982) 63-74.

28. D. C. McCain, J. L. Markley: The Solution Conformation of Sucrose: Concentration and Temperature Dependance. *Carbohydr. Res.* **152** (1986) 73-80.

29. D. C. McCain, J. L. Markley: Rotational Spectral Density Functions for Aqueous Sucrose: Experimental Determination Using ^{13}C-NMR. *J. Am. Chem. Soc.* **108** (1986) 4259-4264.

30. C. Christofides, D. B. Davies: Cooperative and Competitive Hydrogen Bonding in Sucrose Determined by SIMPLE ^{1}H-NMR Spectroscopy. *J. Chem. Soc., Chem. Commun.* **1985**, 1533-1534.

31. D. B. Davies, J. C. Christofides: Comparison of Intramolecular Hydrogen-bonding Conformations of Sucrose-containing Oligosaccharides in Solution and in the Solid State. *Carbohydr. Res.* **163** (1987) 269-274.

32. a) V. H. Tran, J. W. Brady: Disaccharide Conformational Flexibility, I. An Adiabatic Potential Energy Map for Sucrose. *Biopolymers* **29** (1990) 961-976.

    b) V. H. Tran, J. W. Brady: Disaccharide Conformational Flexibility, II. Molecular Dynamics Simulations of Sucrose. *Biopolymers* **29** (1990) 977-997.

    c) V. H. Tran, J. W. Brady: Conformational Flexibility of Sucrose. Static and Dynamic Modeling. *Computer Modeling of Carbohydrate Molecules* (A. D. French, J. W. Brady, Eds.) *ACS Symposium Series # 430*, Am. Chem. Soc., Washington DC, 1990, pp. 213-226.

33. C. K. Johnson: ORTEP II: A Fortran Thermal-Ellipsoid Plot Program for Crystal Structure Illustration. Oak Ridge National Laboratory, Report ORNL-3794, Oak Ridge, Tenn., 1965.

34. For a Pertinent Review, see: N. C. Cohen, J. M. Blaney, C. Humblet, P. Gund, D. M. Barry: Molecular Modeling Software and Methods for Medicinal Chemistry. *J. Med. Chem.* **33** (1990) 883-894.

35. W. G. Richards: *Computer-Aided Molecular Design.* IBC Technical Servies Ltd., London, 1989.

36. A. D. French, J. W. Brady (Eds.): *Computer Modeling of Carbohydrate Molecules.* *ACS Symposium Series # 430*, Am. Chem. Soc., Washington DC, 1990.

37. K. B. Lipkowitz, D. B. Boyd: *Reviews in Computational Chemistry,* VCH Verlagsgesellschaft Weinheim, 1990.

38. J. Brickmann, M. Waldherr-Teschner: Molecular Modeling: Wohin geht der Weg? *Labo* (Hoppenstedt Verlag, Darmstadt) **10** (1989) 7-14.

39. Prof. Dr. J. Brickmann, Institut für Physikalische Chemie, Technische Hochschule Darmstadt, Petersenstr. 20, D-6100 Darmstadt, Germany.

40. F. M. Richards: Areas, Volumes, Packing, and Protein Structure. *Ann. Rev. Biophys. Bioeng.* **6** (1977) 151-176.

41. M. L. Connolly: Analytical Molecular Surface Calculation. *J. Appl. Cryst.* **16** (1983) 548-558.

42. M. L. Connolly: Solvent-Accessible Surfaces of Proteins and Nucleic Acids. *Science* **221** (1983) 709-713.

43. B. Lee, F. M. Richards: Interpretation of Protein Structures: Estimation of Static Accessibility. *J. Mol. Biol.* **55** (1971) 379-400.

44. P. K. Weiner, R. Langridge, J. M. Blaney, R. Schaefer, P. A. Kollman: Electrostatic Potential Molecular Surfaces. *Proc. Natl. Acad. Sci. USA* **79** (1982) 3754-3758.

45. J. J. P. Stewart: MOPAC: A Semiempirical Molecular Orbital Program. *Quantum Chem. Prog. Exch.*, Program No. 455 (1983).

46. F. W. Lichtenthaler, S. Rönninger: α-D-Glucosyl-D-fructoses: Distribution of Furanoid and Pyranoid Tautomers in Water, Dimethyl Sulfoxide, and Pyridine. *J. Chem. Soc., Perkin Trans.* 2, **1990**, 1489-1497.

47. P. Furet, A. Sele, N. C. Cohen: 3D-Molecular Lipophilicity Potential Profiles: A New Tool in Molecular Modeling. *J. Mol. Graphics* **6** (1988) 182-189.

48. J.-L. Fauchère, P. Quarendon, L. Kaetterer: Estimating and Representing Hydrophobicity Potential. *J. Mol. Graphics* **6** (1988) 202-206.

49. E. W. Deutsch, C. Hansch: Dependence of Relative Sweetness on Hydrophobic Bonding. *Nature* **211** (1966) 75.

50. R. U. Lemieux: The Binding of Carbohydrate Structures with Antibodies and Lectins. *Frontiers Chem., Plenary Keynote Lect.* 28th IUPAC Congr., 1981 (K. J. Laidler, Ed.), Pergamon, Oxford, U. K., 1982, pp. 3-24.

51. C.-K. Lee: Chemistry and Biochemistry of Sweetness. *Adv. Carbohydr. Chem. Biochem.* **45** (1987) 199-351; pp. 223 ff., in particular.

52. C. Tanford, *The Hydrophobic Effect.* Wiley, New York, 1973.

53. a) A. K. Ghose, G. M. Grippen: Atomic Physicochemical Parameters for Three-Dimensional Structure-Directed Quantititave Structure-Activity Relationships. Partition Coefficients as a Measure of Hydrophobicity. *J. Comput. Chem.* **7** (1986) 565-577.

    b) A. K. Ghose, A. Pritchett, G. M. Grippen: Atomic Physicochemical Parameters for Three-Dimensional Structure-Directed Quantitative Structure-Acitivity Relationships. Modeling Hydrophobic Interactions. *J. Comput. Chem.* **9** (1988) 80-90.

54. a) L. Hough, R. Khan: Intensification of Sweetness. *Trends Biol. Sci.* **3** (1978) 61-63.

    b) G. Jackson, M. R. Jenner, R. A. Khan, C. K. Lee, K. S. Mufti, G. D. Patel, E. B. Rathbone (Tate & Lyle PLC): 4'-Halo-substituted sucrose derivatives. *Brit. Pat.* 2,088,855 (1982); *Eur. Pat. Appl.* EP 73,093 (1983); *Chem. Abstr.* **99** (1983) 54127j.

    c) L. Hough, R. Khan: Enhancement of the Sweetness of Sucrose by Conversion into Chloro-deoxy Derivatives. *Progress in Sweeteners* (T. H. Grenby, Ed.), Elsevier Appl. Science, London, 1989, pp. 102 ff.

55. R. S. Shallenberger, T. E. Acree: Molecular Theory of Sweet Taste. *Nature* **216** (1967) 480-482; *J. Agric. Food Chem.* **17** (1969) 701-703.

56. L. B. Kier: Molecular Theory of Sweet Taste. *J. Pharm. Sci.* **61** (1972) 1394-1397.

57. W. Dreissig, P. Luger: Die Strukturbestimmung der Isomaltulose. *Acta Crystallogr., Sect. B* **29** (1973) 514-521.

58. J. A. Kanters, W. P. J. Gaykema, G. Roelofson: Conformation and Hydrogen Bonding of Disaccharides. The Crystal and Molecular Structure of Turanose. *Acta Crystallogr., Sect. B* **34** (1978) 1873-1880.

59. M. E. Gress, G. A. Jeffrey: A Neutron Diffraction Refinement of the Crystal Structure of β-Maltose Monohydrate. *Acta Crystallogr., Sect. B* **33** (1977) 2490-2495.

60. S. S. C. Chu, G. A. Jeffrey: The Crystal Structure of Methyl β-Maltopyranoside. *Acta Crystallogr.* **23** (1967) 1038-1049.

61. F. Takusagawa, R. A. Jacobson: The Crystal and Molecular Structure of α-Maltose. *Acta Crystallogr., Sect. B* **34** (1978) 213-218.

62. I. Tanaka, N. Tanaka, T. Ashida, M. Kakudo: The Crystal and Molecular Structure of Phenyl α-Maltoside. *Acta Crystallogr., Sect. B* **32** (1976) 155-160.

63. J. A. Kanters, R. L. Scherrenberg, B. R. Leeflang, J. Kroon, M. Mathlouthi: The Crystal and Molecular Structure of an Intensely Sweet Chlorodeoxysucrose: 4,1',6'-Trichloro-4,1',6'-trideoxy-*galacto*-sucrose. *Carbohydr. Res.* **180** (1988) 175-182.

64. J. C. Christofides, D. B. Davies, J. A. Martin, E. B. Rathbone: Intramolecular Hydrogen Bonding in 1'-Sucrose Derivatives Determined by SIMPLE ^{1}H-NMR Spectroscopy. *J. Am. Chem. Soc.* **108** (1986) 5738-5743.

65. L. Hough, E. O'Brien: α-D-Allopyranosyl β-D-Fructofuranosid (*allo*-sucrose) and its Derivatives. *Carbohydr. Res.* **84** (1980) 95-102.

66. H. Schiweck, M. Munir, K. M. Rapp, B. Schneider, M. Vogel: New Developments in the Use of Sucrose as an Industrial Bulk Chemical. *This Monograph,* pp. 57 ff.

67. D. Schwengers: Leucrose, a Ketodisaccharide of Industrial Design. *This Monograph,* pp. 183 ff.

68. J. Thiem, M. Kleeberg, K.-H. Klaska: Kristallstruktur der Leucrose. *Carbohydr. Res.* **189** (1989) 65-77.

69. H. J. Lindner, F. W. Lichtenthaler: Extended Zigzag Conformation of 1-*O*-α-D-Glucopyranosyl-D-mannitol. *Carbohydr. Res.* **93** (1981) 135-140.

70. F. W. Lichtenthaler, H. J. Lindner: The Preferred Conformations of Glycosyl-alditols. *Liebigs Ann. Chem.* **1981**, 2372-2383.

71. M. Munir, B. Schneider, H. Schiweck: 1-*O*-α-D-Glucopyranosyl-D-fructose. Darstellung aus Saccharose und ihre Reduktion zu 1-*O*-α-D-Glucopyranosyl-D-glucitol. *Carbohydr. Res.* **164** (1987) 477-485.

2

Applications of the Chemistry of Sucrose

Leslie Hough

Department of Chemistry, King's College London
University of London, Campden Hill Road, London W8, 7AH, U.K.

Summary. Sucrose is an attractive feedstock for chemical exploitation. Glycosides of D-fructose are readily obtained by alcoholysis, such as the 2-chloroethyl β-pyranoside from which novel spiro-acetals can be derived. Selective oxidations of sucrose can give mono- and di-carboxy derivatives, and glucosyl and fructosyl morpholines. Differences in the rates of the reaction of the eight hydroxyls in sucrose has been exploited to give partially substituted derivatives. Thus, esterification yields products ranging from the mono- to the octa-esters. Applications of sucrose mono-esters of fatty acids include surfactants and emulsifiers in food products, and preservative coatings for fruits and vegetables. Higher sucrose esters, containing 5 - 8 fatty acid groups, have been developed for use in low caloric fats and oils. The sweetness of sucrose is intensified, up to 2200 times, by replacement of specific hydroxyls by chloro groups.

Introduction

Efforts to extend applications of sucrose to areas other than the traditional sweetener, food and drink markets were made by the Sugar Research Foundation from 1954 onwards.[1] They identified large volume markets such as surfactants, plastics, and polymers, for study with the obvious advantage that sucrose (**1**), is a low-cost, pure, and readily available material with few storage or transportation problems when compared with the inevitable decline in the supply of petrochemicals and their environmental hazards. The Foundation stimulated technical and fundamental studies on the chemistry of sucrose, termed sucrochemistry, which continued apace over the past decade considerably enhancing scientific knowledge of this unique non-reducing disaccharide with the emergence of hundreds of new derivatives of commercial importance.[2,3]

On the European scene the economo-political emphasis on self-sufficiency of sugar from beet crops has transformed the E.E.C. from a nett importer of sugar to an exporter of that which exceeds its requirements as a food. The excess sugar can be utilized at a lower price for chemical production, a considerable incentive for sucrochemicals that are targetted towards the needs of identified markets. Furthermore, many new sugar based products have the advantage that they are "green", namely non-toxic, biodegradable, and generally conducive to the environment.

Hydrolysis and Alcoholysis

The acid lability of sucrose, giving D-glucose (**3**) and D-fructose (**5**), coupled with its insolubility in many organic solvents, usually limits its chemical reactions to either melts or aqueous alkaline and pyridine solutions.

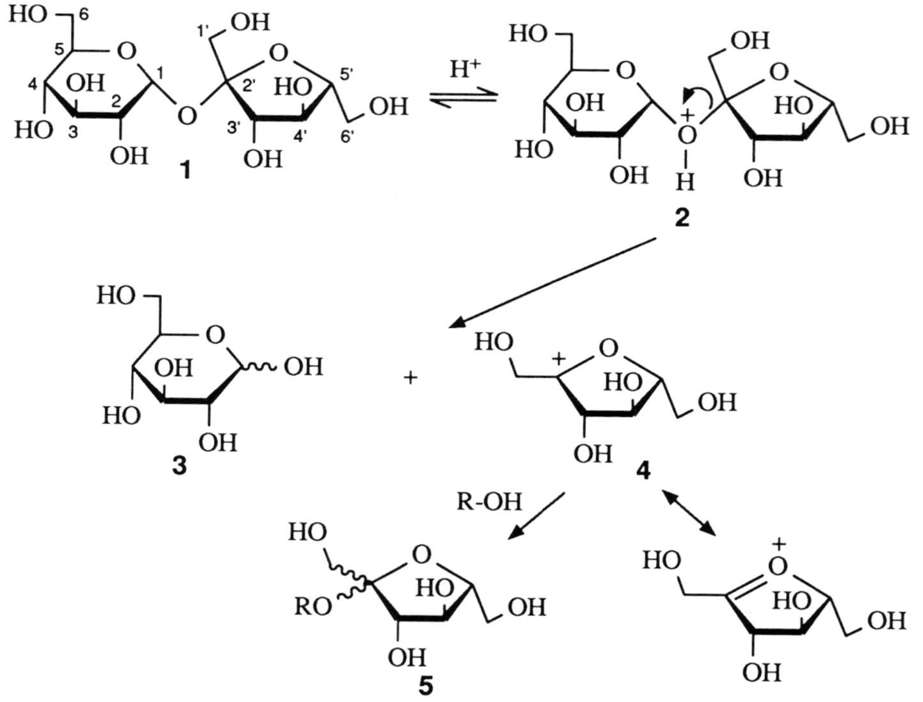

In 0.1 % methanolic hydrogen chloride at 20 °C sucrose (**1**) is protonated giving the oxonium ion **2** which undergoes cleavage between the C-2'-to-O bond

giving glucose (3) and the fructofuranosyl carbocation **4** and then the latter reacts with methanol with complete conversion to methyl D-fructofuranosides (**5**) within 30 minutes.[4]

A similar reaction of sucrose in 2-chloro-ethanol afforded the 2'-chloroethyl β-D-fructopyranoside (**6**), crystallizing out of the reaction mixture.

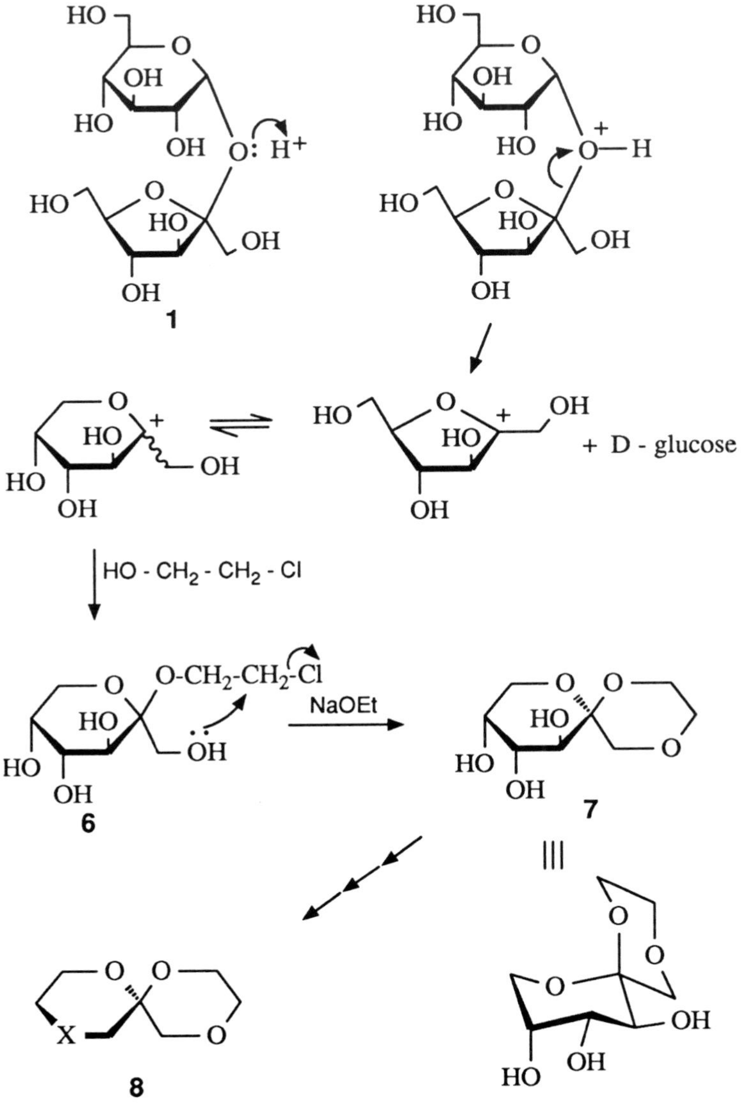

Treatment of the latter (**6**) with base caused internal cyclization, departure of the chloro group resulting from intramolecular attack by the 1-hydroxyl group, giving the 1',2-spiro-anhydride, also a spiro-acetal **7**, incidentally, a tasteless water soluble derivative.[5] From this spiro-product **7** we generated the (*R*)-spirobi-1,4-dioxan (**8**, X= O), its nitrogen analogue morpholino-2-spiro-1,4-dioxan (**8**, X= N), and a thia-analogue (**8**, X= S), novel compounds in which the chirality is due solely to the spiro-ring-junction.

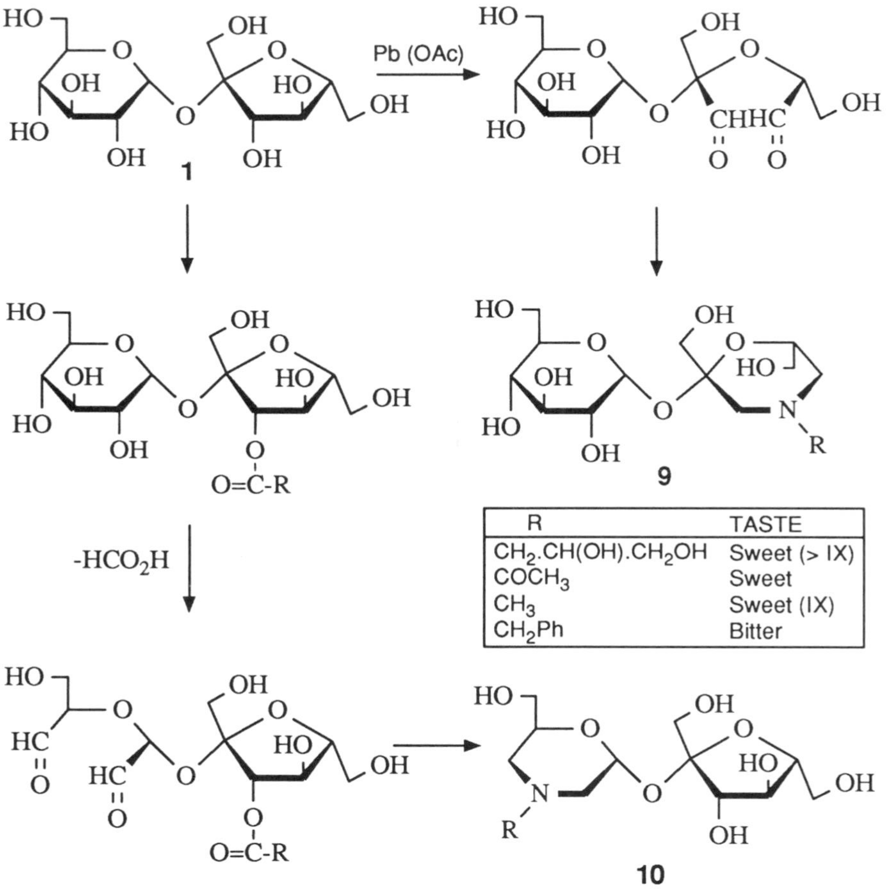

R	TASTE
CH$_2$.CH(OH).CH$_2$OH	Sweet (> IX)
COCH$_3$	Sweet
CH$_3$	Sweet (IX)
CH$_2$Ph	Bitter

α-D-Glucopyranosyl morpholines (**9**) and β-D-fructofuranosyl morpholines (**10**) have been synthesized from sucrose and its 3'-*O*-esters respectively by selective oxidative cleavage of the fructosyl ring with lead tetra-acetate and the glucosyl ring with periodate, followed by reductive amination.[6,7] Some of the products are sweet.[6]

The selective oxidation of the primary hydroxyl groups of sucrose to carboxylic acids has been extensively studied using oxygen with a platinum or palladium catalyst at pH 7 - 9. Edye and Richards[8] have observed that the oxidation with platinum (10 % Pt on C), maintained at pH 7.0, was predominant at C-6 and C-6', yielding a mixture of the 6-monocarboxy **11**, 6'-monocarboxy **12** and 6,6'-dicarboxy **13** derivatives. The two latter sucronic acid derivatives are competitive inhibitors of invertase but the 6-carboxy derivative **11** was hydrolysed by this enzyme to yield fructose and glucuronic acid.

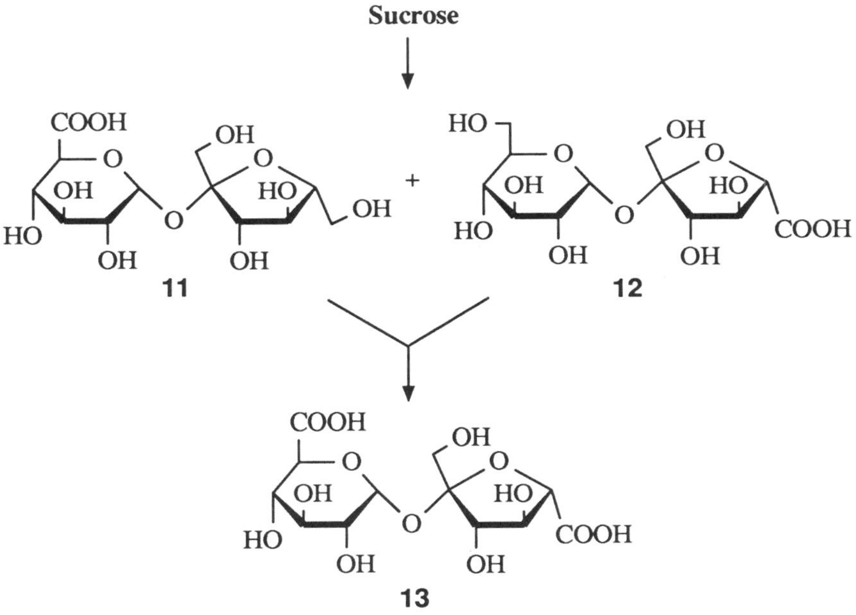

Sucrose

Esterifications of Sucrose

Esterification of sucrose with a variety of acylating reagents can under carefully controlled conditions exhibit selectivity due to the small but significant differences in the rates of reactivity of the hydroxyl groups, thus making partially esterified products available.[9] In general the primary hydroxyls at C-6 and C-6' would show higher reactivity yielding a mixture of 6- and 6'-mono-esters (**14** and **15**) and the 6,6'-diester **16**. Reaction at the remaining 1'-hydroxyl group, a neopentyl and hindered substituent, follows, resulting in the 6,1',6'-triester **17**. Of the remaining five, slower reacting secondary hydroxyl groups, those in close proximity to the anomeric centres at C-1 and C-2', namely the 2-hydroxyl and 3'-hydroxyl, show the higher reactivity; the hindered hydroxyls at C-4 and C-4' are the least reactive.

Thus, Khan and Mufti[10] isolated 6-*O*-acetylsucrose (**14**, R= CH$_3$) in 40 % yield by reaction of sucrose with acetic anhydride in pyridine at -40 °C followed by column chromatography. Esterification has a dramatic effect upon sweetness since the 6-monoacetate is only slightly sweet whereas the 6-monobenzoate is tasteless and the octa-acetate is extremely bitter. Trimolar benzoylation of sucrose with benzoyl chloride in pyridine affords the 6,1',6'-tribenzoates, (**17**, R= Ph) as the major product, also a bitter compound but detailed analysis of the products revealed that the reactivity decreases according to 6-OH > 1'-OH and 6'-OH.[11]

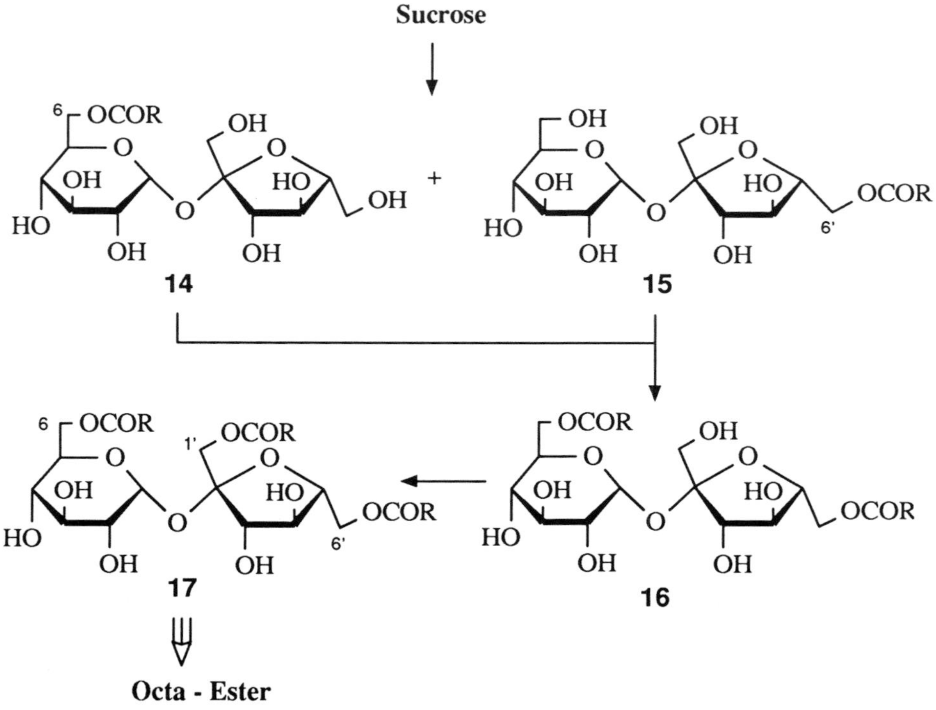

Sucrose

14   **15**

17   **16**

Octa - Ester

As expected, greater selectivity in esterification reactions was observed with pivaloyl chloride (2,2-dimethylpropionyl chloride), a bulky sterically hindered reagent.[12] Thus, pivaloylation under a variety of conditions led to the isolation of 6,6'-dipivalate   **18**,   6,1',6'-tripivalate   **19**,   6,1',4',6'-tetrapivalate   **20**,

2,6,1',4',6'-pentapivalate **21**, and the 4-hydroxy-heptapivalate **22** amongst other pivalates.

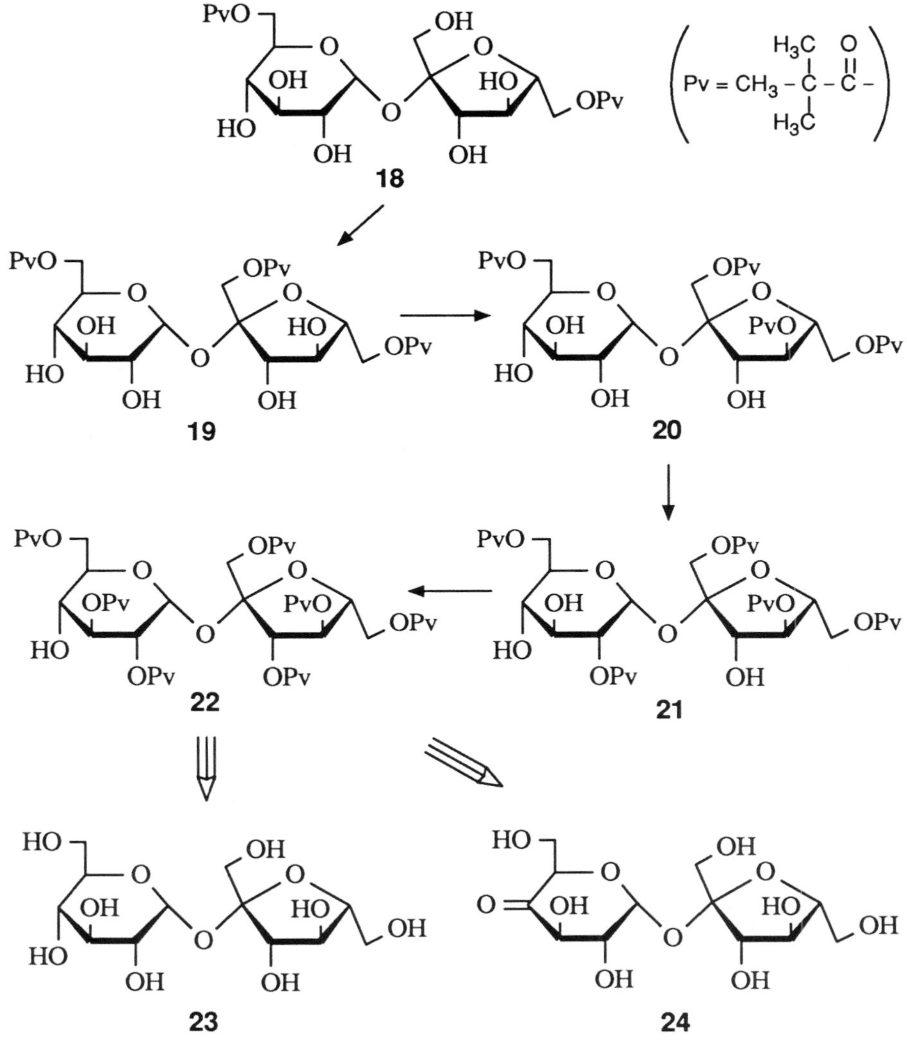

The latter (**22**) was isolated crystalline in 50 % yield and was a useful intermediate in the synthesis of *galacto*-sucrose (**23**), 4-deoxysucrose and 4-ketosucrose (**24**).[13] Interestingly, the former (**23**) is tasteless whereas an equatorial 4-hydroxyl group is not essential for sweetness in sucrose.[14]

The regioselective esterification of sucrose can be influenced by organometallics and transition metals. Thus, Ogawa and Matsui[15] found that three equivalents of bis(tributylstannyl)-oxide and 6 moles of benzoyl chloride gave 2,3,6,1',6'-penta-*O*-benzoylsucrose (**25**) in 87 % yield. Avela et al.[16] found that monoesters were obtained when sucrose was treated with sodium hydride, $CoCl_2$ and acetic anhydride in DMF. Analogous reactions in pyridine for both acetylation and butyroylation gave the 3'-*O*-esters **26** in 60 % yield.[7]

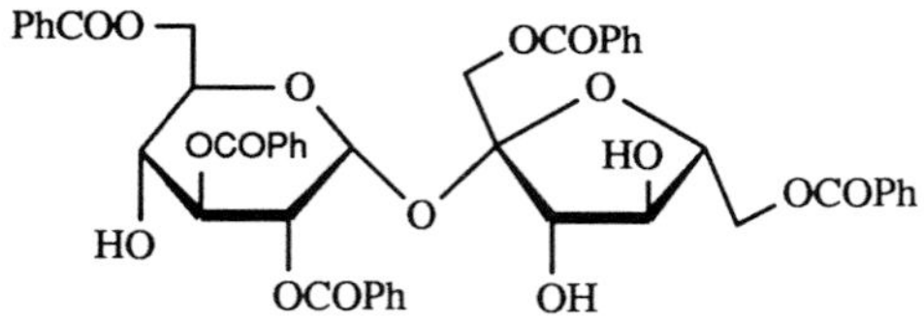

PhCOO — O
OCOPh
HO
OCOPh
OCOPh
O
HO
OCOPh
OH
25

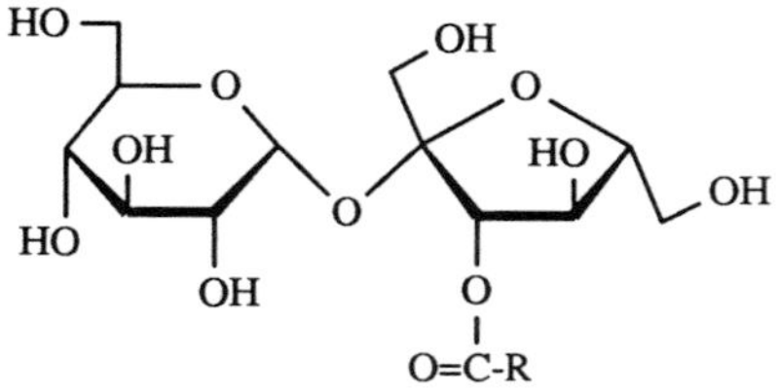

HO — O
OH
HO
OH
HO
OH
O
HO
OH
O
|
O=C-R
26

The synthesis of monoesters of sucrose with long chain fatty acids was a major achievement of the Sugar Research Foundation,[1] quickly approved in Japan for use as a food additive in 1959 and subsequently finding world wide approval for application as non-ionic surfactants and emulsifiers in food products with the major advantage of total metabolism and biodegradability. There are several commercial routes to these monoesters of fatty acids **27**. The original Hass-Snell process[2,17] involves transesterification of a triglyceride fat or oil (**28**) with sucrose using a base catalyst at 90 °C in DMF as solvent but later replaced by dimethyl sulphoxide, as a safer and less expensive solvent. The product contains > 50 % monoesters (**27**) and some di- (**29**) and higher-esters (> 10 %), unreacted sucrose and triglyceride.

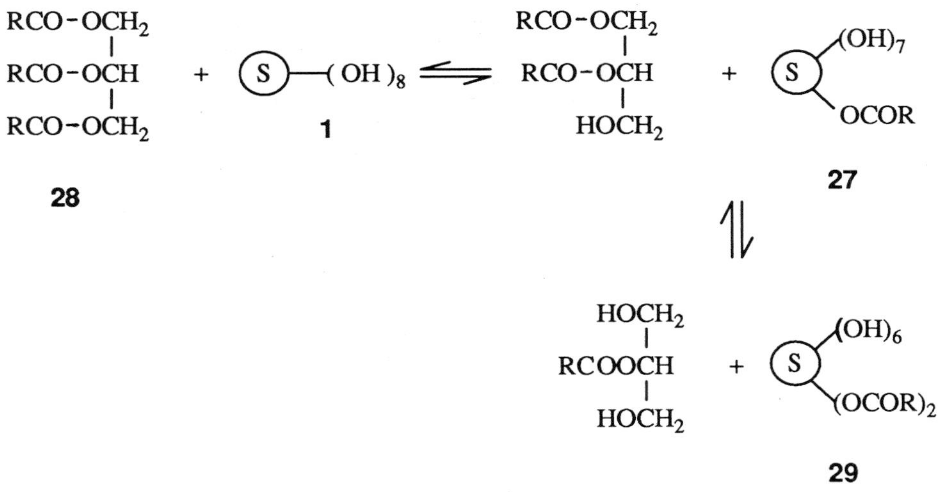

RCO-OCH$_2$ RCO-OCH$_2$
| |
RCO-OCH   + (S)—(OH)$_8$ $\rightleftharpoons$ RCO-OCH   + (S)⟨$\begin{array}{l}\text{(OH)}_7\\\text{OCOR}\end{array}$

RCO-OCH$_2$ **1**   HOCH$_2$

28   **27**

$\updownarrow$

HOCH$_2$
|
RCOOCH   + (S)⟨$\begin{array}{l}\text{(OH)}_6\\\text{(OCOR)}_2\end{array}$
|
HOCH$_2$

29

The use of methyl esters of fatty acids in transesterification reactions with sucrose has its advantages because the methanol can be distilled off as it is formed, thus forcing the equilibrium in favour of the sucrose ester and hence a higher yield of the desired product.

CH$_3$OCOR   +   (S)—(OH)$_8$ $\rightleftharpoons$ CH$_3$OH ($\uparrow$) +   (S)⟨$\begin{array}{l}\text{OCOR}\\\text{(OH)}_7\end{array}$

A solventless process was developed by Tate & Lyle[18] which uses a melt or slurry of sucrose, triglyceride (or methyl ester) and base catalyst (potassium carbonate or potassium soap) at 130 °C. The crude reaction product finds some use in detergent formulations.

Sucrose monoesters including the stearate, behenate (tallow), oleate, palmitate, and myristate are manufactured in Japan by the Mitsubishi Food Corporation and usually contain 70 % monoester (**27**) and 30 % di- (**29**), tri- and poly-esters. In addition sugar monoesters such as the monolaurate, inhibit the growth of *Escherichia coli* and other bacteria[19] with obvious advantages in food and drink products.

Sempernova (U.K.) has exploited research into non-toxic food and vegetable coatings based on sucrose monoesters; the coating is both edible and biodegradable.

An aqueous dispersion of the monoester (Semperfresh®) applied to the fruit creates a semi-permeable membrane which retards the ripening process, reducing costly losses (30 - 40 %) from rotting during storage and transportation.

Since fat has twice the calories of sugar or carbohydrate, the need to develop a low calorie fat or oil has been targetted as a high priority for the slimmers market with the added medical role as a potential agent for lowering cholesterol levels in humans. Procter & Gamble[20] have developed a group of fatty acid esters of sucrose, termed polyesters (SPE, **30**) that are neither absorbed or metabolised – they are not hydrolysed by pancreatic lipase – and the Company is poised to market them under the brand name "Olestra®". The polyesters **30** contain 6 - 8 esters per molecule and are made by the solventless transesterification process for example by treating the ethyl ester of the fatty acid(s) with sodium ethoxide in a melt with sucrose at 100 - 180 °C for 14 hours. Unreacted fatty acid esters and sucrose esters containing four or less ester residues can be removed by enzymic hydrolysis with lipase; SPE with 5 - 8 ester groups is resistant to lipase activity.[21] The fatty acid ester constituent(s) determines the physical properties of the resulting low calorie product varying from a solid fat to a liquid oil at room temperature and indistinguishable in characteristics from the natural oils and fats. Since fat consumption in the USA is 6×10^9 tons per annum, only a few percent of SPE would raise sugar use by more than a million tons.[3]

$$\text{(S)}-\text{(OH)}_8 \quad + \quad \text{EtOCOR} \quad \underset{\longleftarrow}{\overset{\text{NaOEt}}{\rightleftharpoons}} \quad \text{(S)}\begin{matrix}\diagup \text{(OH)}_{0-2} \\ \diagdown \text{(OCOR)}_{6-8}\end{matrix} \quad + \quad \text{EtOH}$$

S.P.E. (OLESTRA)

30

Two other sucrose esters have been manufactured for many years. A fully esterified but mixed sucrose acetate isobutyrate (SAIB) (**31**) is made by Tennessee Eastman Co. (USA) by reaction of sucrose with acetic and isobutyric anhydrides for use in plastics, lacquers and inks.[22] A unique aluminium salt of sucrose octasulphate ("Sucralphate®") is manufactured by esterification of sucrose with either sulfur trioxide-DMF-pyridine or chlorosulphonic acid, to give the octasulphate ester **32** which is then converted into its aluminium salt for pharmaceutical use in the treatment of gastric ulcers since it inhibits peptidase and acts as a buffer.[2]

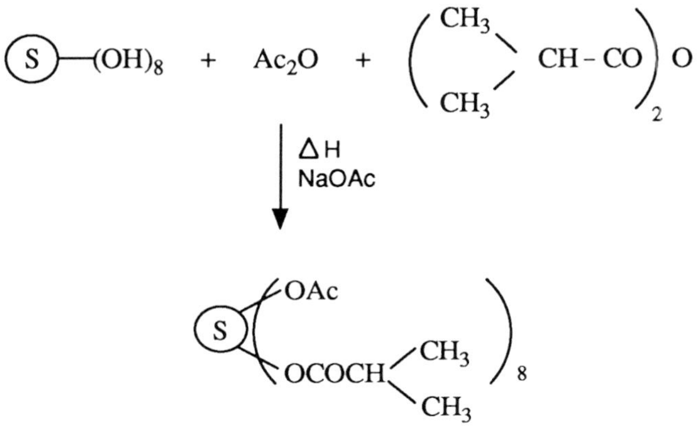

SAIB **31**

32

SUCRALFATE

An alternative route to partially esterified sucrose derivatives is to commence with the fully subtituted octa-ester and then carry out partial, often selective, hydrolysis with mild base.[23,24] Thus, treatment of sucrose octa-acetate with basic alumina, using chloroform as eluant has yielded, after chromatographic separation, hepta-acetates with the 6-OH, 6'-OH, 4-OH, or 4'-OH free. Franzkowiak and Thiem[24] utilized the 4'-hydroxy hepta-acetate **33** in the synthesis of Agrocinopine A, 4'-*O*-sucrose 2-*O*-L-arabinosyl phosphate (**34**). More extensive de-esterification occurs when potassium carbonate is incorporated into alumina. Thus, using methanol as eluant, Čapek et al.[25] have isolated penta-acetates containing 3',4',6-hydroxy, 1',3',4'-hydroxy and 2,3',4'-hydroxy groups with de-esterification predominating on the fructofuranose ring. This group has also isolated a hexa-acetate (**35**, R= H) with 3',4'-hydroxy groups which on conversion to the 3',4'-tosylate (**35**, R= SO$_2$C$_7$H$_7$) followed by treatment with sodium methoxide gave a 3',4'-epoxide, namely α-D-glucopyranosyl 3,4-epoxy-β–D-*lyxo*-hexulofuranoside (**36**).[26]

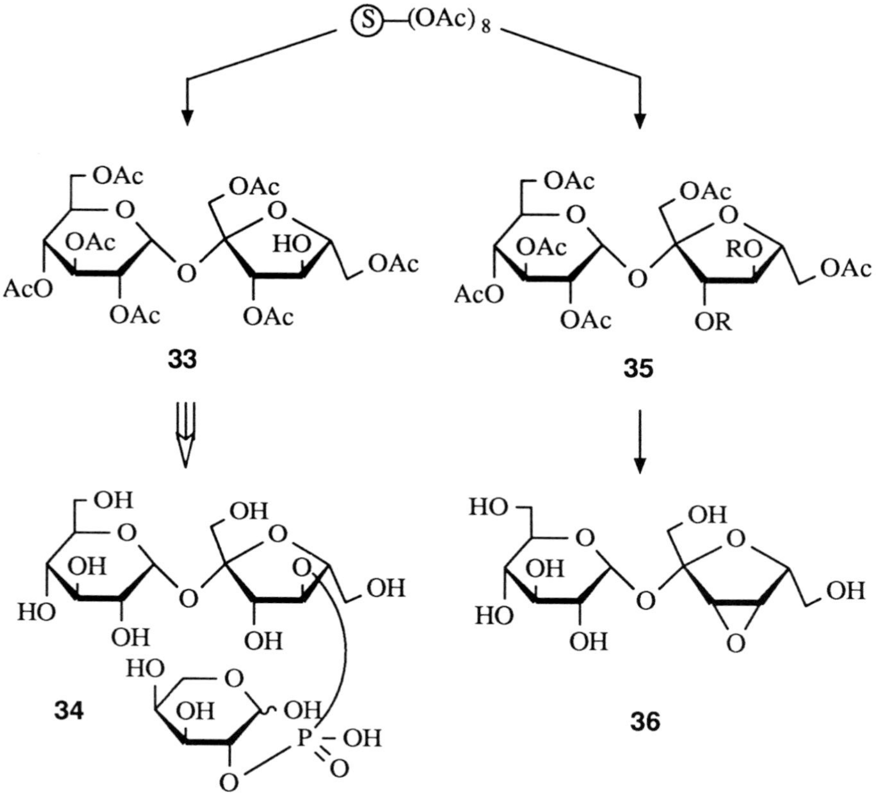

Polyurethanes are manufactured by the reaction of di-isocyanates, such as toluene di-isocyanate **37** and a range of polyols, ethane-1,2-diol for example, the choice of polyol depending upon the type of polymer or polyurethane **38** required. A low functionality polyol is used for flexible and elastomeric polyurethanes, whilst high functionality polyols, such as sucrose derivatives, are used for rigid polyurethanes. Unfortunately, sucrose itself yields brittle polymers but its poly(hydroxypropyl)ether **39**, made by reaction of sucrose with propylene oxide, yields a strong polyurethane foam.[27]

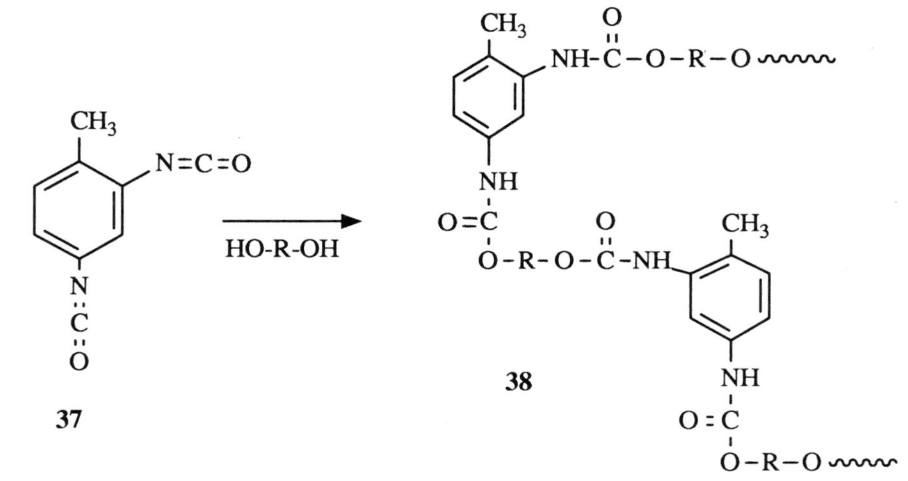

Enhancing the Sweetness of Sucrose

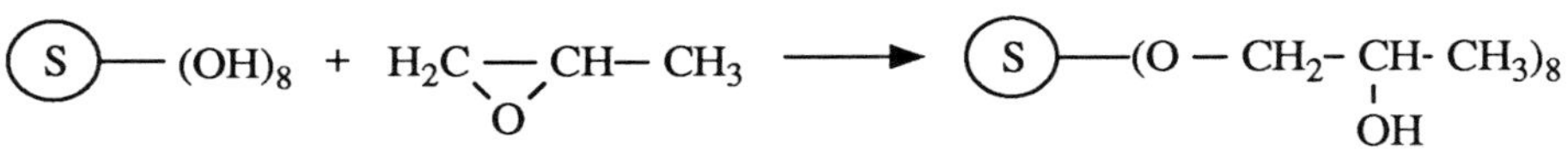

Another target was to enhance the natural sweetness of sucrose in the same way that the greater activity of pharmacologically active substances has been achieved, thereby entering the profitable market of high intensity sweeteners, hopefully with a safe product with a good taste profile similar to that of its parent, sucrose. All high intensity sweeteners, such as saccharin (400 - 500x; **40**), cyclamate (50x; **41**), acesulfam-K (150x; **42**), neohesperidin dihydrochalcone (2000x; **43**) and aspartame (100 - 200x; **44**) are more lipophilic than water-loving sucrose (1x).[28] Unlike sucrose, however, they have unusual and varying taste profiles and lack body in their response to the taste buds.

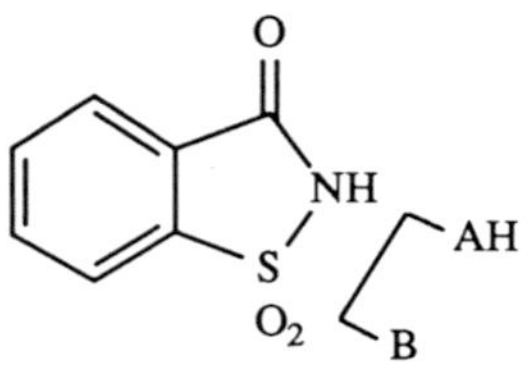

40
Saccharine (200-700x)

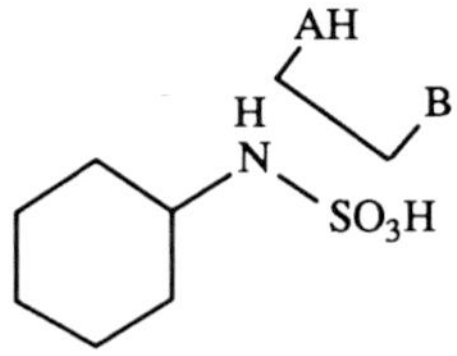

41
Cyclamate (30-80x)

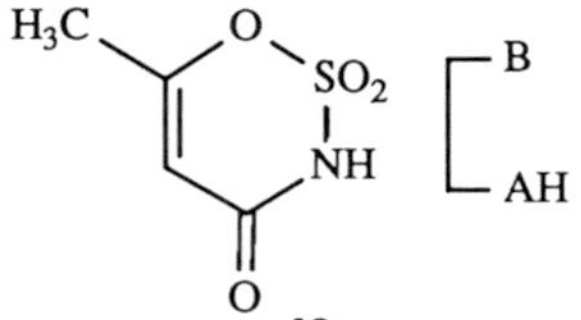

42
Acesulpham - K (150x)

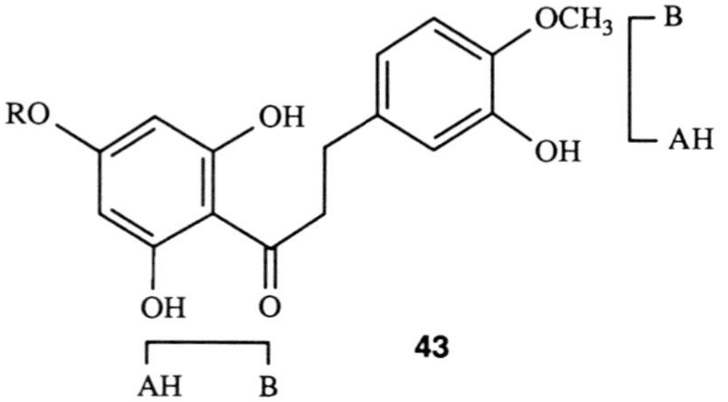

43
Neohesperidin dihydrochalcone (2000x)

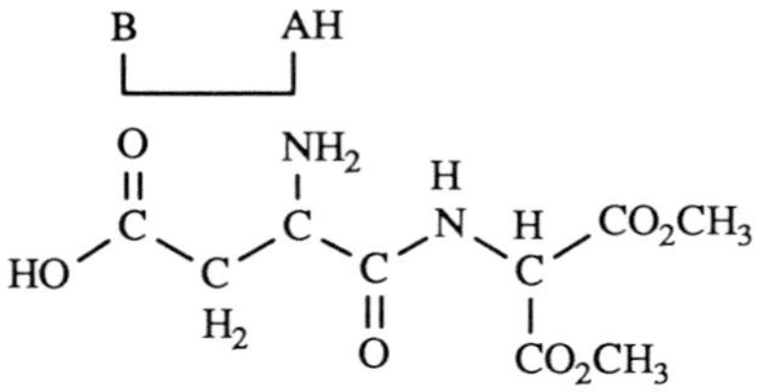

44
L-aspartyl-L-phenylalanine
methyl ester (100-200)
Aspartame: Nutrasweet

Amongst their diverse chemical structures, all high intensity sweeteners have a common feature consisting of two electronegative atoms, designated **A** and **B**, separated by 2.5 to 4.0 Å, with a hydrogen atom attached to **A** thus giving a saporophoric **AH / B** unit.[29] This unit, which arises in sucrose from a hydroxyl group (**AH**) and the oxygen atom (**B**) of another hydroxyl group, interacts with a related feature on the protein of the taste buds, such as a hydroxyl or imino group as **AH** and a carbonyl group **B** of amino acid constituents such as serine or threonine.[30] The resulting pair of intermolecular hydrogen bonds is believed to give rise to the sweetness sensation. In addition a lipophilic factor **X** acts in harmony with the **AH / B** unit to guide and anchor the sweet molecules onto the receptor site (**45**), the more lipophilic molecules being sweeter by virtue of their stronger attachments to the taste bud receptors.

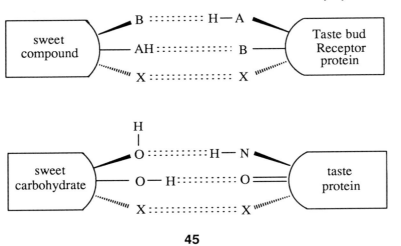

45

When the **AH / B / X** resides within a chiral molecule such as a sugar or amino acid there appears to be a further requirement in that the elements of the sweetness triad are in a clockwise arrangement when viewed from the point of contact with the receptor (**46**).[14,28]

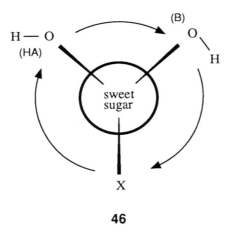

46

The sweetness of sucrose[14,28] has been attributed to the 2-hydroxyl (**B**) of the glucosyl unit and either the 1'- or 3'-hydroxyl (**AH**) of the fructosyl unit, consequently intensification of the sweetness of sucrose requires lipophilic groups to be inserted at other positions in the molecule ideally by replacement or substitution of hydroxyl substituents by groups of a similar size to avoid steric crowding on the receptor site. Significantly, NMR spectroscopy of sucrose solutions has revealed an equilibrium of two different conformations in which each has an intramolecular hydrogen bond, one from the 2-hydroxyl to the 1'-hydroxyl (47) and in the other to the 3'-hydroxyl (48), the former predominating.[31] There is an obvious relationship of these conformations to the **AH / B / X** glucophore and, clearly, the energy involved in hydrogen bonding to the receptor protein will be at a minimum apparently involving the simple transfer of the intra-hydrogen bond (48) to one that is intermolecular (45).[14]

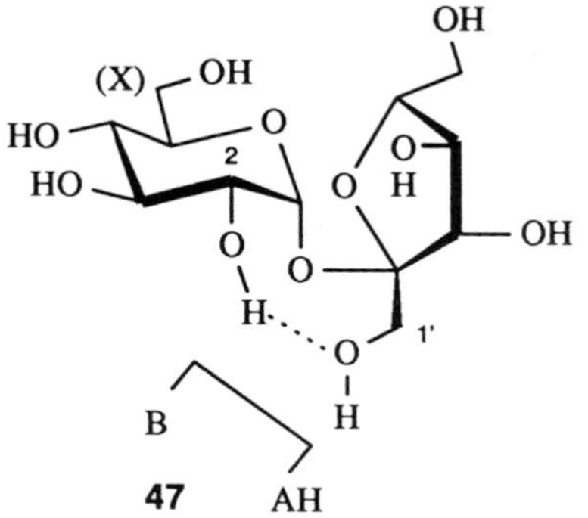

An opportunity to synthesize such lipophilic sucrose derivatives arose from chemical studies on the direct replacement of hydroxyl groups by chloro substituents, a reaction utilizing sulphuryl chloride-pyridine that was discovered in 1925 by Helferich et al.[32] Using this reagent they found that methyl α-D-glucopyranoside (49) gave a methyl 4,6-dichloro-4,6-dideoxyhexopyranoside 2,3-sulphate (50). More detailed investigations by Jones and his colleagues[33] revealed that, with minimal pyridine and chloroform as co-solvent, cyclic sulphate formation could be avoided. At -70 °C the first product was the 2,3,4,6-tetra(chlorosulphate) 51 which at 0 °C gave the 6-chloro-6-deoxy-2,3,4-tris(chlorosulphate) 52, then finally at room temperature 4-substitution occurred to give methyl 4,6-dichloro-4,6-dideoxy-α-D-galactopyranoside 2,3-bis(chlorosulphate) 53. The chlorosulphate groups were conveniently removed by treatment with methanolic sodium iodide yielding the methyl 4,6-dichloro-4,6-dideoxy-α-galactopyranoside (54).

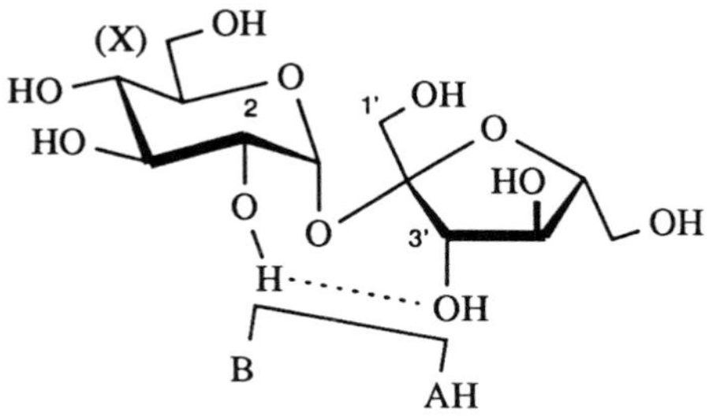

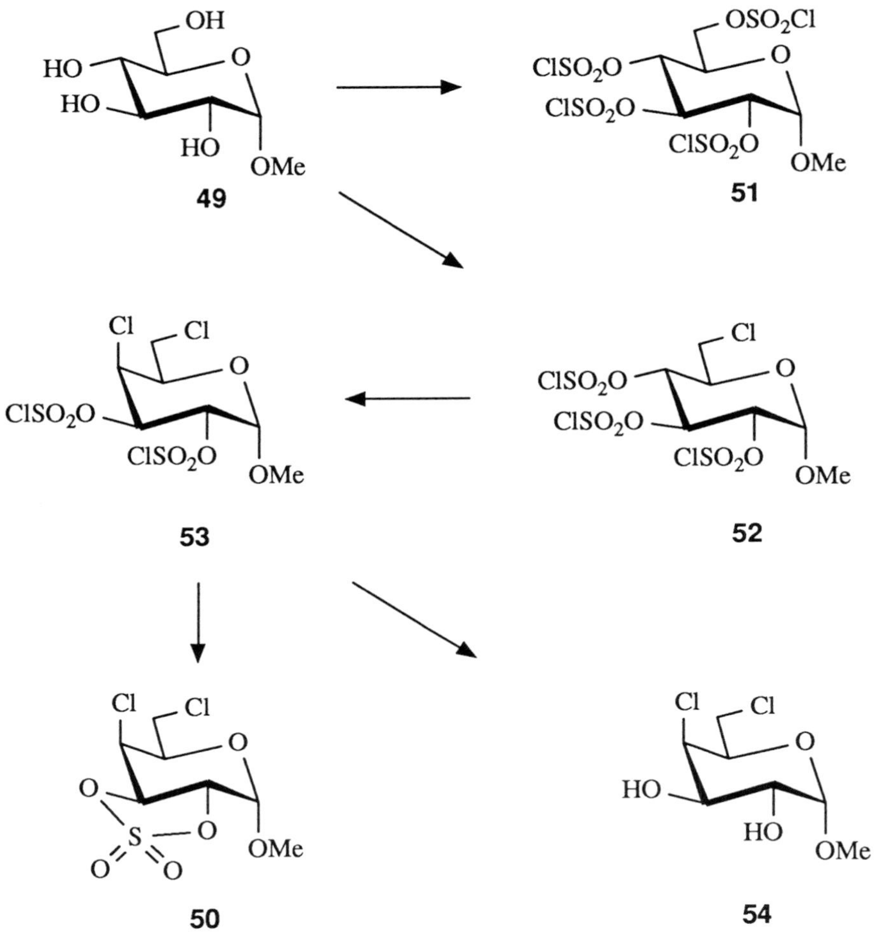

Application of these reactions to sucrose progressed in stages via the initial octa-chlorosulphate to the 6'-monochloride **55**, then the 6,6'-dichloride **56**, followed by the 4,6,6'-trichloride **57** and the 4,6,1',6'-tetrachloride **58** of *galacto*-sucrose.[9] Thus the regio-selectivity of substitution of chlorosulphonyloxy substituents in sucrose octa-chlorosulphate proceeds in the sequence C-6' > C-6 > C-4 > C-1'. Under more forcing conditions (50 - 70 °C) the 4,6,1',4',6'-pentachloride of *galacto*-sucrose (**59**) is obtained probably via the D-*lyxo*-3',4'-epoxide **60**.[34]

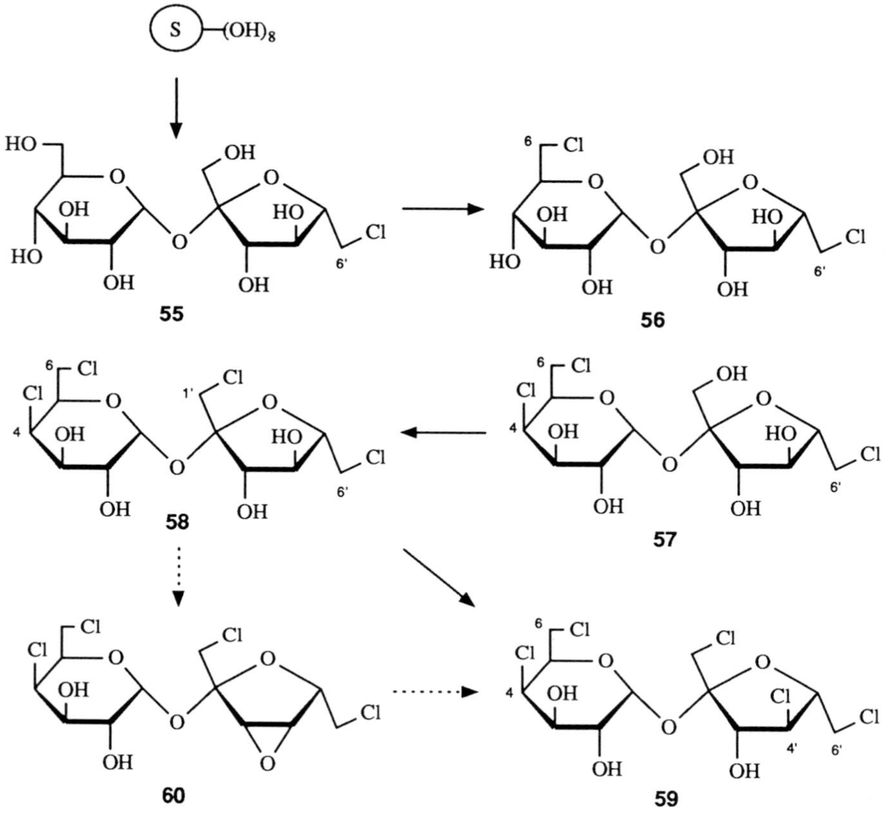

Taste studies then revealed that 4-chloro (**61**), 6'-chloro (**62**), and 1'-chloro (**63**) derivatives were 5x, 20x, and 20x sweeter, respectively, than their parent sucrose, thus suggesting that the lipophilic centres (**X**) had to be located on the upper face of the molecule to enhance its sweetness.[14] The sweetness of 1'-deoxysucrose and its 1'-chloro derivative **63** clearly favours the **AH** group of sucrose at the 3'-hydroxyl[14] which is supported by the lack of sweetness of 3-acetyl sucrose (**26**, R= CH$_3$) where it is blocked.[7]

A combination of two or three chloro groups at C-4, C-1' and C-6' had the effect of further increasing the lipophilicity of the molecule resulting in a dramatic increase in its sweetness. The 1',6'-dichloride **64** and 1',4-dichloride **65** were 70x and 120x as sweet as sucrose, respectively, and the 4,1',6'-trichloro-4,1'6'-trideoxy-*galacto*sucrose (**66**) was even greater at 650x.[14,35]

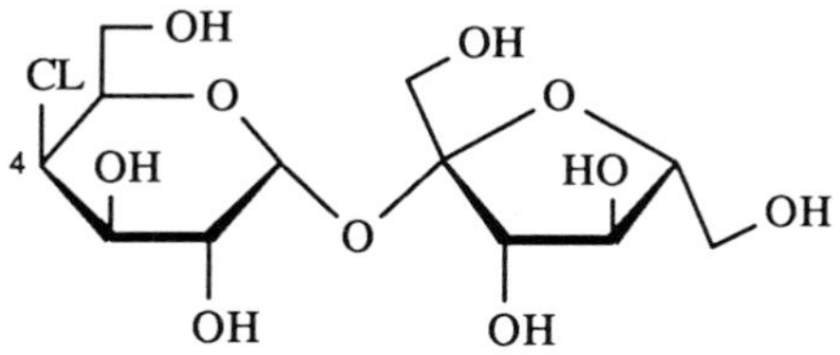

61

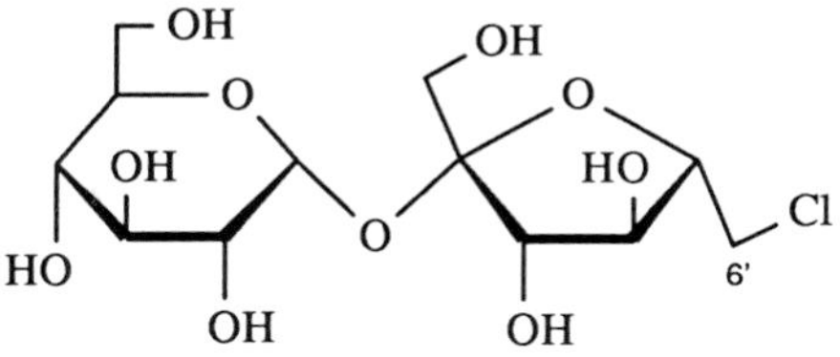

62

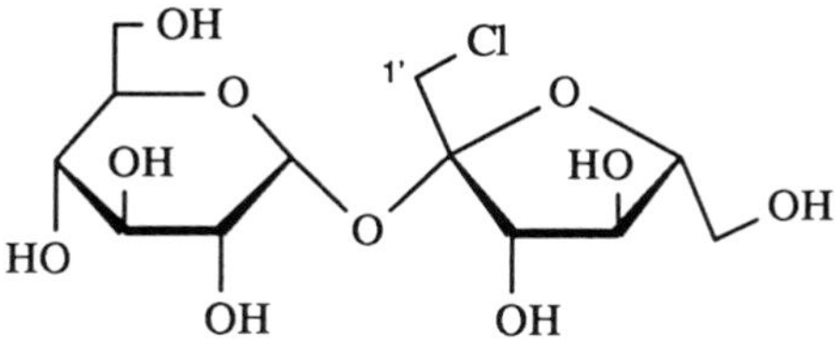

63

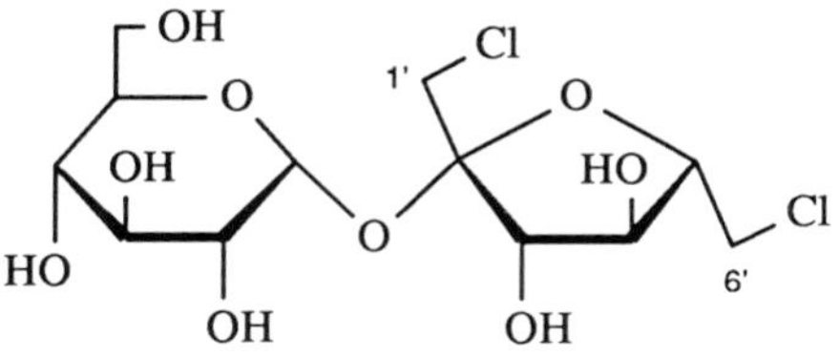

64

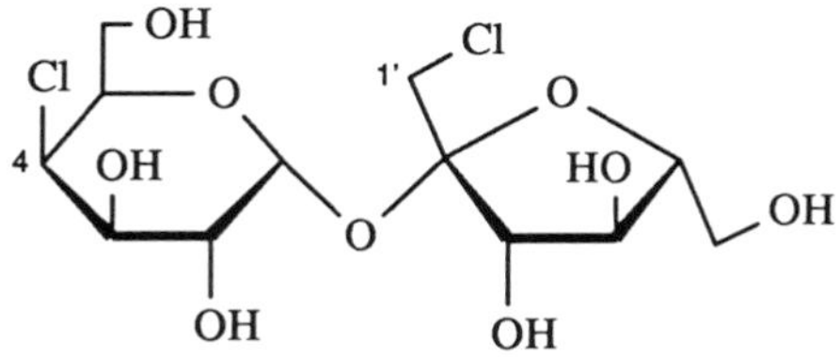

65

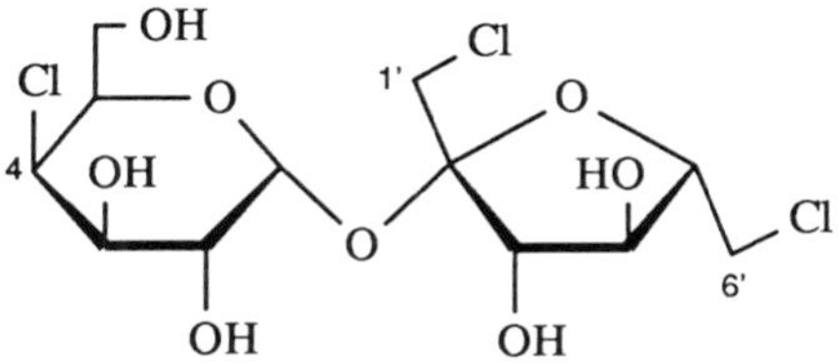

Sucralose   **66**

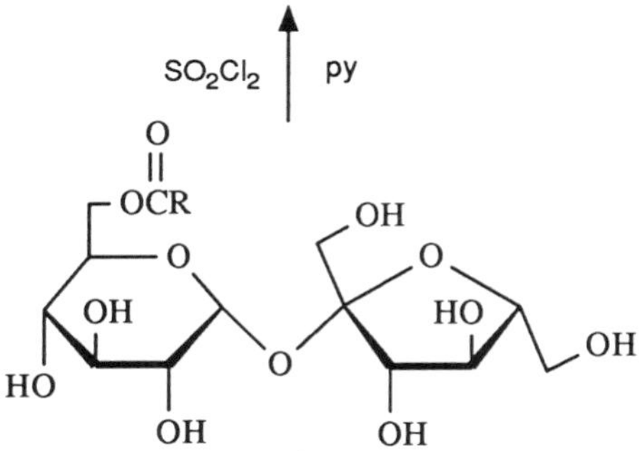

$$SO_2Cl_2 \quad py$$

4

Once this very sweet sucrose derivative **66** was found to have a similar taste profile to sucrose, to be non-toxic, non-nutritive, and 60 times more stable to acid hydrolysis than sucrose, it was promoted by Tate & Lyle (U.K.) and Johnson & Johnson (U.S.A.) as a high intensity sweetener termed "sucralose", and currently awaiting approval as a food and drink additive from the FDA and other Health Authorities.[36] Some idea of the size of market for this sweetener can be judged from the world wide sales of aspartame which jumped from $ 11 million in 1981 to $ 700 million in 1985.

Substitution of the 6-position of sucrose by chloride has an adverse effect upon sweetness as revealed by lack of taste of the 6,6'-dichloride 56 and the low sweetness of the 6,1',6'-trichloride 67, only 25x sweeter than sucrose.[14] On the other hand, an additional chloro group on the upper face of the molecule at the 4'-position, i.e. tetrachloride 68 resulted in a four-fold increase to 2200x sucrose.[34]

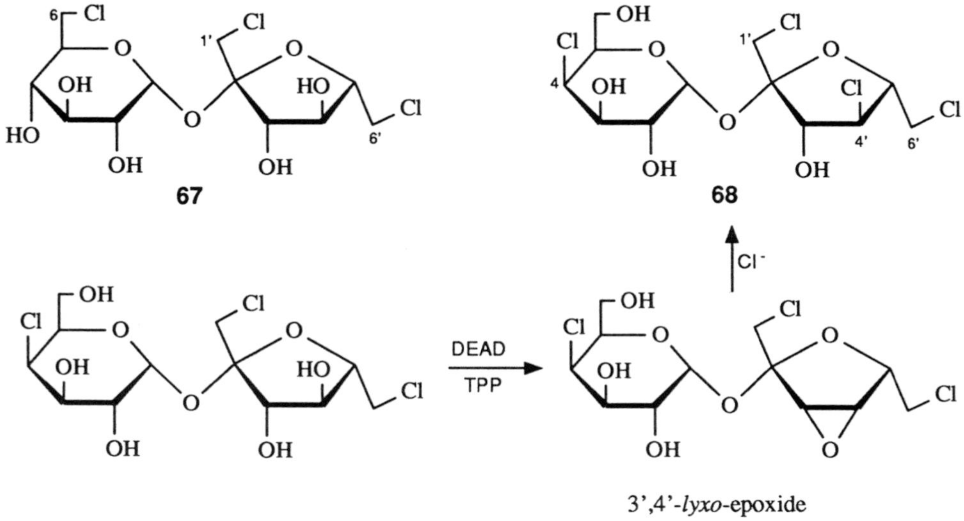

67

68

3',4'-*lyxo*-epoxide

Some Conclusions and Prospects

Chemical modification of the molecular structure of sucrose has revealed a fascinating arrange of new products, many with commercial applications. Those derivatives which have enhanced its natural sweetness, the monoesters that are utilized as emulsifiers and the polyesters that provide new fat and oil substitutes, each comprise a range of new products varying in physical and sensory properties which will stimulate further food and nutritional studies, building on sugar's major role namely, to improve and enhance the quality of food and drink, with or without the calories.

Acknowledgement. This lecture is dedicated to my loyal friend and colleague Dr. John L. HICKSON who as Vice-President and Director of Research of the I.S.R.F., stimulated and initiated this adventure into the chemistry of sucrose.

References

1. V. Kollinitsch (Ed.): Sucrose Chemicals. The International Sugar Research Foundation, Inc., Washington D.C., U.S.A. (1970).

2. J. L. Hickson (Ed.): Sucrochemistry. *A.C.S. Symposium Series No. 41*, Am. Chem. Soc., Washington, 1977.

3. M. Clark, Sugar y azucar **1989**, 24.

4. C. B. Purves, C. S. Hudson: Behaviour of Sucrose in Methyl Alcohol Containing Hydrogen Chloride. *J. Amer. Chem. Soc.* **56** (1934) 1973-1977.

5. J. Y. C. Chan, L. Hough, A. C. Richardson: The Synthesis of (*R*)- and (*S*)-Spirobi-1,4-dioxane and Related Spirobicycles from D-Fructose. *J. Chem. Soc. Perkin Trans. 1*, **1985**, 1457-1462.

6. K. J. Hale, L. Hough, A. C. Richardson: Morpholino-glucosides: New Potential Sweeteners Derived from Sucrose. *Chem. and Ind.* **1988**, 268-269; *Chem. Abstr.* **109** (1988) 36785t.

7. L. Hough, C. E. James, A. C. Richardson, unpublished results.

8. L. A. Edye, G. N. Richards, *Carbohydr. Res.* forthcoming publication.

9. C. E. James, L. Hough, R. Khan: Sucrose and its Derivatives. *Progr. Chem. Org. Natural products*, Springer-Verlag, Vienna / New York, **55** (1989) 117-184.

10. R. Khan, R. S. Mufti (Tate & Lyle): 4,1',6'-Trichloro-4,1',6'-trideoxy-*galacto*sucrose. *U.K. Pat.* 2,079,749 (1980); *Chem. Abstr.* **96** (1982) 163112j.

11. D. M. Clode, D. McHale, J. B. Sheridan, C. G. Birch, E. B. Rathbone: Partial Benzoylation of Sucrose. *Carbohydr. Res.* **139** (1985) 141-146.

12. M. S. Chowdhary, L. Hough, A. C. Richardson: Selective Esterification of Sucrose Using Pivaloyl Chloride. *J. Chem. Soc., Chem. Commun.* **1978**, 664-665.

13. M. S. Chowdhary, L. Hough, A. C. Richardson: Sucrochemistry, Part 33. The Selective Pivaloylation of Sucrose. *J. Chem. Soc., Perkin Trans. 1* **1984**, 419-427.

14. L. Hough, R. Khan: Enhancement of the Sweetness of Sucrose by Conversion into Chlorodeoxy Derivatives. *Progress in Sweeteners* (T. H. Grenby, Ed.), Elsevier Applied Science, London / New York, **1989**, 97-120.

15. T. Ogawa, M. Matsui: A New Approach to Regioselective Acylation of Polyhydroxy Compounds. *Carbohydr. Res.* **56** (1977) C1-C6.

16. E. Avela: Selective Substitution of Carbohydrate Hydroxyl Groups via Metal Chelates. *La Sucrerie Belge* **92** (1973) 337-344; *Chem. Abstr.* **80** (1974) 48248r.

17. H. B. Hass, F. D. Snell, W. I. C. York, L. I. Osipow (Sugar Research Foundation, Inc.): Sugar Esters. *U.S. Pat.* 2,893,990 (1959); *Chem. Abstr.* **53** (1959) 19422c.

18. W. J. Parker, R. A. Khan, K. S. Mufti (Tate & Lyle, Ltd.): Surface-active Product. *Brit. Pat.* 1,399,053 (1973); *Chem. Abstr.* **82** (1975) 100608r.

19. Y. Ando, H. Sunagawa, T. Tsuzuki, K. Kameyama: Effects of Sucrose Esters of Fatty Acids on the Growth of Spores of *Clostridium botulinum* and *Clostridium perfringens*. *Report of the Hokkaido Institute of Hygiene* **33** (1983) 1-7; *Chem. Abstr.* **100** (1984) 188595h.

20 F. H. Mattson, R. A. Volpenheim (Procter & Gamble Co.): Low-Calorie Fat-Containing Food Compositions. *U.S. Pat.* 3,600,186 (1968); *Chem. Abstr.* **75** (1971) 139614v.

21. M. Marek, K. Čapek, P. Musil, M. Ranny: Enzymic Purification of Non-Caloric Fat – Fatty Acid Esters and Sucrose Transesterification Products. *Abstracts Eurocarb V*, Prague (1989) D-20.

22. G. P. Touey, H. E. Davis (Eastman Kodak Co.): Mixed Esters of Glucose and Sucrose. *U.S. Pat.* 2,931,802 (1960); *Chem. Abstr.* **54** (1960) 16401c.

23. J. M. Ballard, L. Hough, A. C. Richardson: Sucrochemistry, Part IV. A Direct Preparation of Sucrose-2,3,4,6,1',3',4'-hepta-acetate. *Carbohydr. Res.* **24** (1972) 152-153.

24. L. Franzkowiak, J. Thiem: Synthesen von Agrocinopin A und B. *Liebigs Ann. Chem.* **1987**, 1065-1071.

25. a) K. Čapek, T. Vydra, M. Ranny, P. Sedmera: Structures of Hexa-*O*-acetyl-sucroses Formed by Deacetylation of Sucrose Octa-acetate. *Coll. Czech. Chem. Commun.* **50** (1985) 2191-2200.

    b) K. Čapek, T. Vydra, P. Sedmera: Structure of Penta-*O*-acetyl-sucroses Formed by Deacetylation of Octa-*O*-acetyl-sucrose. Relation of 2,3,4,6,6'-Penta-*O*-acetyl-sucrose. *Coll. Czech. Chem. Commun.* **53** (1988) 1317-1330.

    c) K. Čapek, T. Vydra: Oxirane-oxetane-1,4-dioxan Anhydro-ring Migration in Sucrose Derivatives. *Carbohydr. Res.* **168** (1987) C1-C4.

26. P. Musil, K. Čapek, M. Marek, P. Sedmera: Enzymic Hydrolysis of α-D-Glucopyranosyl-3,4-anhydro-β-D-*lyxo*-hexulofuranoside. *Abstracts of Eurocarb V*, Prague 1989, C-59.

27. a) K. C. Frisch, J. E. Kresta: An Overview of Sugars in Urethanes. In reference 2, p. 238.

b) A. R. Meath, L. D. Booth: Sucrose and Modified Polyols in Rigid Urethane Foam. In reference 2, p. 257.

28. L. Hough: Sucrose, Sweetness and Sucralose. *Int. Sugar J.* **91** (1989) 23-31, 35, 37; *Chem. Abstr.* **110** (1989) 191307c.

29. R. S. Shallenberger: Sweetness and Sweeteners (G. G. Birch, Ed.). Applied Science Publishers, London, 1971, 47.

30. T. Suami: Synthetic Ventures in Pseudo-sugar Chemistry. *Pure Appl. Chem.* **59** (1987) 1509-1520.

31. D. B. Davis, J. C. Christophides: Comparison of Intramolecular Hydrogen-bonding Conformations of Sucrose-containing in Solution and the Solid State. *Carbohydr. Res.* **163** (1987) 269-274.

32. B. Helferich, G. Sprock, E. Besler: Über ein D-Glucose-5,6-dichlorohydrin. *Ber. Dtsch. Chem. Ges.* **58** (1925) 886-891.

33. W. Szarek: Deoxyhalogeno Sugars. *Adv. Carbohydr. Chem.* **28** (1973) 225-306.

34. C. K. Lee: Synthesis of an Intensely Sweet Chlorodeoxy-sucrose: Mechanism of 4'-Chlorination of Sucrose by Sulfurylchloride. *Carbohydr. Res.* **162** (1987) 53-63.

35. L. Hough, S. P. Phadnis, R. Khan, M. R. Jenner: Chlorine Derivatives of Sucrose. *Brit. Pat.* 1,543,167 (1977); *Chem. Abstr.* **87** (1977) 202019v.

36. M.R. Jenner: Sucralose; Unveiling its Properties and Applications. *Progress in Sweeteners* (T.H. Grenby, Ed.), Elsevier Applied Science, London / New York, **1989**, 121-141.

3

New Developments in the Use of Sucrose as an Industrial Bulk Chemical

Hubert Schiweck, Mohammed Munir, Knut M. Rapp,
Bernd Schneider, and Manfred Vogel

Zentrallaboratorium, Südzucker AG Mannheim / Ochsenfurt,
D-6718 Grünstadt, Germany

Summary. Isomaltulose (Palatinose[®]) and Isomalt (Palatinit[®]) are not only versatile food ingredients, they can also be used as chemical feed stocks for speciality products. Palatinose can be derived directly by enzymatic conversion of sucrose using an immobilized enzyme as biocatalyst. Hydrogenation of isomaltulose leads directly to isomalt.

Fatty acid esters of sucrose, previously known as sucrose polyesters (SPE), have emerged as a potential non-adsorbable substitute for fats and oils in food. Research has shown that, unlike triglycerides, sucrose esters cannot be hydrolyzed by the human corporeal enzymes. Fructo-oligosaccharides comprising mainly 1-kestose, nystose, and 1-fructofuranosyl-nystose can be manufactured by enzymatic action of fructosyltransferase on sucrose. Treatment of inulin-containing roots and tubers, e.g. chicory or Jerusalem artichoke with endo-inulinase leads to soluble fructo-oligosaccharides with a degree of polymerization up to 7. Both products possess very similar properties. They are low caloric and are promoters of bifidobacteria. They have therefore a beneficial effect on humans and animals.

5-Hydroxymethyl-furfural (HMF) is a sugar derivative, which can compete with chemicals based on petrochemistry. As a key substance between carbohydrate chemistry and today's industrial organic chemistry, HMF may widen the field of non-food uses of sugars. It is best prepared via fructose either from sucrose or inulin.

Levoglucosan, one of the simplest derivatives of glucose, is prepared from starch in a pyrolysis process followed by ion exchange chromatography and a crystallization step. It can be transformed into tailor-made dextran derivatives or into complex biologically active carbohydrates.

1. Production of Isomaltulose and Isomalt (Palatinit®)

In 1957 Weidenhagen and Lorenz[1,2] reported on the conversion of sucrose to an unknown reducing disaccharide through the action of a bacterium they had isolated from sugar beet raw juice.[3] This disaccharide was identified as isomaltulose and given the trivial name palatinose® (derived from "Palatinum", the Latin name of the German province where the disaccharide was found).[1,2] The first process for the production of isomaltulose was subsequently patented for Süddeutsche Zucker AG, Mannheim in 1959.[4]

The bacterium effecting the enzymatic transformation of sucrose to isomaltulose was identified as *Protaminobacter rubrum* (den Dooren de Jong) by Windisch:[5]

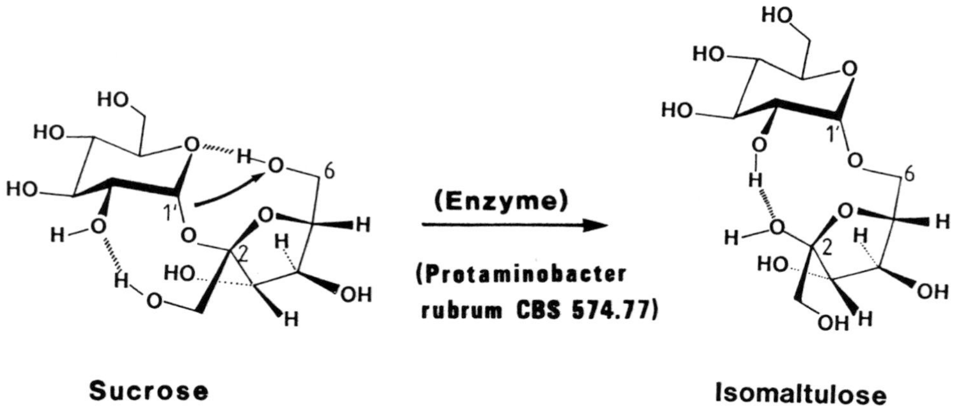

(Enzyme)

(Protaminobacter rubrum CBS 574.77)

Sucrose   **Isomaltulose**

Some other microorganisms have also been reported to be capable of transforming sucrose into isomaltulose. The most important of these are: *Leuconostoc mesenteroides*,[6] *Serratia plymuthica*,[7] *Serratia marcescens*,[7] *Erwinia carotovera*,[7,8] and *Erwinia rhapontici*.[9]

The development of a continuous fermentation process for *Protaminobacter rubrum* CBS 574.7712[10] in 1976, comprising a single-stage simultaneous cell-propagation and substrate-conversion, opened the possibility of an industrial scale production of isomaltulose. The availability of advanced immobilization technology made it possible to develop processes for the conversion of sucrose to isomaltulose with the help of immobilized cells of isomaltulose-forming microorganisms.[11-15] At least two of these processes are already being employed on an industrial scale.

From the microorganisms reported to be capable of converting sucrose to isomaltulose,[4,6-15] only *Protaminobacter rubrum* CBS 574.77, the strain found originally by Weidenhagen and Lorenz,[1-3] seems to have reached industrial scale use.[16,17] Processes using living and propagating free microorganism cells[10,18] have the disadvantage of higher product purification cost and lower yield.

Production of Isomaltulose

Food-grade isomaltulose as well as that suitable for conversion to isomalt can be produced in a process comprising essentially of an enzymatic conversion of a sucrose-containing solution to a solution containing mainly isomaltulose followed by a purification step. Such a production process is schematically described in Fig. 1.

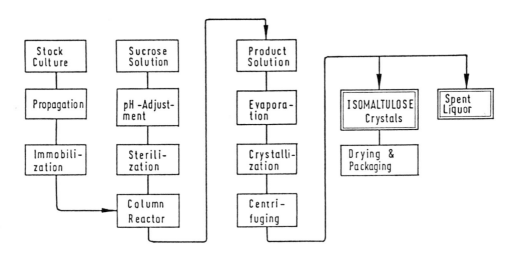

Fig. 1. Flow sheet of the production of isomaltulose from sucrose using immobilized cells.

Stock Culture: The culture maintenance of *Protaminobacter rubrum* CBS 574.77 can be carried out either by cryopreservation or by lyophilization. Subcultures can be maintained in the laboratory at 4 °C on agar slants containing beet or cane thick juice (an intermediate product from a sugar factory). These subcultures have to be tested and renewed regularly at intervals of about 4 weeks.

Propagation: A growth medium based either on thick juice supplemented with corn steep liquor or on molasses with an additional nitrogen and phosphate source e.g. $(NH_4)_2HPO_4$, adjusted to a concentration of 5 % total solids and pH of 7.2 is heat sterilized and used for the propagation of *Protaminobacter rubrum* cells.

Immobilization: Essentially all of the cell immobilization methods described in the literature can be used for the immobilization of whole cells of *Protaminobacter rubrum* or other isomaltulose producing microorganisms. It must however be taken into consideration that the enzyme sucrose $\rightarrow$ isomaltulose mutase is sensitive towards glutaraldehyde. Therefore practically only two methods, working under mild conditions, have so far been employed for industrial scale production of immobilized *P. rubrum* CBS 574.77 cells.[13,19] Although the isolation and immobilization of the enzyme itself, rather than whole cells is also possible,[14,15] it has not been found to be feasible yet, the immobilization of whole cells being more economical.

Sucrose Solution: The literature gives a wide range of concentrations and purities for the sucrose solution to be used as substrate. Thus it would be a matter of individual choice to use the concentration and purity as that most suited to individual needs. The sugar solution as well as the biocatalyst being excellent nutrients, the reaction has to be carried out under conditions which would exclude the contamination with foreign microorganisms.

Column Reactors: To ensure aseptic operation the column reactors must be designed, built and run according to guidelines applicable to fermenters. Since the density of the immobilized cells is rarely going to be very different from that of the sucrose solution both upflow and downflow operation is possible. In actual practice however, downflow operation has been found to give more stable results.

Product Solution: If the column reactors are run so as to convert almost all of the sucrose supplied, then the average composition of the product solution running from the reactors is as follows:

Table 1. Composition of product solution

Fructose	2.5 – 3.5 %	on total solids		
Glucose	2.0 – 2.5 %	"	"	"
Sucrose	0.5 – 1.0 %	"	"	"
Isomaltulose	79.0 – 84.5 %	"	"	"
Trehalulose	9.0 – 11.0 %	"	"	"
Isomaltose	0.8 – 1.5 %	"	"	"
Higher homologues	0.7 – 1.5 %	"	"	"

The generation of sugars other than isomaltulose does not seem to be a question of impurities in the immobilized cells; even the purified enzyme gives a product spectrum very similar to the one given above. On the other hand operating parametes like substrate concentration, temperature, mean residence time, residence time distribution and, to some extent, the method of immobilization do influence the product composition. The extent of byproduct generation determines not only the yield but also the extent of purification required.

Evaporation and Crystallization: Crystallization from aqueous solutions is the most efficient method for the isolation of pure isomaltulose from the product solution coming from the column reactors. Although conventional evaporation technology can be applied to concentrate isomaltulose solutions, one has to bear in mind that due to its fructose moiety isomaltulose is rather heat sensitive. Conventional equipment and technology available from crystallization in motion can also be applied for the crystallzation of isomaltulose. Separation of isomaltulose crystal-magmas on wire-basket centrifuges yields isomaltulose crystals and spent liquor. The isomaltulose can be dried in conventional drying equipment, e.g. rotary drum dryers. The drying and packaging of isomaltulose is however necessary only if it is to be stored prior to the further use. If the isomaltulose is to be processed to isomalt on the spot, then the moist crystals from the centrifuges can be used directly.

Properties of Isomaltulose: Isomaltulose is a free-flowing, non-hygroscopic crystalline substance. It crystallizes easily from aqueous solutions with 1 mol water of crystallization so that the moisture content ranges from 5.0 to 5.05 %. The following table shows a comparison of the properties of isomaltulose with those of sucrose.

Table 2. Comparison of the properties of sucrose and isomaltulose

	Sucrose	Isomaltulose
Sweetening power	100	42
Sweetening character	round, balanced	neutral
Melting point (range)	160-185 °C	123-124 °C
$[\alpha]_D^{20}$	+ 66.5°	+ 103°
Solution enthalpy	- 18.2 kJ/kg	- 21.7 kJ/kg
Cooling effect on dissolving	none	none
Solubility at 20 °C	~ 2 g/g water	0.49 g/g water
Hygroscopicity in powder	low	very low
Viscosity in solution	low	low
Browning reactions	+	+
Calories / g DS	4	4
Suitability for diabetics	-	+
Suitability for teeth	-	+

Isomaltulose shows a neutral sweetness without any after-taste, its sweetening power being about 42 % of that of sucrose. With -21.7 kJ/kg the solution enthalpy is of the same order as that of sucrose and alike it doesn't cause a sensation of cold on dissolving in the mouth. The solubility at 20 °C is about one fourth of that of sucrose, but the viscosity of isomaltulose and sucrose solutions is of the same order.

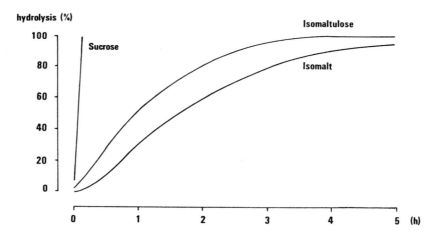

Fig. 2. Hydrolysis of Sucrose, Isomaltulose, and Isomalt in 1 % HCl at 100 °C.

Isomaltulose is hydrolyzed and absorbed in the small intestine; thus its energetic value of 4 kcal/g equals that of sucrose. On the other hand, the absorption rate in the small intestine is lower compared to sucrose, it may thus be regarded suitable at least for some types of diabetics. Since isomaltulose can neither be converted to acids nor polymerized by the mouth flora, it is a suitable, tooth-sparing sweetener. Due to its fructose moiety being free at the anomeric center, isomaltulose exhibits stronger browning reactions than sucrose, yet as shown in Fig. 2, is more stable towards acid hydrolysis.

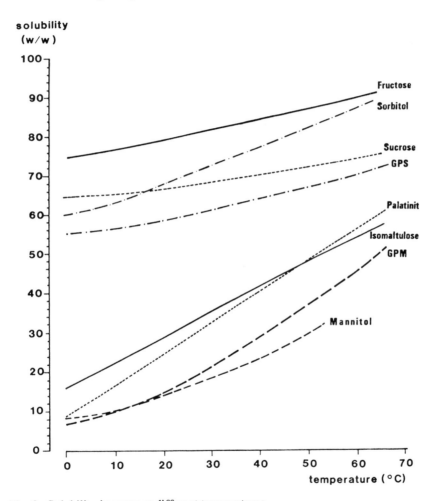

Fig. 3. Solubility in water at different temperatures

The solubility of a saccharide is one of the major parameters important not only for the production process but more so for its applications. The saccharides differ vastly in solubility but till now no explanation as to a possible relationship between structure and solubility behaviour has been put forward.

Perhaps a number of tautomers in which a saccharide may exist in, is related to its solubility. Thus, as shown in Fig. 3, fructose with five tautomeric forms has a substantially higher solubility than isomaltulose which can only form three tautomers. The hydrogenation products of isomaltulose, i.e. GPM and GPS (cf. below) are not capable of forming tautomers and, interestingly, exhibit no marked solubility differences as compared to mannitol and sorbitol.

Isomaltulose is a sugar with special properties; it can be used as a bulk chemical, and it is being used as feed-stock for the production of isomalt.

Production of Isomalt

With the hydrogenation of isomaltulose to isomalt, Südzucker[16] could produce a low-caloric sweetening agent with the bodying and texture-giving properties of sucrose, with enhanced stability, and without such drawbacks as cariogenicity and insulin dependancy. Isomalt (synonym: Palatinit®) is an equimolar mixture of 6-O-(α-D-glucopyranosyl)-D-sorbitol ("GPS") and 1-O-(α-D-glucopyranosyl)-D-mannitol ("GPM"):

1-O-(α-D-glucopyranosyl)-D-mannitol
(" GPM ")

6-O-(α-D-glucopyranosyl)-D-sorbitol
(" GPS ")

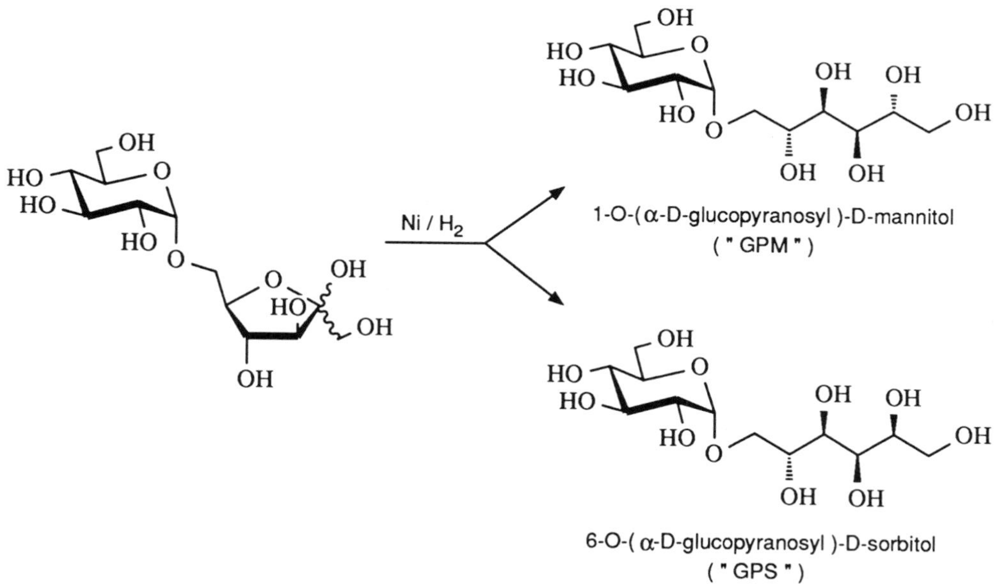

It is prepared by hydrogenation of isomaltulose according to the following process:

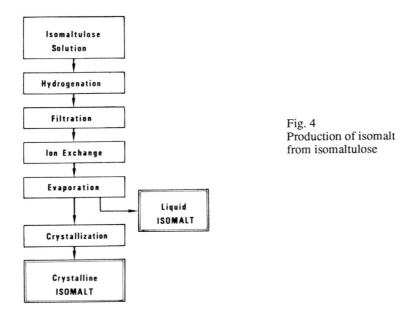

Fig. 4
Production of isomalt
from isomaltulose

Hydrogenation: The isomaltulose solution can be hydrogenated easily at elevated temperature and pressure on any of the usually employed hydrogenation catalysts. A suspension of Raney-Nickel is quite suitable for this purpose, although this catalyst has also been employed in pelletized form in a fixed-bed reactor.[20] This latter process is specially suited for continuous operation. The hydrogenation can be carried out at essentially neutral pH or under mildly alkaline conditions. Theoretically the reduction of isomaltulose should yield an equimolar mixture of 1-*O*-(α-D-glucopyranosyl)-D-mannitol and 6-*O*-(α-D-glucopyranosyl)-D-sorbitol. Depending on the conditions of hydrogenation the proportion of each component, however, can vary between 43-57 %.

Filtration and Ion Exchange: Traces of nickel may go into solution during the hydrogenation as some degree of acid formation cannot be totally excluded. This nickel together with the anions must be removed completely from the solution prior to crystallization. This purification can be carried out through an ion exchange treatment with strong acid cation and medium to strong base anion exchangers.

Evaporation: The isomalt solution coming from the ion exchange purification can either be sold as liquid isomalt or sent to the crystallization step. In either case it has first to be evaporated to the appropriate concentration (for solubility data cf. ref.[18]).

The evaporation is at best carried out in a multi-stage falling film evaporator operating at reduced pressure.

Crystallization: Isomalt crystallizes readily from super-saturated aqueous solutions. The GPM part of isomalt crystallizes with 2 mols of crystal water[21] whereas GPS crystals are anhydrous.[22] This results in a water content of the order of 5 % in crystalline isomalt. Since the crystal water of GPM is integrated into the molecule via hydrogen bridges (Fig. 5), it is difficult to produce anhydrous GPM. For the technical crystallization of isomalt a special equipment, combining the functions of a vacuum evaporator, crystallizer, and a vacuum dryer, is used.

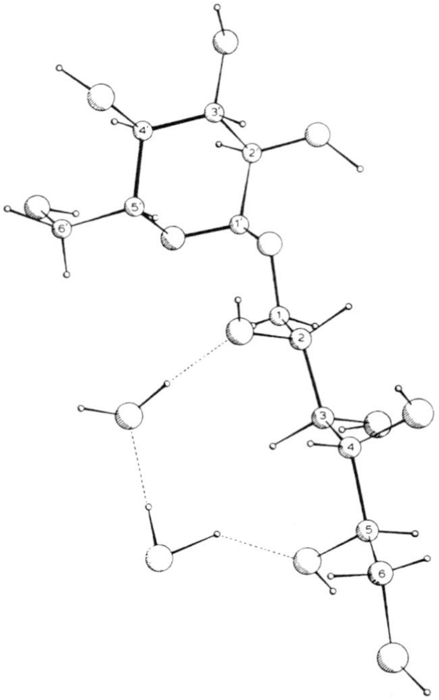

Fig. 5
X-Ray crystal structure
of GPM

Properties of Isomalt: Isomalt is a low-calorie, tooth-sparing alternative sweetener, suitable for diabetics. It compares very favourably with sucrose. The main features of comparison are listed in Table 3.

Table 3. Comparison of the properties of sucrose and isomalt

	Sucrose	Isomalt
Sweetening power	100	45-60
Sweetening character	round, balanced	neutral
Melting point (range)	160-185 °C	145-150 °C
Solution enthalpy	- 18.2 kJ/kg	- 39.4 kJ/kg
Cooling effect on dissolving	none	low
Solubility at 20 °C	~ 2 g/g water	0.33 g/g water
Hygroscopicity in powder	low	very low
Viscosity in solution	low	low
Browning reactions	+	-
Calories / g DS	4	2
Suitability for diabetics	-	+
Suitability for teeth	-	+

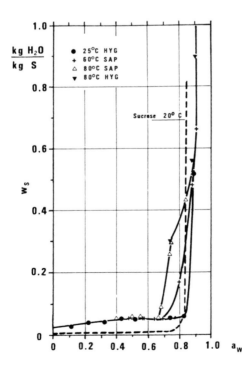

Fig. 6. Adsorption isotherms of isomalt at different temperatures

The sorption isotherms in Fig. 6 show that isomalt is practically non-hygroscopic. Therefore isomalt is the ideal feedstock for the production of hard boiled candy and for coating surfaces.

Many isomalt sweetened products, particularly sweets and chocolates are being produced today in Japan and Switzerland. In Germany, the nation-wide introduction of isomalt products is scheduled for the summer of 1990.

2. Fatty Acid Esters of Sucrose

Since 1948 basic research has been done by Mattson and Volpenhein[23,24] on the absorption of fats. Thinking that esters of alcohols above three carbon atoms might be more easily digested, they analyzed the ability of intenstinal lipases to cleave fatty acid esters of various carbohydrates. They found, however, that hydrolysis stopped when fully esterified polyols reached six carbon atoms. Even at C-4 there is only a low rate of hydrolysis, as shown in Fig. 7.

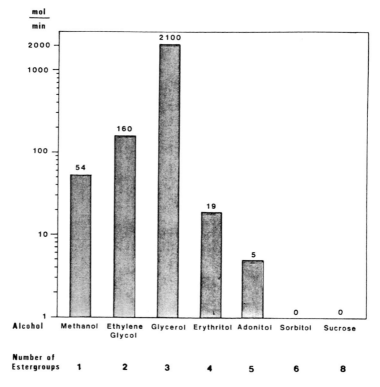

Fig. 7. Hydrolysis of fatty acid esters[23,24]

It can be seen that corporeal enzymes almost selectively attack triglycerides as the most common fat components occuring in Nature. Less than 10 % of that activity is found on the hydrolysis of esters of glycol or methanol. Esters of sorbitol and sucrose for instance are not hydrolyzed at all. Fatty acid esters of sucrose, commonly known as sucrose polyesters (SPE), are represented by the formula:

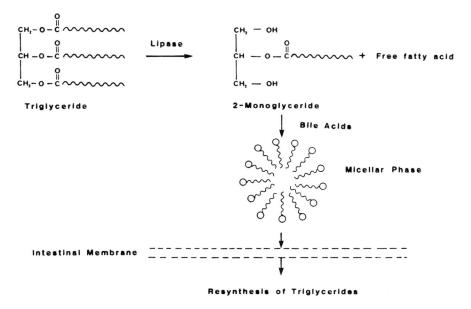

Sucrose polyesters (SPE)

R = H
or
$R = CH_3 - (CH_2)_n - C -$
(n = 6 to 20)

(molecules contain 6, 7 or 8 fatty acid residues)

Sucrose esters of this type have been given the brand name Olestra® by Procter & Gamble in 1986. To understand the reason why SPE are not absorbed, it is necessary to give a short review about the mechanism of triglyceride absorption in the small intestine.

Triglyceride

Lipase

2-Monoglyceride + Free fatty acid

Bile Acids

Micellar Phase

Intestinal Membrane

Resynthesis of Triglycerides

Fig. 8. Mechanism of triglyceride absorption[25-27]

It is well known today that triglycerides must first be hydrolyzed by pancreatic lipase to fatty acids and 2-monoglycerides, before they can be absorbed, as is schematically depicted in Fig. 8.

After partial hydrolysis, these two compounds form a micellar phase with the bile acids. After diffusion through the cell wall of the intestinal mucosa a re-synthesis into triglycerides is possible, followed by the transport to the liver.[25-27] The important thing is that both the triglyceride and SPE molecules, are excluded from the micellar phase because they are bulkier and more lipophilic.

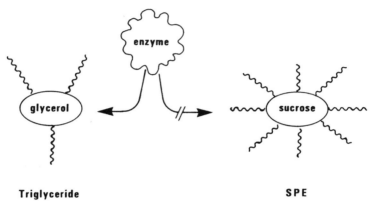

Triglyceride   SPE

Fig. 9. Why SPE is not hydrolyzed

Fig. 9 illustrates by a simple model, that a triglyceride allows an attack by the enzyme from the side of its classic Y-configuration and an SPE does not so due to its bulky appearance. So, due to the fact that SPE are not hydrolyzed by lipases they are not absorbed and provide no fat and zero calories to the diet. Similar to the physical properties of triglycerides which are controlled by the nature of the fatty acid molecules, particularly chain length and degree of unsaturation, the properties of SPE can be tailored during preparation, so that taste, appearance, aroma and lubricity are nearly indistinguishable from certain triglycerides. Especially the rheology of fats at body temperature is of consierable importance for the mouth feeling when used, e.g. as spread or in chocolates. This is shown in Table 4, where the melting ranges of some natural fats are given.[28]

Table 4. Melting ranges of some natural fats

Natural fat	M.P. (°C)
Coconut oil	20 - 28
Palm kernel oil	25 - 30
Cocoa butter	32 - 36
Palm oil	27 - 43
Butterfat	28 - 38
Beef tallow	40 - 50
Mutton tallow	44 - 55
Lard	28 - 40
Goose fat	32 - 34
Chicken fat	30 - 32
Butter	33.5
Margarine	27 / 29 / 31 / 35

Only those fats melting below body temperature and not being liquid at room temperature impart a pleasant perception to the tongue. Tallows with melting points a few degrees higher are not very much preferred by the human taste.

For instance a great deal of know-how is necessary for the correct handling of chocolate mass, where pre-crystallization of higher melting compounds by defined tempering and cooling processes is of great significance for the quality, especially for a brilliant surface and a soft melting on the tongue. As an example of the composition, Procter & Gamble patented a cocoa butter substitute made from sucrose containing about 35 % lauric acid, about 60 % palmitic acid, and about 5 % other fatty acids, preferably capric, myristic and/or stearic acid.

SPEs are prepared by a number of patented processes starting with edible triglycerides or fatty acids and sucrose. Many of the procedures described use organic solvents, such as dimethylformamide (DMF), dimethylsulfoxide (DMSO), various glycols, esters, ketones, and alcohols. But there are also a few solventless processes, for instance in the U.S. the USDA process and a method developed in Germany, known as the *Zimmer* method, which works with sucrose in the molten state. The method used by Procter & Gamble to prepare Olestra® is shown in Fig. 10. The triglyceride is converted into fatty acid methyl esters through alkali-catalyzed transesterification in methanol. These methyl esters are used to esterify sucrose, thereby liberating methanol.

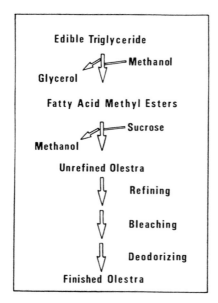

Fig. 10
Preparation of Olestra®
(Procter & Gamble)

The resulting Olestra® is then purified by refining, bleaching, and deodorizing in the same way as full-caloric fats and oils. As further prospective manufacturers of SPE Unilever, Mitsubishi, and Rhóne-Poulenc should be mentioned.[29,30] The main advantage of SPE, besides their application for calorie-reduced foods, is their depressing effect on the cholesterol level so that they could reduce the risk of coronary heart disease. This results from the fact that cholesterol is one of the most lipid-soluble materials in the diet and is carried out of the body by dissolving in the non digestible sucrose esters. Procter & Gamble has applied for FDA approval of Olestra® for use as a food additive in the U.S. and in the U.K.. In 27 clinical tests sucrose esters have been tested by about 1 800 persons. As the main disadvantages researchers have noted some gastro-intestinal side effects on larger intakes and a loss of vitamin E and other fat-soluble vitamins.

As a result of further development on the composition, gastro-intestinal side effects have now been diminished and nutritional concerns have been removed with vitamin E supplementation. At this time, Olestra® still awaits FDA regulatory approval, which is estimated to take at least two more years.

3. Fructo-oligosaccharides

Fructo-oligosaccharides belong to the group of fructans, which are storage polymers of fructose synthesized by at least 36 000 species of higher plants representing 10 families.[31] Three types of fructans can be defined: levan, branched fructans, and inulin. Fructo-oligosaccharides are found in Nature, e.g. in asparagus, onions, Jerusalem artichoke, and in chicory. Today large scale production of mixtures of fructo-oligosaccharides has become possible using one of the two processes known. They shall be described here.

In the first process fructo-oligosaccharides are synthesized by the enzymatic action of fructosyltransferase on sucrose. The product from this process, which was developed first by the Japanese company Meiji Seika, carries the trade name "Neosugar". This is a mixture of glucose, sucrose, and $\beta(2\rightarrow1)$-linked fructo-oligosaccharides with a terminal glucose unit.

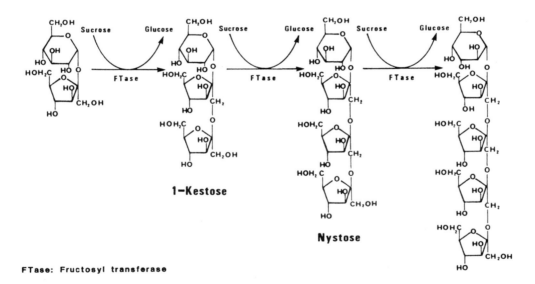

FTase: Fructosyl transferase

Fig. 11. Enzymatic preparation of fructo-oligosaccharides by fructosyl-transfer to sucrose

Fig. 11 shoes that the enzyme fructosyltransferase transfers the fructosyl part of the sucrose molecule on to the acceptor sucrose molecule. As 1-kestose and nystose also act as acceptor molecules, the final product contains 1-kestose, nystose, and fructosyl-nystose. Oligosaccharides with longer chains than DP 5 are produced in minor amounts only. For large scale production immobilized fructosyl-transferase is used, i.e. cells of *Aureobasidium pullulans var. melanigenum* or *Aspergillus niger* containing this enzyme are propagated and immobilized in calcium alginate as described under isomaltulose. The transfer reaction is performed at 50-60 °C with a 60 % (w/v) sucrose solution at a pH of 5.5 to 6.0.[32] A chromatogram of a typical unrefined product solution is shown in Fig. 12.

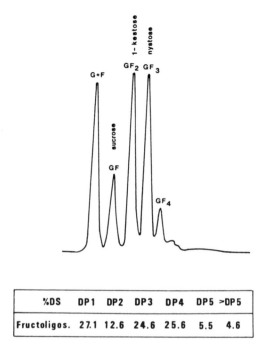

%DS	DP1	DP2	DP3	DP4	DP5	>DP5
Fructoligos.	27.1	12.6	24.6	25.6	5.5	4.6

Fig. 12. Fructo-oligosaccharide mixture: composition of an unrefined product

The fructo-oligosaccharides 1-kestose and nystose at about 25 % on total solids each, are the main components whereas more than 27 % of dry substance is composed of monosaccharides, mainly glucose originating from sucrose. The solution contains nearly 13 % residual sucrose and minor amounts of fructosyl nystose and higher molecular weight components. The unrefined product is subjected to the usual decolourization, demineralization, and concentration process, to obtain the so-called Neosugar G. To get a product with a higher amount of fructo-oligosaccharides the glucose and sucrose is removed by ion exchange chromatography similar to that described for the fructose syrup production.[33]

Another process for the production of fructo-oligosaccharides starts with inulin. In this case the linear $\beta(2\rightarrow1)$-linked fructose polymers built on a sucrose residue are partially hydrolyzed by the enzyme endo-inulinase. Fig. 13 shows, that the enzymatic inulin hydrolysis yields two types of fructo-oligosaccharides: hetero-oligomers, identical with those mentioned above for the Neosugar and homo-oligomers composed of $\beta(2\rightarrow1)$-linked-fructose molecules.

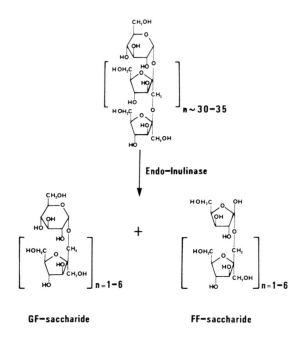

Fig. 13. Enzymatic preparation of inulo-oligosaccharides by splitting inulin

As shown by chromatographic separation, in Fig. 14, the product by this process contains more fructo-oligosaccharides with longer chains (up to DP 7) than the Neosugar product; the amount of the mono- and disaccharides is also lower and inulobiose, a component of interest, is included besides sucrose.

The large scale production starts from chicory roots or Jerusalem artichoke tubers. There are two possibilities of processing the plant material. In the classical manner the inulin is extracted out of the plant material and submitted to a juice purification very similar to that used for the sucrose production from sugar beet.

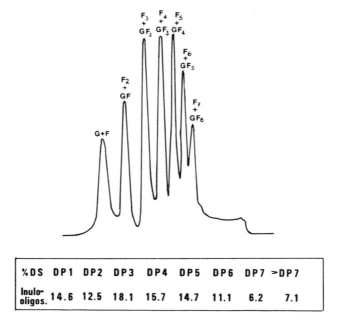

%DS	DP1	DP2	DP3	DP4	DP5	DP6	DP7	>DP7
Inulo- oligos.	14.6	12.5	18.1	15.7	14.7	11.1	6.2	7.1

Fig. 14. Inulo-oligosaccharide mixture: composition of an unrefined product

In a simpler process, developed by Südzucker (patent pending), the plant material is submitted directly after a maceration and pasteurization step to the endo-inulinase treatment at 56 °C and a pH of 5.4. The amount of enzyme and the reaction time depends on the desired product composition. For the case shown in Fig. 14 the enzyme dosage was 2 units/g inulin and the reaction was stopped after 16 hours by adjusting the pH to 10.7 with milk of lime.

After separation from the solids the liquid phase was pre-concentrated, decolourized, demineralized, and concentrated finally to a syrup which could be spray-dried. A product free of glucose, fructose, and sucrose can be obtained by ion exchange chromatography either directly or following a specific enzymatic splitting of the sucrose, the second way giving higher yields.

The most important characteristics of fructo-oligosaccharides, apart of their natural occurance, are that they are

- sweet
- non-digestible in the stomach and small gut
- fibre soluble
- fermentable by intestinal microorganisms
- selectively utilizable by bifidobacteria
- suppressing production of intestinal putrefactive substances.

Inulin, for example, is consumed to about 6 g/d in southern European countries, possessing a sweetness about 0.4 to 0.6 times of that of sugar. Furthermore fructo-oligosaccharides are scarcely digestible in the alimentary tract.[34] In vitro studies also showed that they are not hydrolyzed by the glucoamylase / maltase- and saccharase / isomaltase-complex isolated from the mucosa of humans, pigs, and rats.[35,36] Identical results were obtained for in vivo studies with rats and humans.[37,38] Therefore, fructo-oligosaccharides can be regarded as a kind of soluble "dietary fibre" with reduced caloric value.

In the large intestine fructo-oligosaccharides are fermented to carbon dioxide, methane, and volatile fatty acids preferentially by the bifidobacteria. This, so called bifidus factor, means, that these fructo-oligosaccharides have a positive effect on the colonization of the large intestine with bifidobacteria eliminating other noxious bacteria like *Clostridium* and *Staphylococcus sp.* which are responsible for the production of putrefactive substances.[39] Based on these properties the fructo-oligomers can be applied as a healthy, low-caloric ingredient in food.

In addition these oligosaccharides have a positive effect as an ingredient in feed. Experiments with young pigs and male broilers showed, that amounts of 0.25-0.5 % (on total solids) of fructo-oligosaccharides to the diet lead to an improvement in feed conversion and body weight gain of the animals.[40,41]

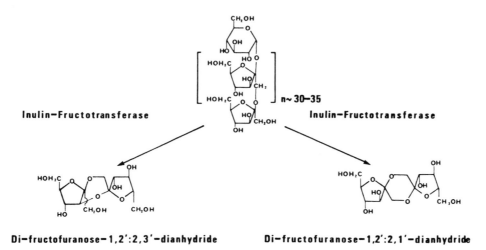

Fig. 15. Enzymatic generation of difructose dianhydrides by splitting of inulin

Fig. 15 shows another application of inulin. In this case, two types of difructose-dianhydrides can be produced by an enzymatic treatment of this polymer with specific inulin-fructotransferases to obtain in one case the di-fructofuranose-1,2':2,1'-dianhydride and, in the other case, di-fructofuranose-1,2':2,3'-di-anhydride.[42,43] The corresponding enzymes are isolated in both cases from different *Arthrobacter* species. Both difructose-dianhydrides are being discussed as alternative low-calorie sweeteners.

4. 5-Hydroxymethyl-furfural (HMF)

HMF is a common dehydration product of all hexuloses and, as such,
- an aldehyde
- an alcohol
- an aromatic compound
- a cisoid dien
- a difunctional furan
- available from biomass (fructose) in a simple dehydration reaction
- convertible in di- and tetrahydrofuran derivatives
- convertible in benzene-, pyridazine-, pyridine- etc. derivatives.

It is produced by an acid-catalyzed intramolecular elimination of 3 moles of water.[44] Basically, three methods of HMF-production can be distinguished:

(i) reaction of a ketose with an acid in an organic solvent,

(ii) reaction of a ketose with an acid in aqueous solution and extraction with the aid of an organic solvent, and

(iii) reaction of a saccharide with an acid in water and purification without any organic solvent.

This last methodology seems to be the most economical one in spite of a lower degree of transformation.[45]

The Südzucker-process for HMF Manufacture

Fructose, derived either from sucrose or polyfructans like inulin, is the most suitable raw material for industrial scale HMF production. A typical reaction batch composed of 100 kg fructose-dry substance (DS), e.g. a mixture of mother liquor from a fructose crystallization step with a purity of ~90 %, 34 kg dried chicory pellets and 1.34 kg sulfuric acid is made up to 360 kg with water and heated to 150 °C for 2 h in a stirred pressure tank. The addition of chicory pellets helps to prevent the formation of deposits on the reactor walls. Since chicory contains inulin, it can be preferably used for this purpose.

After the reaction, the mixture is cooled to 20 °C and the pH raised from 1.8 to neutral with calcium hydroxide or calcium carbonate. The resulting calcium sulfate is removed by filtration along with any solid residues from the reaction. The HMF-containing filtrate is then subjected to a chromatographic separation on calcium-loaded strong acid cation exchanger resin of low cross-linkage using demineralized water as eluent.

It is important that the cation used for neutralization is identical with the cation of the resin used for chromatography. On the other hand, it is also possible to carry out the chromatographic separation on H⁺-loaded cation exchanger resin without prior neutralization.

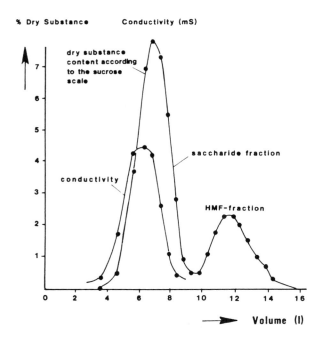

Fig. 16. Chromatographic separation of HMF

The chromatographic separation process has sometimes been regarded as too expensive and unsuitable for large scale operation. But it is well known that products like fructose syrups or 2nd generation HFCS (High Fructose Corn Syrup[33]) are being produced today at a scale of some millions of tons per year using this process. Therefore large scale recovery of HMF with chromatographic separation is feasible, too.

A typical separation diagram for a saccharide / HMF mixture is shown in Fig. 16. First to be eluted is ash, followed by unreacted saccharides whereas HMF emerges as the last component. The purity of the HMF fraction is high enough to permit its crystallization without further treatment.

The cooling crystallization process has to be adapted to the high solubility of HMF in water and its low melting point of only 28 °C. However by utilizing proper conditions the difficulties can easily be surmounted. The reaction yield of HMF is about 43 mol % corresponding to 30 % by weight.

Applications of HMF and its Derivatives

Hydroxymethyl furfural is a versatile sugar derivative which can be regarded as a key substance between carbohydrate chemistry and mineral oil based industrial organic chemistry. Regretfully the development of its chemistry and its broad application have, till today, been hampered by its prohibitive selling price at presently ca. 40,-- DM/g (Merck 90/91) and its non-availability on a larger scale. Some of the possible applications are shown in Fig. 17.

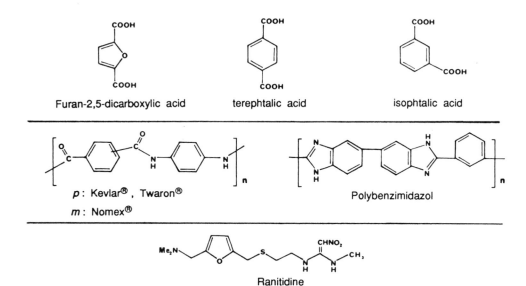

Fig. 17. HMF as raw material for pharmaceuticals and polymers

Terephthalic or isophthalic acid are bulk products which could be replaced by furan-2,5-dicarboxylic acid. Polymers, especially polyamides have been prepared using this HMF-derivative instead of terephthalic acid.[46a-c]

Ranitidine or Zantac®, an H_2-receptor antagonist being used as a gastric acid anti-secretory agent, includes structural features of HMF.[47a-c] This antiulcer drug from Glaxo was the first prescription drug to exceed $ 1 bn in annual sales, raising Glaxo from the 16th to the 7th place in the global pharmaceutical sales league.[47d]

Special electronical and optical effects of the furan nucleus versus the phenyl ring are shown in dicyanovinyl substituted furans derived from HMF (pathway 8, Fig. 18), which could be used in opto-electronical devices, e.g. in information processing.[55]

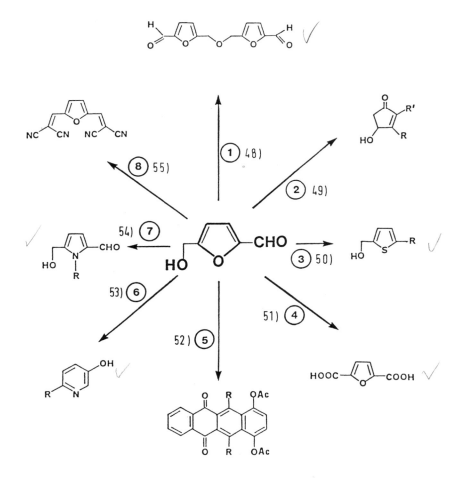

Fig. 18 Conversions of HMF(-derivatives)

Due to the extraordinary versatility of its reactions, HMF can be transformed into carbocyclic and heterocyclic compounds, e.g. pyridine derivatives which are important intermediates for pharmaceuticals or into cyclopentenones, important for the production of pharmaceuticals or plant protection agents. Some of the many structures which can be derived from HMF are shown in Fig. 18 (the relevant references are shown by the numbers 48 to 55 attached to the individual pathways).

Finally, hydroxymethyl-furfural and its derivatives can play a role in almost every part of applied chemistry making this sugar derivative, which is based on renewable resources, a unique compound indeed.

5. Levoglucosan

Levoglucosan, being an intramolecular anhydride or glycoside, is one of the simplest derivatives of glucose. Due to the $\beta(1\rightarrow6)$-bond, the number of free OH-groups is reduced and more selective reactions at the three remaining OH-groups are possible. Another characteristic feature is the reversed chair conformation respective to glucose, so that all the OH-groups are in the energetically unfavourable axial position.

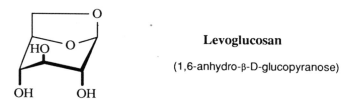

Levoglucosan

(1,6-anhydro-β-D-glucopyranose)

Levoglucosan was first prepared by Tanret in 1894.[56] It is a non-reducing glucose derivative, which is not fermented by yeast. It is stable towards alkali, but is hydrolyzed by acids to yield glucose, oligo- and polysaccharides.[57] It is a product of thermal degradation of cellulose or cellulosic material, e.g. wood, and it is connected directly with the development and spreading of human culture, at least since the utilization of fire. Levoglucosan is therefore surely one of those "chemicals", with which human beings are in contact for a long time.[58]

Methods of Preparation.   There are two general methods for the preparation of levoglucosan. 1. Pyrolysis of oligo- or polysaccharides and 2. Conversion of glucose derivatives. The last method requires chemicals, solvents, and expenditure of work, moreover the chemicals for derivatization have to be disposed of. Besides, an alkali catalyzed elimination from phenyl-[59] or pentachlorophenyl-glucosides[60] or from glucosyl fluoride,[61] an acid-catalyzed elimination from 1,2,3,4-tetra-*O*-acetyl-6-*O*-triphenyl-methyl-β-D-glucopyranose[62] has also been described.

Pyrolysis of polysaccharides, as shown in the ensuing formula scheme, leads directly to levoglucosan. The advantage of this method lies in the availability of the raw material, e.g. starch or cellulose. The process can do without protecting groups, chemicals, and solvents:

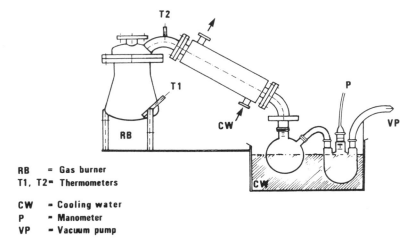

The practicability of producing crystalline levoglucosan by this method and using water as the only solvent, has been demonstrated on a pilot plant scale by Südzucker.[63] A specially designed stainless steel equipment comprising a gas-heated pyrolysis vessel fitted with a condenser as shown schematically in Fig. 19 was used for the pyrolysis of commercial starch.

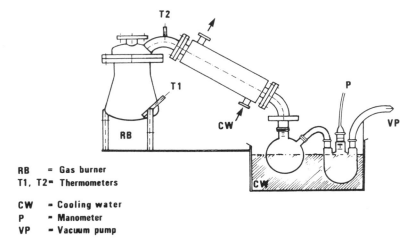

RB   = Gas burner
T1, T2 = Thermometers

CW   = Cooling water
P   = Manometer
VP   = Vacuum pump

Fig. 19. Pyrolysis apparatus for the production of levoglucosan

The layer of the raw material in the pyrolysis vessel has to have an optimum thickness as otherwise the heat transfer from the reactor wall into the starch layer may become a problem. Mixing an appropriate amount of metal turnings to the raw material may help to improve the heat transfer.

The measurement of the temperature profile of the reaction (Fig. 20) shows that the temperature at the condenser-inlet drops towards the end of the reaction, whereas the temperature of the residue is still rising. The crude distillate is treated with calcium carbonate and filtered. The resulting aqueous solution, which has a dry substance content of ~40 %, is subjected directly to a chromatographic separation step using demineralized water as eluent.

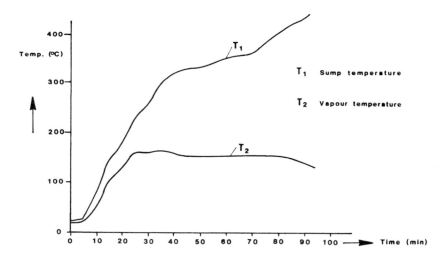

Fig. 20. Temperature / time diagram of starch pyrolysis

A typical diagram for such a chromatographic separation is shown in Fig. 21. As indicated by the curves for absorption at 420 nm and electrical conductivity, ash, and colour are the first to be eluted. They are followed by levoglucosan as shown not only by the line representing dry substance content, measured as refractive index (n_D^{20}), but also by measuring the optical rotation expressed as polarization.

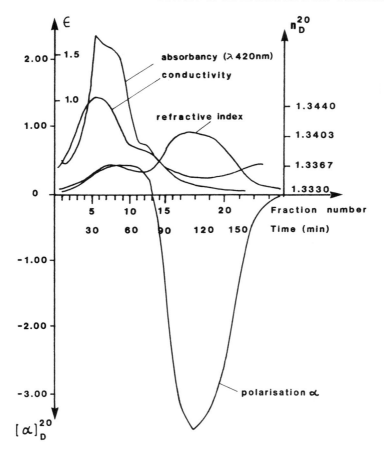

Fig. 21. Chromatographic separation of levoglucosan

The purity of the levoglucosan fraction is high enough to allow direct crystallization from water by evaporation or cooling.

Reactions with Levoglucosan

Some of the reactions which can be carried out with levoglucosan are shown schematically in the ensuing formulae. Relevant literature citations are shown by references[64-69] attached to the individual reaction pathways:

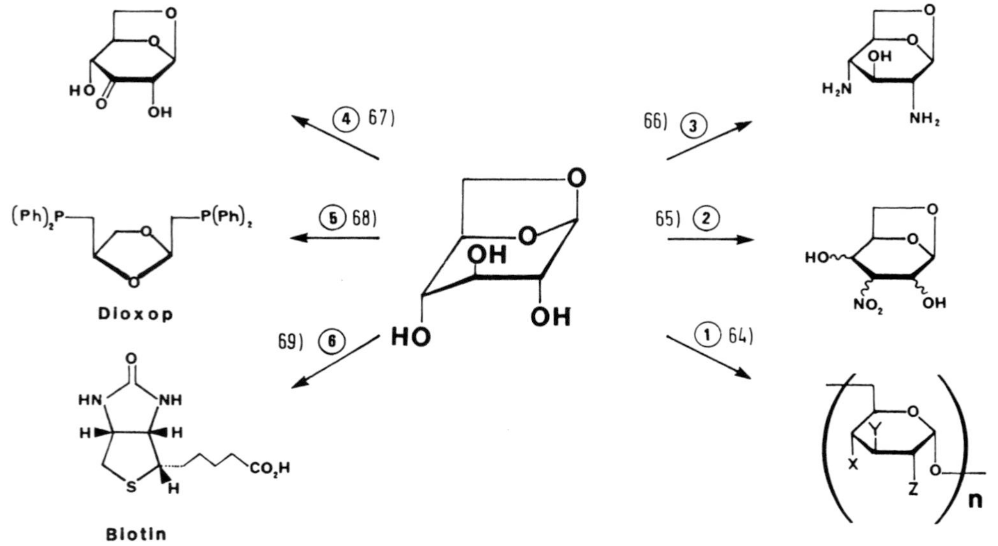

The polymerization with ring opening (route ①) yields dextrans which possess a highly linear structure in comparison to the branched-chain structure of natural dextran.[64] Sulfated dextran exhibits anticoagulant activity like heparin.

The polymerization of levoglucosan derivatives like tri-*O*-benzyl levoglucosan alone or with other compounds may lead to new homo- or hetero-polysaccharides. The glycol cleavage of levoglucosan, followed by addition of nitromethane to the resulting dialdehyde yields the 3-nitro derivatives of 1,6-anhydrohexoses (pathway ②), from which the respective 3-amino- and 2,3,4-triamino-compounds are readily accessible.[65]

Selective tosylation at *O*-2 and *O*-4 and reaction of the azide ion with the intermediately formed epoxide yields the diamino sugar of pathway ③.[66] Catalytic oxidation is selectively performed at C-3 as shown in reaction ④.[67] Cleavage of the triol part of levoglucosan, followed by reduction and substitution with diphenyl phosphide yields the chiral ligand Dioxop (⑤), which constitutes a part of a metal complex used for asymmetric hydrogenations.[68] Due to its easy accessibility and its special chirality levoglucosan can be used in the preparation of a wide variety of complex compounds.

References

1. R. Weidenhagen, S. Lorenz: Ein neues bakterielles Umwandlungsprodukt der Saccharose. *Angew. Chem.* **69** (1957) 641.

2. R. Weidenhagen, S. Lorenz: Palatinose (6-(α-Glucopyranosido)-fructofuranose), ein neues bakterielles Umwandlungsprodukt der Saccharose. *Z. Zuckerind.* **7** (1957) 533-534.

3. S. Lorenz: Über bakteriologische Betriebskontrolle in Zuckerfabriken und papierchromatographische Studien zu Saccharoseumwandlung der dabei gefundenen Bakterienstämme. *Z. Zuckerind.* **8** (1958) 490-494 and 535-541.

4. R. Weidenhagen, S. Lorenz (Süddeutsche Zucker AG): Process for the Production of Palatinose. *Ger. Pat.* DE 1,049,800 (1957); *Chem. Abstr.* **55** (1961) 2030b.

5. S. Windisch: Über einen Farbstoffbildner von Zuckerrüben, *Protaminobacter rubrum* (den Dooren de Jong). *Z. Zuckerind.* **14** (1958) 446.

6. F. H. Stodola, E. S. Sharp, H. J. Koepsell: The Preparation, Properties and Structure of the Disaccharide Leucrose. *J. Am. Chem. Soc.* **78** (1956) 2514-2518.

7. S. Schmidt-Berg-Lorenz, W. Mauch: Ein weiterer Isomaltulose bildender Bakterienstamm. *Z. Zuckerind.* **14** (1964) 625-627.

8. B. M. Lund, G. M. Wyatt: The Nature of Reducing Compounds Formed from Sucrose by *Erwinia carotovora var. atroseptica. J. Gen. Microbiol.* **78** (1973) 331-336.

9. P. S. J. Cheetham, C. E. Imber, J. Isherwood: The Formation of Isomaltulose by Immobilized *Erwinia rhapontici. Nature* **299** (1982) 628-631.

10. W. Crueger, L. Drath, M. Munir (Bayer AG): Process for the Continuous Isomerization of Sucrose to Isomaltulose with Microorganisms. *Ger. Offen.* DE 2,741,197 (1977) and *Ger. Offen.* DE 2,806,216 (1978); *Eur. Pat.* 1,099 (1978); *Chem. Abstr.* **90** (1979) 184932t and **91** (1979) 156068t.

11. C. Bucke, P. S. J. Cheetham (Tate & Lyle): Production of Isomaltulose. *Brit. Pat. Appl.* 2,063,268 (1979); *Eur. Pat. Appl.* 28,900 (1981); *Chem. Abstr.* **95** (1981) 95468g.

12. M. Munir (Süddeutsche Zucker AG): Process for the Production of Isomaltulose with Immobilized Bacterial Cells. *Ger. Offen.* DE 3,038,219 (1980) and 3,213,107 (1982); *Chem. Abstr.* **97** (1982) 4680x and **100** (1984) 66611g.

13. I. Shimizu, K. Suzuki, Y. Nakajima (Mitsui Sugar Co.): Palatinose Production Using Immobilized α-Glucosyltransferase. *Jap. Pat. Appl.* 55-113,982 (1980); *Chem. Abstr.* **97** (1982) 53994m.

14. C. Kutzbach, G. Schmidt-Kastner, H. Schutt (Bayer AG): Sucrose-Mutase, Immobilized Sucrose-Mutase and Use of Immobilized Sucrose-Mutase for the Production of Isomaltulose. *Eur. Pat. Appl.* 49,801 (1981); *Chem. Abstr.* **97** (1982) 90444c.

15. P. Egerer, W. Haese, H. Perrey, G. Schmidt-Kastner (Bayer AG): Immobilization of *Protaminobacter rubrum* and Use of Immobilized Product for Conversion of Sucrose to Isomaltulose. *Eur. Pat. Appl.* 160,253 (1985) and 160,260 (1985); *Chem. Abstr.* **104** (1986) 49828f and **104** (1986) 87147r.

16. Y. Nakajima: Palatinose Production by Immobilized α-Glucosyl-transferase. *Proc. Res. Soc. Japan Sugar Refineries Technol.* **33** (1984) 55-63; *Chem. Abstr.* **103** (1985) 213357d.

17. T. Kaga, T. Mizutani: Applications of Palatinose for Foods. *Proc. Res. Soc. Japan Sugar Refineries Technol.* **34** (1985) 45-57; *Chem. Abstr.* **105** (1986) 224782f.

18. H. Schiweck, G. Steinle, L. Haberl (Süddeutsche Zucker AG): Process for the Production of α-D-Glucopyranosyl(1-6)sorbitol (Isomaltitol). *Ger. Pat.* DE 2,217,628 (1972); *Chem. Abstr.* **80** (1974) 83527c.

19. O. J. Lantero (Miles Laboratories Inc.): Immobilization of the Sucrose Mutase in Whole Cells of *Protaminobacter rubrum*. *U.S. Pat.* 4,390,627 (1983); *Chem. Abstr.* **99** (1983) 35243c.

20. G. Darsow, W. Biedermann (Bayer AG): Process for the Production of a Mixture of α-D-Glucopyranosyl-1,6-mannitol and α-D-Glucopyranosyl-1,6-sorbitol from α-D-Glucopyranosyl-1,6-fructose. *Ger. Offen.* DE 3,403,973 (1984); *Chem. Abstr.* **104** (1986) 130220y.

21. H.-J. Lindner, F. W. Lichtenthaler: Extended Zig-Zag Conformation of 1-*O*-α-D-Glucopyranosyl-D-mannitol. *Carbohydr. Res.* **93** (1981) 135-140.

22. F. W. Lichtenthaler, H.-J. Lindner: The Preferred Conformations of Glucosylalditols. *Liebigs Ann. Chem.* **1981**, 2372-2383.

23. F. H. Mattson, G. A. Nolen: Absorbability by Rats of Compounds Containing from One to Eight Ester Groups. *J. Nutr.* **102** (1972) 1171-1175; *Chem. Abstr.* **77** (1972) 137821c.

24. F. H. Mattson, R. A. Volpenhein: Rate and Extent of Absorption of the Fatty Acids of Fully Esterified Glycerol, Erythritol, Xylitol, and Sucrose as Measured in Thoracic Duct Cannulated Rats. *J. Nutr.* **102** (1972) 1177-1180; *Chem. Abstr.* **77** (1972) 137850m.

25. F. H. Mattson, R. A. Volpenhein: The Digestion and Absorption of Triglycerides. *J. Biol. Chem.* **239** (1964) 2772-2777.

26. A. F. Hofmann: Function of Bile Salts in Fat Digestion: Behavior and Solubility of Saturated Fatty Acids in Bile Salt Solutions. *Chem. Phys. Appl. Surface Active Subst., Proc. 4th Int. Congr.*, 1964 (Publ. 1967), 3, 371-379; *Chem. Abstr.* **72** (1970) 52618p.

27. M. W. Riegler, R. E. Honkanen, J. S. Patton: Visualization by Freeze Fracture, in Vitro and in Vivo, of the Products of Fat Digestion. *J. Lipid. Res.* **27** (1986) 836-857; *Chem. Abstr.* **105** (1986) 223367n.

28. H. Pardun: Principles and Advances in Food Study and Food Technology, Vol. 16: Analysis of Edible Fats. (Parey: Berlin Ger., 1976); *Chem. Abstr.* **86** (1977) 21682a.

29. a) J. Bodor, G. Page (Unilever N. V.): Fatty Acid Esters of Sugars and Sugar Alcohols. EP 235,836 (1987); *Chem. Abstr.* **108** (1988) 118976z.

    b) P. van der Plank, A. Rozendaal (Unilever N. V.): Process for Preparing Polyol Fatty Acid Polyesters. EP 256,585 (1988); *Chem. Abstr.* **110** (1989) 58002h.

    c) G. J. van Lookeren (Unilever N. V.): Method of Purifying Crude Polyol Fatty Acid Polyesters. EP 319,091, EP 319,092 (1989); *Chem. Abstr.* **111** (1989) 195325a.

    d) G. W. M. Willemse (Unilever N. V.): Process for the Synthesis of Polyol Fatty Acid Polyesters. EP 322,971 (1989); *Chem. Abstr.* **111** (1989) 214883z.

30. a) C. A. Bernhardt (Procter and Gamble Co.): Method for Making Better Tasting Low Calorie Fat Materials. EP 233,856 (1987); *Chem. Abstr.* **107** (1987) 196781a.

    b) C. A. Bernhardt (Procter and Gamble Co.): Low Calorie Fat Materials that Eliminate Laxative Side Effects. EP 236,288 (1987); *Chem. Abstr.* **107** (1987) 174789g.

    c) S. A. McCoy, P. M. Self, B. L. Madison (Procter and Gamble Co.): Cocoa Butter Substitute Made From Sucrose Polyester. EP 271,951 (1988); *Chem. Abstr.* **109** (1988) 127539g.

    d) J. L. Y. Kong-Chen (Procter and Gamble Co.): Calorie and Fat-low Chocolate Confectionery Compositions. EP 285,187 (1988); *Chem. Abstr.* **111** (1989) 95934e.

    e) K. Masaoka, Y. Kasori (Mitsubishi Kasei Corp.): Process for Producing Sucrose Fatty Acid Polyester. EP 346,845 (1989); *Chem. Abstr.* **112** (1990) 198975u.

31. N. C. Karpita: Linkage Structure of Fructans and Fructan Oligomers from *Triticum aestivum* and *Festuca arundinacea* Leaves. *J. Plant Physiol.* **134** (1989) 162-168.

32. H. Hidaka: Neosugar – Manufacturing and Properties. *Proceedings of 1st Neosugar Res. Conference*, Tokyo (1982) 3-13.

33. H. Schiweck: Crystalline Fructose: Production, Specifications, and Physiological Properties. *Alimenta* **2** (1989) 31-35.

34. U. Nilsson, J. Bjoerk: Cereal Fructans: In Vitro and in Vivo Studies on Availability in Rats and Humans. *J. Nutr.* **118** (1989) 1325-1330.

35. S. C. Ziesenitz, G. Siebert: In Vitro Assessment of Nystose as a Sugar Substitute. *J. Nutr.* **117** (1987) 846-851.

36. F. Heinz, Medizinische Hochschule Hannover, unpublished results.

37. T. Oku, T. Tukunaga, N. Hosoya: Nondigestability of a New Sweetener, "Neosugar", in the Rat. *J. Nutr.* **114** (1984) 1574-1581.

38. T. Sano, M. Ishikawa, Y. Nozawa, K. Hoshi, K. Someya: Application of Neosugar P on Blood Glucose. *Reports of Neosugar Conference 2nd*, (N. Hosoya, Ed.) (1985) 29-38.

39. H. Hidaka, T. Eida, T. Takizama, T. Tokunaga, Y. Tashiro: Effects of Fructooligosaccharides on Intestinal Flora and Human Health. *Bifidobacteria Microflora* **5** (1986) 37-50.

40. H. Morimoto, H. Noro, H. Ohtaki, H. Yamazaki: Study on Feeding Fructooligosaccharides (Neosugar G) in Suckling Pigs. Japan Scientific Food Association, Report of August 22, 1984.

41. C. Quarles, E. Ammerman, P. V. Twining: Dietary Fructooligosaccharides Increase Feed Efficiency in Floor Pen-Reared Male Broilers. Southern Poultry Science Society Meeting, Atlanta, GA, January (1988) 19-20.

42. K. Seki, K. Haraguchi, M. Kishimoto, S. Kobayashi, K. Kainuma: Production of a Novel Inulin Fructotransferase (DFAI producing) by *Arthrobacter globiformis* S 14-3. *Stärke* **40** (1988) 440-442.

43. T. Uchiyama, K. Tanaka, M. Kawamura (Mitsubishi Kasei Corp.): Process for the Preparation of Difructose Dianhydride III. *Eur. Pat. Appl.* 332,108 (1989).

44. B. F. M. Kuster: 5-Hydroxymethylfurfural (HMF). A Review Foccusing on its Manufacture. *Starch / Stärke* **42** (1990) 314-321.

45. K. M. Rapp (Süddeutsche Zucker AG): Process for Preparing Pure 5-Hydroxymethylfurfuraldehyde. *U.S. Pat.* 4,740,605 (1988); *Chem. Abstr.* **107** (1987) 154231r.

46. a) A. Mitiakoudis, A. Gandini, H. Cheradame: Polyamides Containing Furanic Moieties. *Polym. Commun.* **26** (1985) 246-249.

    b) A. Mitiakoudis, A. Gandini, (Stamicarbon B.V.): Poly-(p-phenylene-2,5-furandicarbonamide) and Anisotropic Solutions Thereof, Their Conversion into Filaments and Films. *Eur. Appl.* 256,606 (1987); *Chem. Abstr.* **109** (1988) 38462j.

c) A. Gandini: Furan Polymers. in *Encycl. Polym. Sci. Engineer.* 2nd ed., Vol. 7, p. 454-473, Interscience Publ., New York, 1984.

d) G. A. Serad, H. Zimmermann: PBI, ein neuer Polymer-Werkstoff z.B. für feuerfeste Fasern mit außergewöhnlichen Eigenschaften. *HOECHST High Chem. Magazin* **5** (1988) 50-53.

47. a) A. A. Garcia, J. L. O. Martinez (Laborotorios Liade, S.A.): 5-[(Dimethylamino)methyl]furfuryl 2-Aminomethyl Sulfide Derivative and its Physiologically Acceptable Salts. *Span.* ES 506,422 (1982); *Chem. Abstr.* **98** (1983) 179197t.

b) J. W. Clitherow (Glaxo Group Ltd.): Process for the Preparation of Ranitidine. US 4,497,961 (1985); *Chem. Abstr.* **102** (1985) 166604z.

c) B. J. Price, J. W. Clitherow, J. Bradshaw (Allen & Hansbury Ltd.): Aminoalkylfuran Derivative. *Patentschrift* CH 640,846 (1984); *Chem. Abstr.* **101** (1984) 23317b.

d) M. Stone: Keeping Pace with Surging Market. *Performance Chem.*, August 1987, p. 50, 52.

48. T. Iseki: Über Di-(furfural-methyl-2)-äther. *Z. physiol. Chem.* **216** (1933) 130-132; *Chem. Abstr.* **27** (1933) 2953b

49. G. Piancatelli, A. Scettri, G. David, M. D'Auria: A New Synthesis of 3-Oxocyclopentenes. *Tetrahedron* **34** (1978) 2775-2778.

50. V. G. Kharchenko, T. I. Gubina, S. P. Voronin, J. A. Markushina: About the Reaction of the Conversion of Furans into Thiophenes and Selenophenes. *Top. Furan. Chem.; Proc. Symp. Furan. Chem. 4th*, Slovak Techn. Univ. Bratislava, Czech., 1983, p. 142-145.

51. a) B. W. Lew (Atlas Chemical Industries, Inc.): Method of Producing Dehydromucic Acid. US 3,326,944 (1967); *Chem. Abstr.* **68** (1968) 49434n.

b) I. Leupold, M. Wiesner, W. Fritsche-Lang: New Applications for the Catalytic Oxidation with Oxygen. *5th Eur. Symp. Carbohydr. Chem.*, Prag, August 1989, Abstract D-1.

52. W. C. Christopfel, L. L. Miller: Synthesis of Polyacenequinones via a Benzo[2,3-c]furan. *Tetrahedron* **43** (1987) 3681-3688; (R= p-*t*-Butylphenyl; with derivatized HMF as starting material compounds with different R-groups should be accessible).

53. N. Elming, N. Clauson-Kaas: Transformation of 2-Hydroxymethyl-5-aminomethyl-furan into 6-Methyl-3-pyridinol. *Acta Chem. Scand.* **10** (1956) 1603-1605; (R= CH_3 or CH_2OH).

54. N. Elming: Dialkoxydihydrofurans and Diacyloxydihydrofurans as Synthetic Intermediates. *Adv. Org. Chem.* **2** (1960) 67-115; (Formal Ways to pyrrols from furans via 2,5-dialkoxytetrahydrofurans).

55. J. Daub, J. Salbeck, T. Knöchel, C. Fischer, H. Kunkeley, K. M. Rapp: Lichtsensitive und elektronentransferreaktive molekulare Bausteine: Synthese und Eigenschaften eines photochemisch schaltbaren, dicyanvinylsubstituierten Furans. *Angew. Chem.* **101** (1989) 1541-1542; *Angew. Chem., Int. Ed. Engl.* **28** (1989) 1494.

56. C. Tanret: *Compt. Rend.* **119** (1894) 158.

57. L. Reichel, H. Schiweck: Einwirkung von konzentrierter Salzsäure auf Lävoglucosan, III. *Liebigs Ann. Chem.* **761** (1972) 182-188.

58. M. Cerny, J. Stanek, Jr.: 1,6-Anhydro Derivatives of Aldohexoses. *Adv. Carbohydr. Chem.* **34** (1977) 23-177.

59. Y.-Z. Lai, D. E. Ontto: Base-catalyzed Formation of 1,6-Anhydro-β-D-glucopyranose from Phenyl α-D-glucopyranoside. *Carbohydr. Res.* **67** (1978) 500-502.

60. I. Fujimaki, Y. Ichikawa, H. Kuzuhara: A Modified Procedure for the Synthesis of 1,6-Anhydro Disaccharides. *Carbohydr. Res.* **101** (1982) 148-151.

61. H. Hardt, D. T. A. Lamport: Hydrogen Fluoride Saccharification of Cellulose and Xylan: Isolation of α-D-Glucopyranosyl Fluoride and α-D-Xylopyranosyl Fluoride Intermediates, and 1,6-Anhydro-β-D-glucopyranose. *Phytochem.* **21** (1982) 2301-2303.

62. M. V. Rao, M. Nagarajan: An Improved Synthesis of 2,3,4-Tri-*O*-acetyl-1,6-anhydro-β-D-glucopyranose (Levoglucosan Triacetate). *Carbohydr. Res.* **162** (1987) 141-144.

63. M. Gander, K. M. Rapp, H. Schiweck (Südzucker AG Mannheim / Ochsenfurt): Verfahren zur Herstellung von 1,6-β-D-Anhydroglucopyranose (Levoglucosan) in hoher Reinheit. *Ger. Offen.* DE 3,803,339 (1988); *Eur. Appl.* 327,920 (1989); *Chem. Abstr.* **111** (1989) 176618g.

64. a) K. Kobayashi, H. Sumitomo: Regioselectively Modified Stereoregular Polysaccharides. VII. Synthesis of (1→6)-α-D-glucopyranans Having Dodecyl Group in Position 3. *Polymer J.* **16** (1984) 297-301.

    b) T. Uryu, M. Yamanaka, M. Henmi, K. Hatanaka, K. Matsuzaki: Ring-opening Polymerization of 1,6-Anhydro-2,4-di-*O*-benzyl-3-*O*-*tert*-butyldimethylsilyl-β-D-glucopyranose and Synthesis of α-(1→3)-Branched Dextrans. *Carbohydr. Res.* **157** (1986) 157-169.

c) M. Okada, H. Sumitomo, T. Hirasawa: Chemical Synthesis of Polysaccharides. VIII. Synthesis and Enzymatic Hydrolysis of (1→6)-α-Linked Heteropolysaccharides Consisting of D-Glucose and 2,3,4-Trideoxy-D,L-*glycero*-hexopyranose Units. *Polymer J.* **19** (1987) 581-591.

d) H. Ichikawa, K. Kobayashi, M. Okada, H. Sumitomo: Regioselectively Modified Stereoregular Polysaccharides. X. Equilibrium Polymerization of 1,6-Anhydro-2-*O*-benzyl-3,4-dideoxy-β-D-*threo*-hexopyranose. *Polymer J.* **19** (1987) 873-880.

e) Lit. cited in N. K. Kochetkov: Synthesis of Polysaccharides with Regular Structure. *Tetrahedron* **43** (1987) 84-97.

65. F. W. Lichtenthaler: Cyclisierung von Dialdehyden mit Nitromethan. *Angew. Chem.* **76** (1964) 84-97; *Angew. Chem., Int. Ed. Engl.* **3** (1964) 211-224.

66. C. Kolar, H.-P. Kraemer (Behringwerke AG): Cis-Platin(II) - Komplexe mit Diaminozucker-Derivaten als Liganden, Verfahren zu ihrer Herstellung und diese enthaltende Arzneimittel. *Ger. Offen.* DE 3,424,,217 (1984); *Eur. Appl.* 167,071; *Chem. Abstr.* **105** (1986) 209336v.

67. K. Heyns, J. Weyer, H. Paulsen: Selektive katalytische Oxidation von 1,6-Anhydro-β-D-hexopyranosen zu 1,6-Anhydro-β-D-hexopyranos-ulosen. *Chem. Ber.* **100** (1967) 2317-2334.

68. a) G. Descotes, D. Lafont, D. Sinou: Asymmetric Hydrogenation of α-Amino Acid Precursors with a New Chiral Diphosphine (dioxop) derived from 1,6-D-Anhydroglucose. *J. Organomet. Chem.* **150** (1978) C14-C16.

b) J. M. Brown, P. A. Chaloner, G. Descotes, R. Glaser, D. Lafont, D. Sinou: Asymmetric Hydrogenation Catalyzed by Rhodium Complexes of (2*R*,4*R*)-Bis(diphenylphosphinomethyl)dioxolan. A Stable Rhodium Dihydride Derived from a Chelating Diphosphine Complex. *J. Chem. Soc., Chem. Commun.* **1979**, 611-613.

c) D. Lafont, D. Sinou, G. Descotes: Hydrogenation Asymetrique de Precurseurs d'Aminoacides a l'Aide de Mono- et Diphosphines Derivees de Sucres. *J. Organomet. Chem.* **169** (1979) 87-95.

d) D. Sinou: Hydrogenation Asymetrique a l'Aide du Complexe Dioxop-Rh(I) Voie Dihydro ou Voie Insaturee. *Tetrahedron Lett.* **22** (1981) 2987-2990.

e) D. Lafont, D. Sinou, G. Descotes, R. Glaser, S. Geresh: Asymmetric Homogenous Hydrogenation of α-Acylaminocinnamic Acids and Esters Catalyzed by a Rhodium(I) Complex of Dioxop. *J. Mol. Catal.* **10** (1981) 305-311.

f) G. Descotes, D. Lafont, D. Sinou, J. M. Brown, P. A. Chaloner, D. Parker:
Further Studies on Asymmetric Hydrogenation by Rhodium Complexes of
(2*R*,4*R*)-Bis(diphenylphosphinomethyl)dioxolane. *Nouv. J. Chem.* **5** (1981)
167-173.

69. T. Ogawa, T. Kawano, M. Matsui: A Biomimetic Synthesis of (+)-Biotin from
D-Glucose. *Carbohydr. Res.* **57** (1977) C31-C35.

4

Reactive Sucrose Derivatives

Heinrich Gruber and Gerd Greber

Institut für Chemische Technologie Organischer Stoffe,
Technische Universität Wien, A-1060 Wien

Summary. New sucrose esters and ethers with functional groups exhibiting reactivities different from the hydroxyl groups are presented:

♦ methacrylic esters of sucrose,

♦ hydrophilic sucrose methacrylate gels for the immobilization of reagents and catalytic active groups, respectively,

♦ sucrose derivatives with carbonic acid amide groups or N-methylated amide groups as condensation components for formaldehyde resins,

♦ sucrose derivatives with photoactive groups for the preparation of photoresists,

♦ sucrose derivatives with primary amino groups and their fatty acid amides exhibiting surface activity, and

♦ water soluble sucrose derivatives with biologically active groups.

Introduction

With regard to the utilization of naturally renewable raw materials we have been engaged for a number of years in the preparation of new reactive sucrose derivatives for industrial applications.

Our objective was the synthesis of new sucrose esters and ethers with functional groups exhibiting reactivities different from the hydroxyl groups, without change of the basic structure of sucrose. These reactions do not afford clearly defined, singular compounds but mixtures of isomers with average degrees of substitution. These mixtures are difficult to separate and, hence, are of use for applications that do not require uniform substances only.

1. Synthesis of Methacrylic Esters of Sucrose

Polymerizable sucrose methacrylates were prepared by esterification or transesterification of sucrose with methacrylic acid chloride, methacrylic acid anhydride or methyl methacrylate (MMA), respectively:[1]

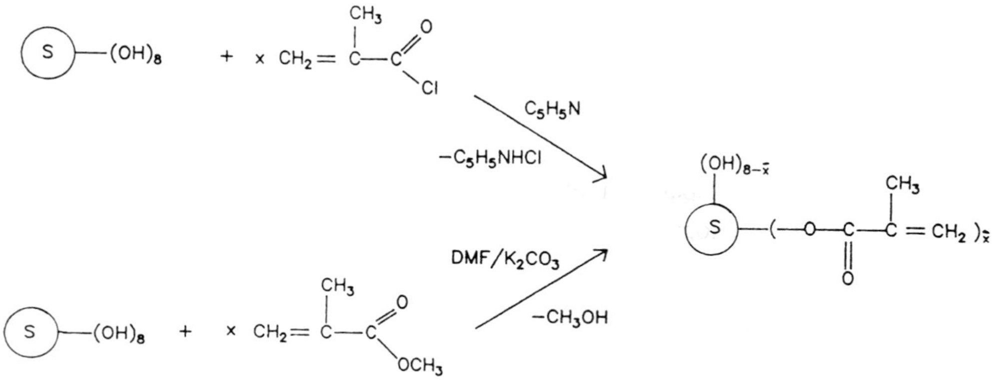

The average degree of substitution (DS) can be regulated by the molar ratio of sucrose to methacrylic acid derivative but even with equimolar amounts of the starting materials, mixtures of about 50 % monoesters and 0.7 % higher substituted esters, together with a lot of unreacted sucrose, were obtained (Table 1). Thus, these mixtures can directly be used for the preparation of crosslinked products.

Table 1. Esterification of sucrose with methacrylic acid anhydride

molar ratio sucrose : methacrylic-acid anhydride	wt.% unreacted sucrose	mono substituted esters	higher substituted esters
1 : 1	47.8	51.5	0.7
1 : 2	21.3	44.2	34.4
1 : 3	12.0	40.8	47.2
1 : 5	4.5	18.1	77.3

From the technical point of view the transesterification of sucrose with MMA is more interesting, affording reaction mixtures containing somewhat more monoesters proportionally to the esters with higher degree of substitution (Table 2).

Table 2. Transesterification of sucrose with methyl methacrylate (MMA)

molar ratio sucrose : methacrylic- acid anhydride	wt.% unreacted sucrose	mono	higher
		substituted esters	
1 : 1	84.6	15.4	< 0.1
1 : 2	37.3	50.4	12.3
1 : 3	25.8	46.6	27.6
1 : 5	21.2	39.9	39.5

The preparative separation of the components is very tedious and can be achieved by chromatographic methods only. Thus, for further reactions, predominantly sucrose methacrylate (SM) mixtures are used without purification.

2. Synthesis and Applications of Sucrose Methacrylate Gels

Radical polymerization of the SM-mixtures yields crosslinked hydrophilic gels:[1]

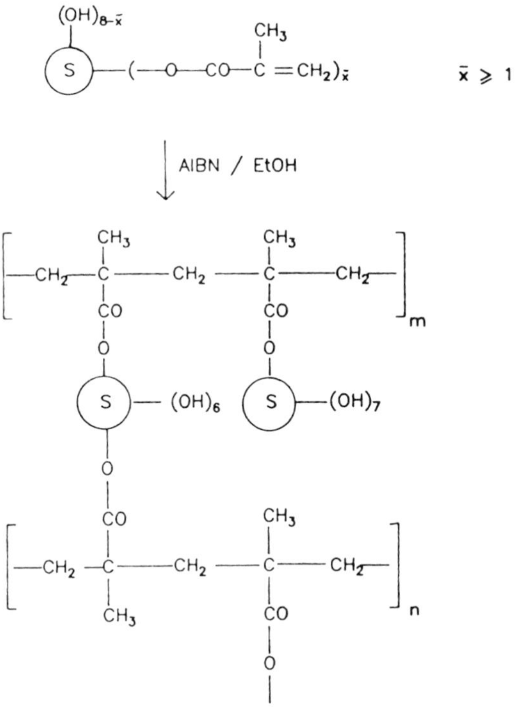

Unreacted sucrose as well as other low molecular substances can easily be removed from the gels by extraction. The swelling degrees of the gels are in the range of 1,1 - 7 depending on the content of the crosslinking agents, which is determined by the concentration of higher substituted SMs in the monomer mixture.

The SM-gels can advantageously be used as hydrophilic carriers for the covalent immobilization of chelating agents for heavy metals, of catalytic active groups, and of Girard-reagents, by reactions of appropriate functional compounds with the free hydroxyl groups of the gels. The resulting gels are hydrolytically stable in the pH range of 1 - 10 and resistant to biochemical degradation, thus exhibiting all the requirements expected from a tough hydrophilic carrier material. The immobilization of reactive groups on unsoluble carriers is a versatile technique applicable in many fields of chemical technology, the main advantages being the easy separation from excess reagents and products, and the regenerability of the immobilized species allowing repeated use without loss of activity by bleeding.

3. Chelating Resins Based on Sucrose Methacrylate

Chelating groups for the complexation of heavy metals can readily be bound to the SM-gels by esterification of the hydroxy-groups with thioglycolic acid[2] or with chlorides of phosphoric acid esters[3] leading to chelating resins.

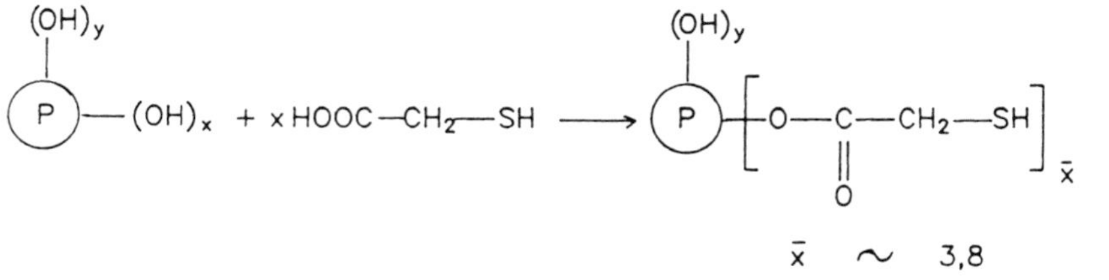

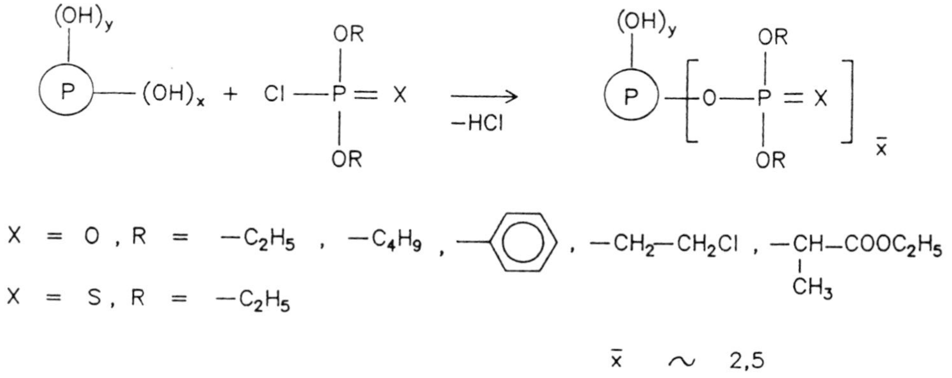

$$\bar{x} \sim 3{,}8$$

$$X = O\ ,\ R = -C_2H_5\ ,\ -C_4H_9\ ,\ \langle\bigcirc\rangle\ ,\ -CH_2-CH_2Cl\ ,\ -CH-COOC_2H_5$$
$$\overset{|}{CH_3}$$

$$X = S\ ,\ R = -C_2H_5$$

$$\bar{x} \sim 2{,}5$$

Esterification of SM-gels with 4-nitrobenzoyl chloride and subsequent reduction with sodium dithionite and diazotation of the amino groups yields gels with diazonium salt residues, providing a variety of immobilization reactions.

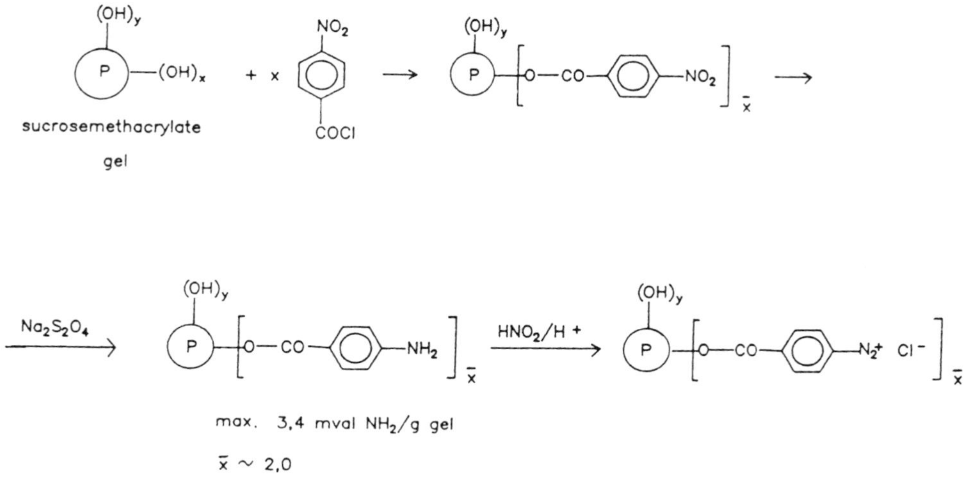

max. 3,4 mval NH$_2$/g gel

$\bar{x} \sim 2,0$

Thus by coupling with 8-hydroxyquinoline, anthranilic acid, salicylic acid, dithiazone[4,5] as well as with sulfur containing reagents like thiols or xanthogenates[2] various chelating groups can be fixed.

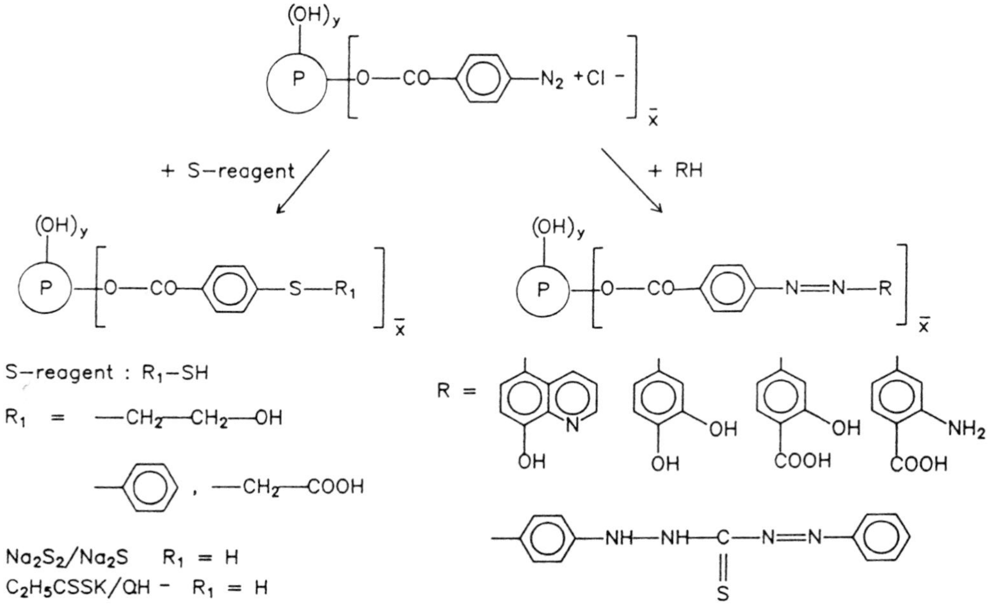

By reaction of the aminophenyl-gels with carbondisulfide or ammonium rhodanide the amino-groups were substituted by xanthogenate groups or thiourea groups:[2]

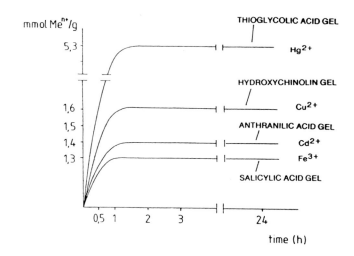

The selectivity of these chelating resins for heavy metals depends on the reactivity of the chelating group. Because of the hydrophilic properties of the carrier the rate of heavy metal removal is very high, in most cases more than 90 % of the final capacity is obtained after half an hour (Fig. 1).

Fig. 1. Capacities of chelating resins based on sucrose methacrylate gels for heavy-metals: 1.3 - 5.9 mmol Me^{n+}/g gel *versus* rate of heavy-metal removal.

Comparable rates are achieved with cellulose ion exchangers only, but contrary to these exchangers, the SM-gels exhibit high capacities for metal ions. Generally the capacities are higher than 1 mval/g, the Hg^{2+} and Ag^+-capacities of the thioglycolate gels were found to be above 5 mval/g. The heavy metals can be recycled from the loaded gels by treatment with diluted acids reversibly. Repeated loading and regeneration cycles cause no loss of capacity.

4. Immobilized Catalysts Based on SM-gels

Immobilized Phase Transfer Catalysts

On account of their capability to complex alkali metal ions, crown ethers and cryptands are the most effective known phase transfer catalysts (PTCs) but, according to their high prices, they are not yet used for industrial purposes. Comparable efficiencies in phase transfer reactions are achieved by polyfunctional molecules having several pendant polyethylenoxide residues, the so-called podands.

A convenient preparation of immobilized podands is accomplished by partial substitution of the end-capped hydroxy-groups of sucrose-ethylenoxide adducts by methacrylic acid residues and subsequent radical polymerization:[6]

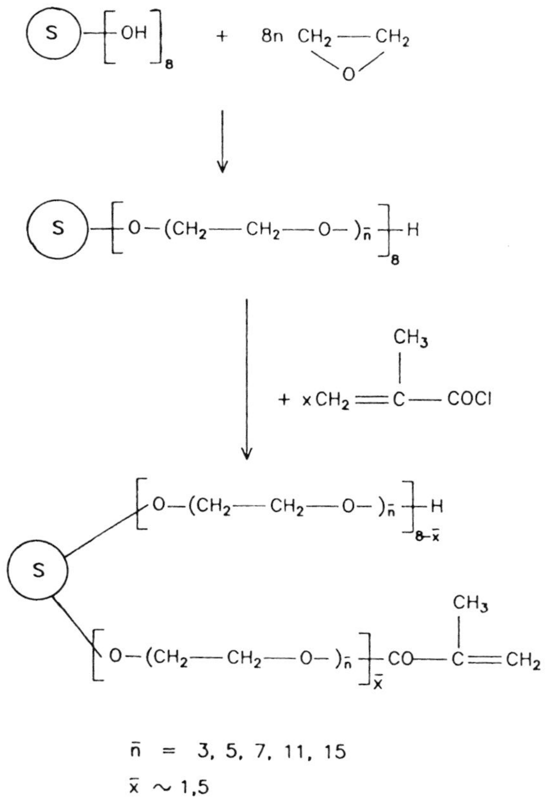

$$S{-}[{-}OH]_8 \quad + \quad 8n\ CH_2{-}CH_2 \underset{O}{\diagdown\diagup}$$

$$\downarrow$$

$$S{-}[O{-}(CH_2{-}CH_2{-}O{-})_{\bar{n}}]_8{-}H$$

$$\downarrow \quad + x\,CH_2{=}\underset{\underset{CH_3}{|}}{C}{-}COCl$$

$$\downarrow$$

$$S\,\Big\langle\,{\begin{array}{l}[O{-}(CH_2{-}CH_2{-}O{-})_{\bar{n}}]_{8-\bar{x}}{-}H \\[2mm] [O{-}(CH_2{-}CH_2{-}O{-})_{\bar{n}}]_{\bar{x}}{-}CO{-}\underset{\underset{CH_3}{|}}{C}{=}CH_2\end{array}}$$

$$\bar{n} \;=\; 3,\ 5,\ 7,\ 11,\ 15$$
$$\bar{x} \sim 1{,}5$$

Assuming regular distribution of the polyethylenoxide residues to the OH-groups of sucrose the length of the "tentacles" of the resulting immobilized podands is determined by the degree of ethoxylation.

These immobilized PTCs exhibit efficiencies higher than crown ethers in many two-phase nucleophilic substitution reactions and oxidations (cf. Fig. 2). Compared with the corresponding not immobilized sucrose ethylene oxide adducts their catalytic activity is somewhat lower, but the immobilized catalysts can be separated from the reaction by filtration easily and reused several times without loss of effectiveness.

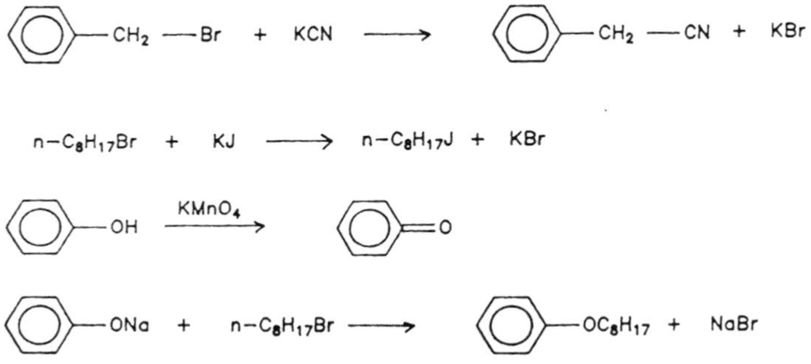

	% conversion
without catalyst	2 - 30
with immobilized sucrose ethyleneoxide adduct	40 - 100
with polyethyleneglycol 10 000	12 - 85
with dibenzo-18-crown-6	4 - 80

Fig. 2. Catalytic activity of immobilized sucrose ethyleneoxide adducts as phase transfer catalysts.

Immobilized Tris(2,2'-bipyridine)ruthenium(II) Complexes (Ru-bipy complexes)

Ru-bipy complexes – being efficient photosensitizers for the light induced water splitting – were immobilized by esterification of the hydroxy-groups with a new bipyridine carbonic acid and subsequent reaction with cis-Dichloro-bis(2,2'-bipyridine)ruthenium.

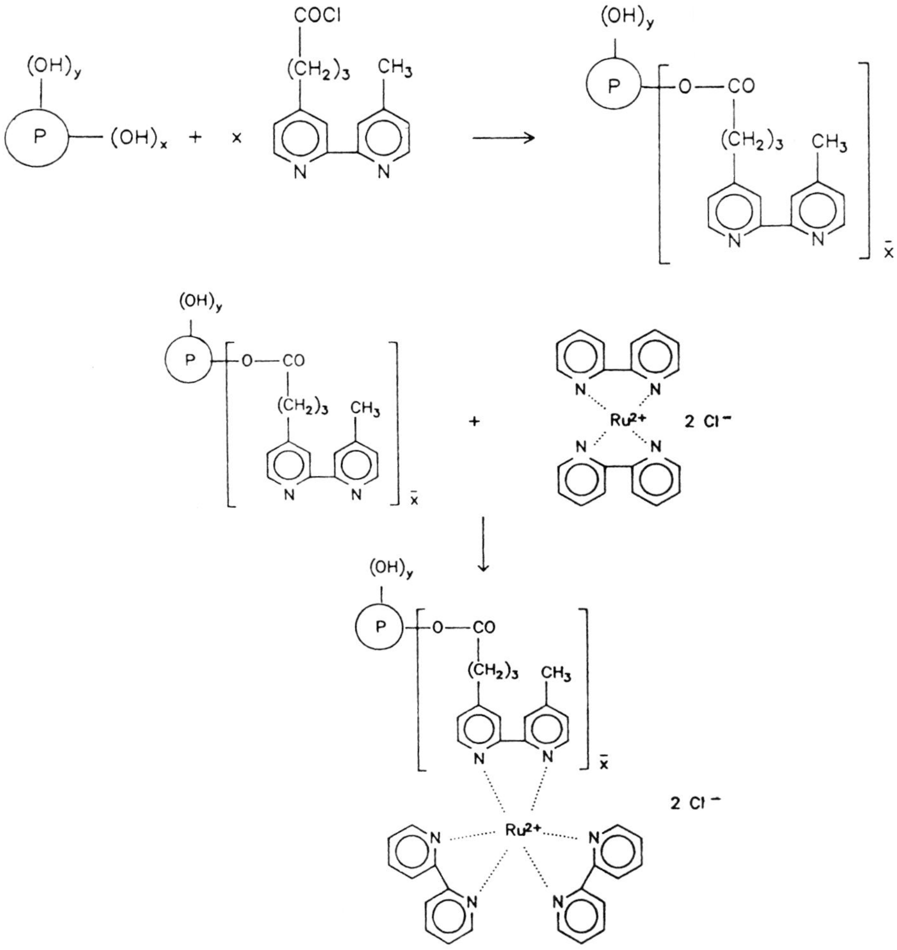

To achieve a high catalytic activity it is essential that the Ru-complex is bonded by a spacer of at least three methylene groups.[7] Photochemical water splitting reactions by the system:

♦ immobilized Ru-bipy complex,
♦ methylviologen,
♦ ethylenediamine tetraacetic acid, and
♦ Pt-catalyst

demonstrated much lower hydrogen production rates than with not immobilized Ru-complexes but the activity of the latter stops after about 40 hours whereas the immobilized complexes still are active after 8 days of irradiation with visible light, thus permitting linear hydrogen production over a longer period (Fig. 3).

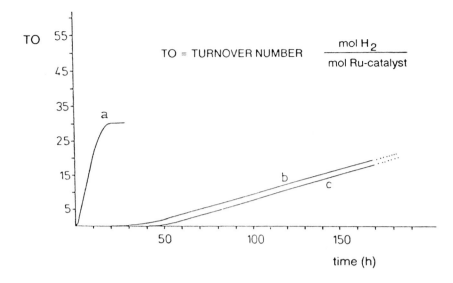

a...... low molecular Ru-bipyridine complex

b, c   immobilized Ru-bipyridine complexes

Fig. 3. Photolytic water splitting by the system Tris-(2,2'-bipyridine)ruthenium (immobilized on sucrose methacrylate gel), methylviologen, ethylene diaminetetraacetic acid and Pt catalyst.

Immobilized Girard Reagents

Girard reagents react with aldehydes and ketones to give hydrazones which can be splitted by acids to regenerate the starting compounds. This very specific reaction enables the detection and separation of carbonyl compounds from complex mixtures even in low concentrations.

For the immobilization of Girard-P reagent 4-vinylpyridine was polymerized in the presence of SM-gels with potassium peroxodisulfate as initiator to yield graft copolymers containing about three vinylpyridine residues per sucrose unit. Quaternization with chloroacetic acid ethyl ester and subsequent treatment with hydrazine gave the desired immobilized reagent.

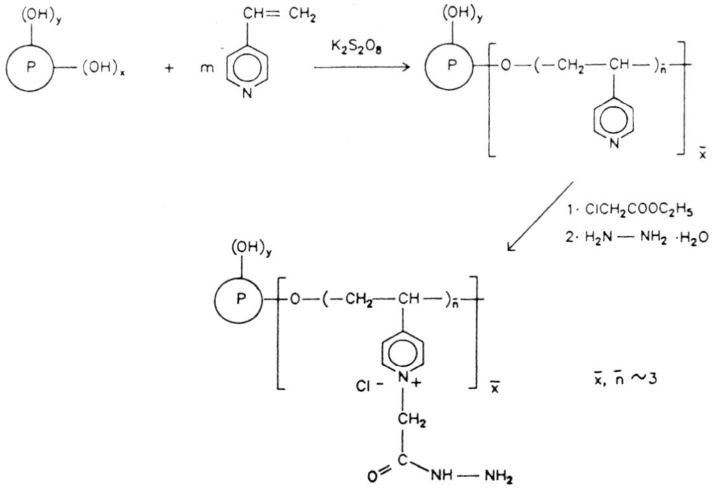

Girard-T groups were bonded to the gel by etherification of the hydroxy-groups with chloroalkylammonium chlorides and subsequent quaternization and hydrazinolysis:

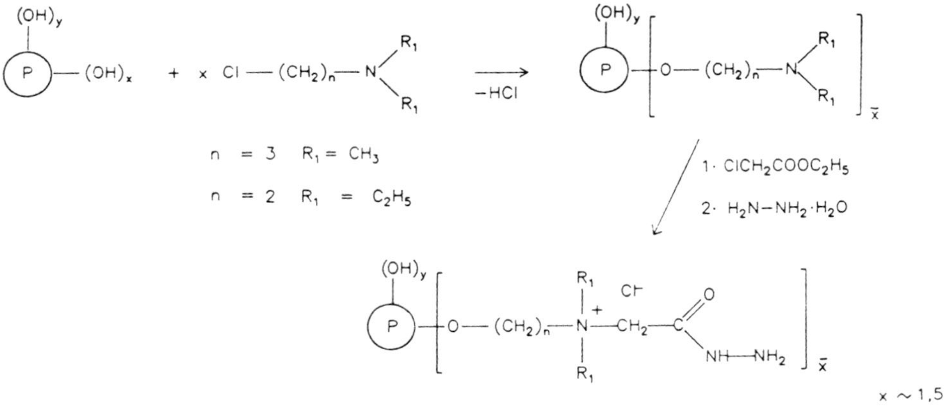

Investigations of the reactivity of the immobilized Girard reagents with benzaldehyde, 4-nitrobenzaldehyde, glycerolaldehyde, and reducing sugars like glucose, maltose, and cellobiose demonstrated much lower rates of hydrazone formation as well as regeneration compared with low molecular reagents which is attributed to steric hindrance.[8]

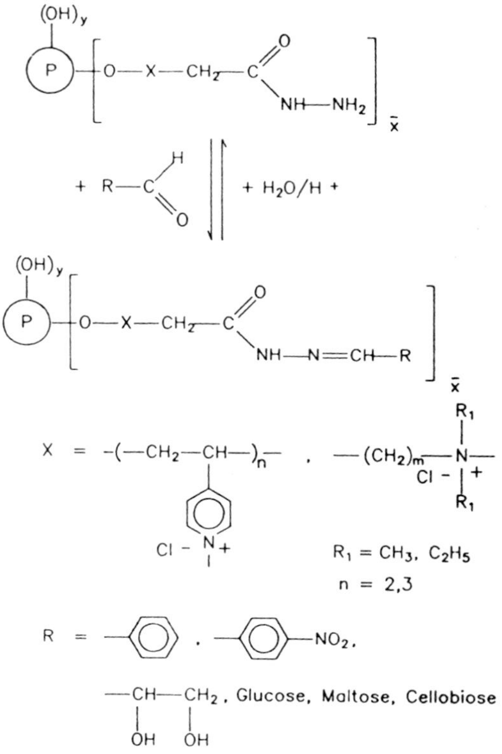

5. Sucrose Derivatives with Carboxylic Acid Amide Groups

Polyfunctional sucrose derivatives containing amide groups capable of reaction with formaldehyde are of special interest with respect to the utilization of sucrose as condensation component in formaldehyde resins. The synthesis of such derivatives by addition of acrylamide to sucrose in aqueous alkaline solution is well known but gives several byproducts which are difficult to separate because of their similar solubility.

Better yields could be obtained by hydrolyzing cyanoethyl sucroses with hydrogen peroxide to yield sucrose-β-amidoethyl ethers with DS corresponding to the cyanoethyl ether.[9] The starting material for this reaction is readily available and the hydrolysis proceeds without side reactions in slightly alkaline solutions, thus preventing further hydrolysis of the amide groups to carboxyl groups. Starting from sucrose and acrylonitrile the two-step reaction can be carried out as a one pot reaction too.

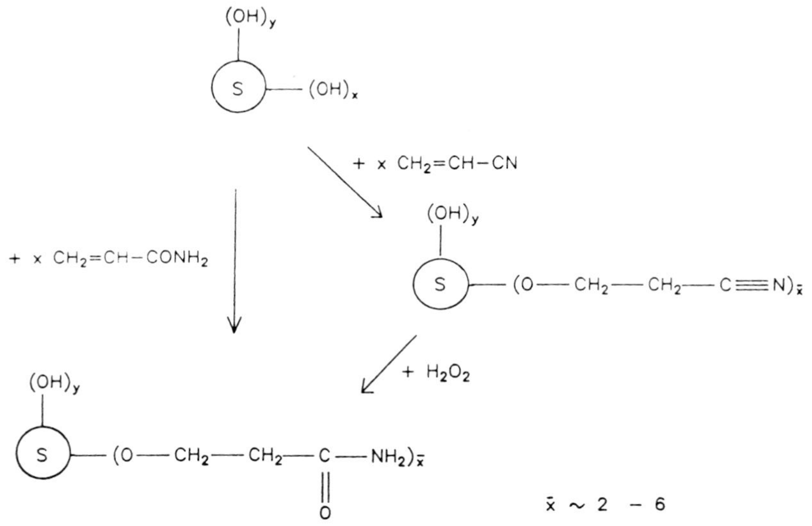

Sucrose-β-amidoethyl ethers can be methyolated with formaldehyde and therefore used as condensation component for commercial formaldehyde resins:

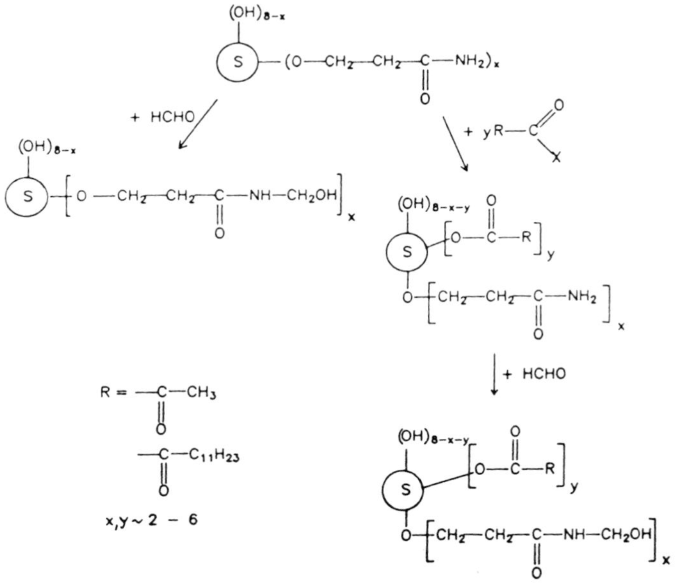

Conveniently they are condensed together with urea, melamine or phenol in the common manner with urotropine as a formaldehyde donor to give hydrophilic resins exhibiting better processibility and resistence as well as minor cracking than comparable not modified resins.[10]

Hydrophobic condensation components were obtained by esterification of the free hydroxy-groups of the sucrose-β-amidoethyl ethers with acetic anhydride or lauric acid chloride.

6. Sucrose Derivatives with Photoactive Groups

Photoresists are of great importance in electronic industries, e.g. for the manufacture of printed circuits. They are prepared by photochemical crosslinking of polymers containing cinnamic acid-, dimethylmaleinimdyl residues or aromatic azide groups, respectively. The former dimerize under UV-irradiation to give cyclobutane-residues by $2 + 2$ cycloaddition reaction whereas aromatic azide groups split off nitrogen to yield highly reactive nitrenes capable of crosslinking polymers by insertion reaction with C-H-bonds or addition to C=C-double bonds.

Sucrose derivatives with photoactive groups might offer various advantages, e.g.

♦ several photoactive groups can be in one molecule due to the polyfunctionality of sucrose, leading to low molecular photohardeners

♦ additional hydrophilic or hydrophobic substituents can be bonded, thus influencing the solubility of the photoresists

♦ adhesion of the photolaquers on various surfaces can be improved, for example by introduction of chelating groups

♦ they can be bonded to polymers by reaction of the free hydroxy-groups easily.

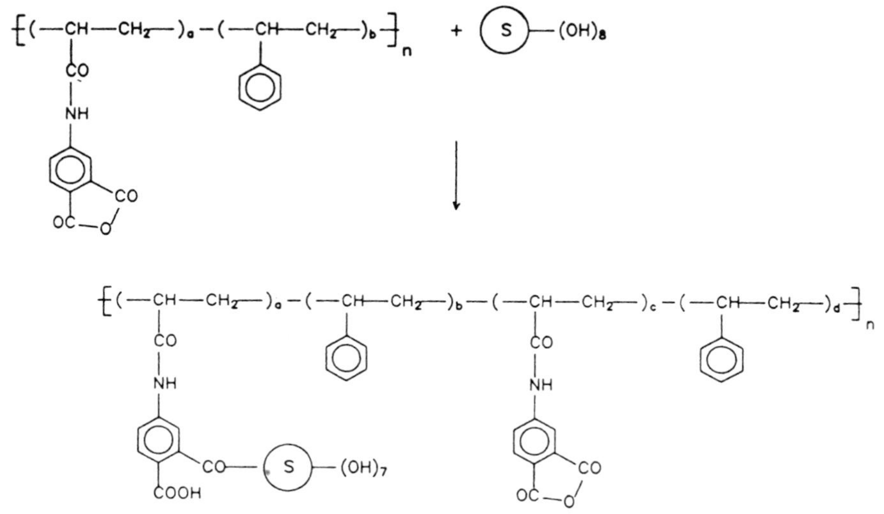

We have prepared photoactive sucrose derivatives by the preceding two methods. First we reacted sucrose with homo- and copolymers of 4-acrylamido-phthalic acid anhydride to give linear, aceton soluble polymers with pendant sucrose residues together with small amounts of insoluble crosslinked products.

Photoactive groups were bonded to the polymers by esterification of the hydroxy-groups with 4-cinnamido-phthalic acid anhydride or 4-dimethyl-maleinimidyl-phthalic acid anhydride, respectively, crosslinkable under UV-irradiation by intermolecular formation of cyclobutane rings to give adhesive films on aluminum plates:[11]

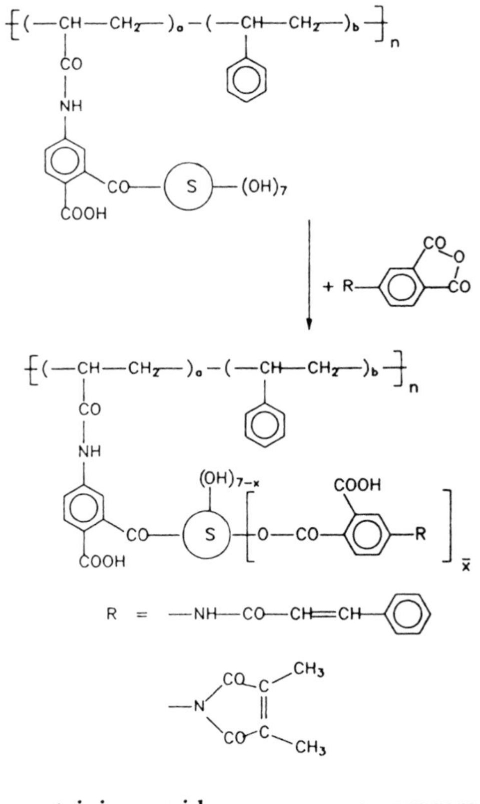

$$\left\{ \left(-CH-CH_2- \right)_a - \left(-CH-CH_2- \right)_b \right\}_n$$

$$+ \ R-\text{(phthalic anhydride)}$$

$$R = \ -NH-CO-CH{=}CH-\text{(phenyl)}$$

Photohardeners containing azide groups we prepared by a method not requiring covalent bonding of sucrose to polymers, thus applicable to all commercial polymers containing C-H- and / or C=C-double bonds. Hydrophilic sucrose-azide derivatives we obtained by esterification of sucrose with 3-azidophtalic acid anhydride:

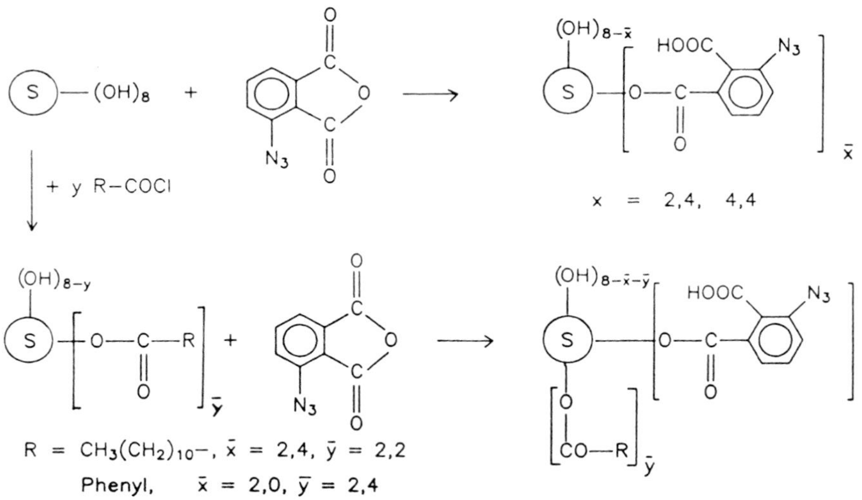

$$R = CH_3(CH_2)_{10}-, \ \bar{x} = 2,4, \ \bar{y} = 2,2$$
$$\text{Phenyl}, \quad \bar{x} = 2,0, \ \bar{y} = 2,4$$

Photocrosslinkable films can be prepared easily from solutions of the polymers and the sucrose derivatives containing about 2 to 6 aromatic azide groups.

By additional substitution of the residual free hydroxy-groups with laurinic acid or benzoyl residues hydrophobic derivatives exhibiting good solubility in organic solvents were prepared, which can be used as photohardeners for polystyrene, polymethylmethacrylate, or polyvinylacetate. The resulting images are characterized by high definition and good adhesion to copper plates.[12]

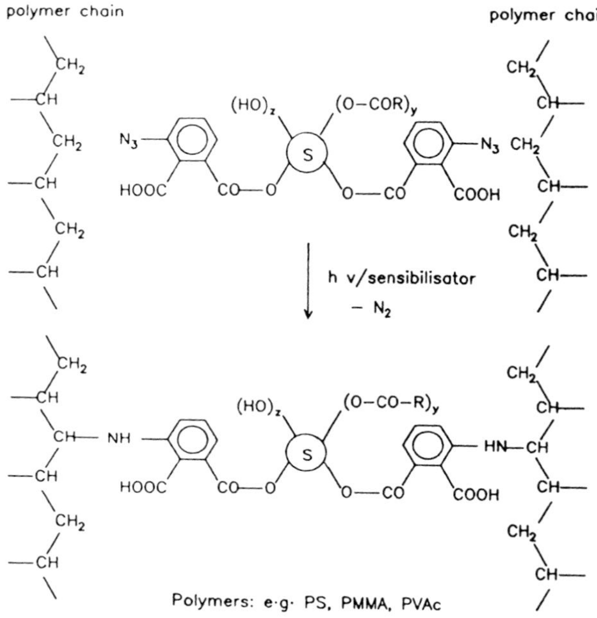

Polymers: e.g. PS, PMMA, PVAc

7. Sucrose Derivatives with Primary Amino Groups

Sucrose derivatives with primary amino groups are of special interest because of their enhanced reactivity which enables new and selective reaction of sucrose.

Although there are some patents concerning the catalytic hydrogenation of cyanoethyl-ethers of mono- and polyhydric alcohols, attempts to hydrogenate cyanoethyl sucrose to prepare sucrose-γ-aminopropylethers were not successful, probably due to the rough reaction conditions requiring temperatures higher than 100 °C in aqueous or liquid ammonia. $LiAlH_4$ or $NaBH_4/AlCl_3$ were not suitable as reduction reagents of the free OH-groups of sucrose. Successful reduction was accomplished by the complex $BH_3 \cdot (CH_3)_2S$ affording the desired γ-aminopropyl ethers in good yields:[13]

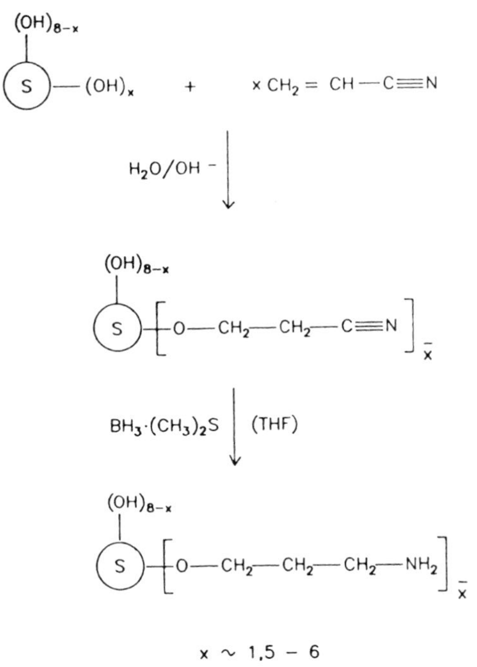

$$x \sim 1.5 - 6$$

A technically more interesting method for the preparation of sucrose-β-aminoalkylethers was accomplished by hydrogenation of sucrose-β-nitroalkylethers with common catalysts like Pt/carbon or Raney-Ni. The nitroalkyl ethers can be prepared readily by Michael addition of 1- or 2-nitroolefines to sucrose.[14]

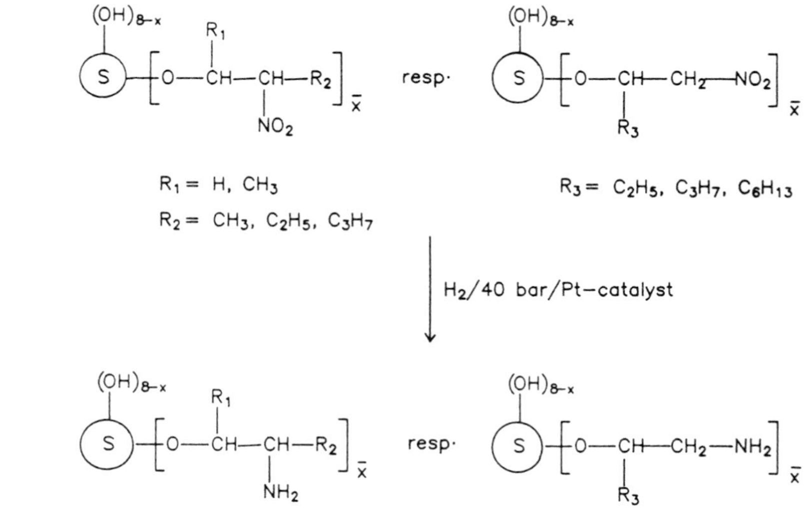

$(OH)_{8-x}$

$$S \left[O-\underset{\underset{NO_2}{|}}{CH}-\underset{\underset{}{|}}{CH}-R_2 \right]_{\bar{x}} \quad \text{resp.}$$

with R_1 substituent

$R_1 = H, CH_3$

$R_2 = CH_3, C_2H_5, C_3H_7$

$(OH)_{8-x}$

$$S \left[O-CH-\underset{\underset{R_3}{|}}{CH_2}-NO_2 \right]_{\bar{x}}$$

$R_3 = C_2H_5, C_3H_7, C_6H_{13}$

H_2/40 bar/Pt–catalyst

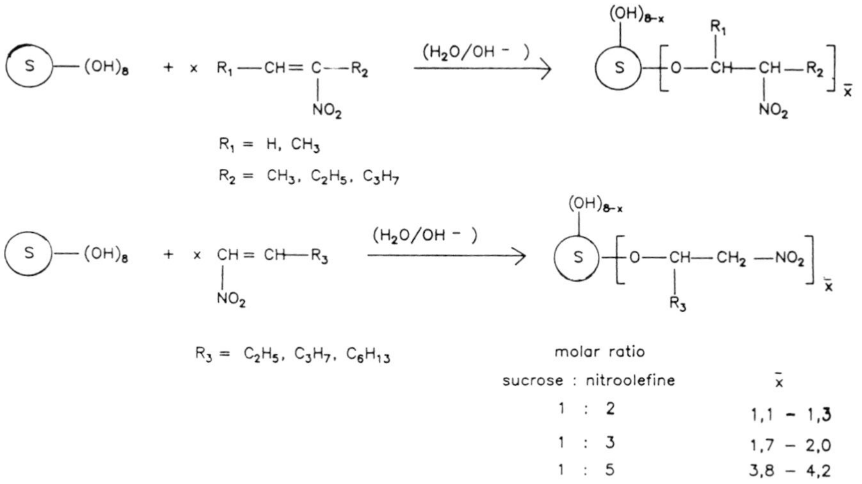

$(OH)_{8-x}$

$$S \left[O-\underset{\underset{NH_2}{|}}{CH}-CH-R_2 \right]_{\bar{x}} \quad \text{resp.}$$

with R_1 substituent

$(OH)_{8-x}$

$$S \left[O-CH-\underset{\underset{R_3}{|}}{CH_2}-NH_2 \right]_{\bar{x}}$$

$$S-(OH)_8 + x \; R_1-CH=\underset{\underset{NO_2}{|}}{C}-R_2 \xrightarrow{(H_2O/OH^-)} S \left[O-\underset{\underset{}{|}}{CH}-\underset{\underset{NO_2}{|}}{CH}-R_2 \right]_{\bar{x}}$$

with $(OH)_{8-x}$ and R_1

$R_1 = H, CH_3$

$R_2 = CH_3, C_2H_5, C_3H_7$

$$S-(OH)_8 + x \; CH=\underset{\underset{NO_2}{|}}{CH}-R_3 \xrightarrow{(H_2O/OH^-)} S \left[O-CH-\underset{\underset{R_3}{|}}{CH_2}-NO_2 \right]_{\bar{x}}$$

with $(OH)_{8-x}$

$R_3 = C_2H_5, C_3H_7, C_6H_{13}$

molar ratio sucrose : nitroolefine	$\bar{x}$
1 : 2	1,1 – 1,3
1 : 3	1,7 – 2,0
1 : 5	3,8 – 4,2

Due to the electron withdrawing effect of nitro groups, the nitroolefines are excellent Michael acceptors, which can be prepared by reaction of nitroalkanes with aldehydes to give nitroalcohols, acetylation and subsequent alkaline treatment to split off acetic acid. Avantageously the nitroolefines are generated in situ by reaction of sucrose with nitroalkylacetates in slightly alkaline aqueous solutions.[14]

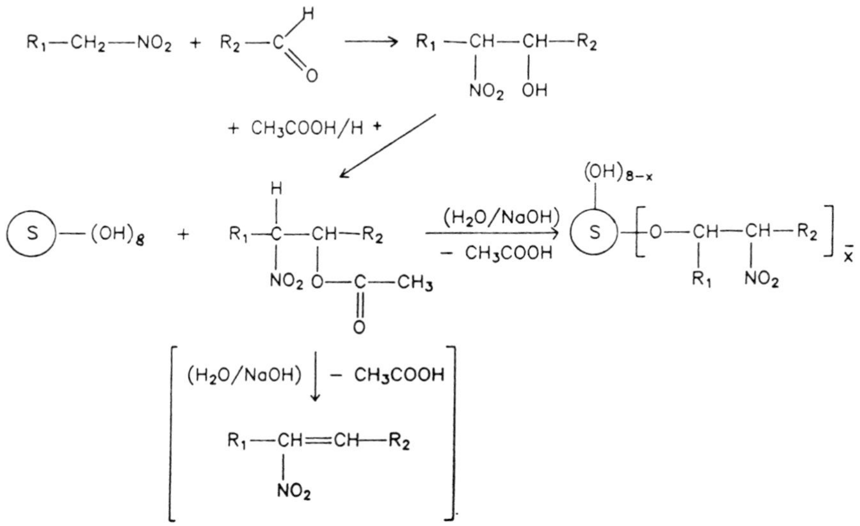

$R_1-CH_2-NO_2$ + $R_2-C{\overset{H}{\underset{O}{}}}$ $\longrightarrow$ $R_1-\underset{NO_2}{CH}-\underset{OH}{CH}-R_2$

+ CH_3COOH/H^+

$\boxed{S}-(OH)_8$ + $R_1-\underset{NO_2}{\overset{H}{C}}-\underset{O-C-CH_3}{\underset{\underset{O}{\parallel}}{CH}}-R_2$ $\xrightarrow[-\,CH_3COOH]{(H_2O/NaOH)}$ $\boxed{S}\left[O-\underset{R_1}{\underset{|}{CH}}-\underset{NO_2}{CH}-R_2\right]_x^{(OH)_{8-x}}$

$\left[\begin{array}{c}(H_2O/NaOH)\downarrow\ -\ CH_3COOH\\[2mm] R_1-CH=\underset{NO_2}{CH}-R_2\end{array}\right]$

Reaction of sucrose-β-aminoalkylethers with long chain aliphatic carbonic acid chlorides affords carbonic acid amides exhibiting surface active properties due to the hydrophilic sucrose residue:[15]

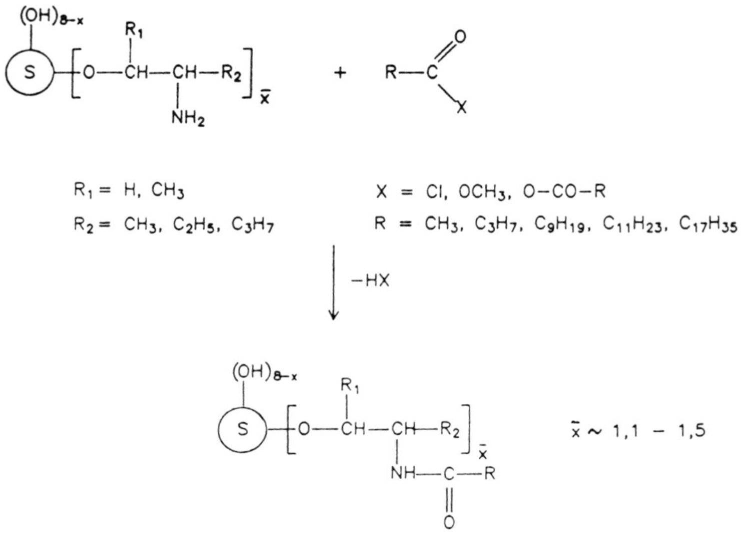

$\boxed{S}\left[O-\underset{\underset{NH_2}{|}}{\overset{R_1}{\overset{|}{CH}}}-CH-R_2\right]_{\bar x}^{(OH)_{8-x}}$ + $R-C{\overset{O}{\underset{X}{}}}$

$R_1 = H,\ CH_3$ $X = Cl,\ OCH_3,\ O-CO-R$
$R_2 = CH_3,\ C_2H_5,\ C_3H_7$ $R = CH_3,\ C_3H_7,\ C_9H_{19},\ C_{11}H_{23},\ C_{17}H_{35}$

$\downarrow\ -HX$

$\boxed{S}\left[O-\underset{\underset{NH-C-R}{\underset{\underset{O}{\parallel}}{|}}}{\overset{R_1}{\overset{|}{CH}}}-CH-R_2\right]_{\bar x}^{(OH)_{8-x}}$ $\bar x \sim 1,1\ -\ 1,5$

Surface active derivatives containing urea groups can be prepared by reaction with alkyl isocyanates in low alcohols or even in water as solvent because isocyanates react generally much faster with amines than with alcohols or water.[16]

$$\underset{\substack{| \\ S}}{(OH)_{8-x}} \left[O - \underset{\substack{| \\ NH_2}}{\overset{\substack{R_1 \\ |}}{CH}} - CH - R_2 \right]_{\bar{x}} \quad + \quad R-N{=}C{=}O$$

$$R_1 = H, CH_3 \qquad\qquad R = C_4H_9, C_{12}H_{25}, C_{18}H_{37}$$
$$R_2 = CH_3, C_2H_5, C_3H_7$$

$$\downarrow$$

$$\underset{\substack{| \\ S}}{(OH)_{8-x}} \left[O - \underset{\substack{| \\ NH-C-NH-R \\ \| \\ O}}{\overset{\substack{R_1 \\ |}}{CH}} - CH - R_2 \right]_{\bar{x}}$$

$$\bar{x} \sim 1{,}1 - 1{,}5$$

With respect to the desired high surface activity the utilization of low substituted sucrose-β-aminoethylethers (DS 1.1 - 1.5) is preferable in these reactions thus yielding amide- or urea derivatives which reduce the surface tension of water to about 28 mN m^{-1} at very low critical micellization concentrations of 0.15 g/l. Contrary to the known surface active fatty acid esters of sucrose the amides and ureas exhibit much better hydrolytical stability. Thus, no change in surface activity could be observed when heating to 90 °C at pH 12 for several hours.

8. Sucrose Derivatives with Biologically Active Groups

The application of biologically active substances in human and veterinary medicine as well as in plant protection frequently is impeded by their poor water solubility, requiring additional solubilizers that may be toxic or give rise to side effects. Drugs with poor water solubility can be administered orally only, thus causing gastrointestinal complaints when used in high dosages. Besides oral application of high dosages generally fails to accumulate the drug in the deceased part of the body. Therefore often intravenous or subcutaneous administration of high dosages on the affected part is desirable.

Sucrose- and glucose-residues seem to be potential candidates for increasing the water solubility of drugs because of their biocompatibility and environmental innocuity, provided that they are bonded to the active substances by easy hydrolizable bonds, e.g. of the ester type. For that purpose we reacted a number of carboxyl group containing drugs with sucrose and glucose, e.g. the analgetic agent acetyl salicylic acid, antiphlogistic agents like ketoprofen, indomethacin, and naproxene, the anticonvulsant valproic acid, the herbicide Dichlorprop and the growth regulator naphthylacetic acid:[17]

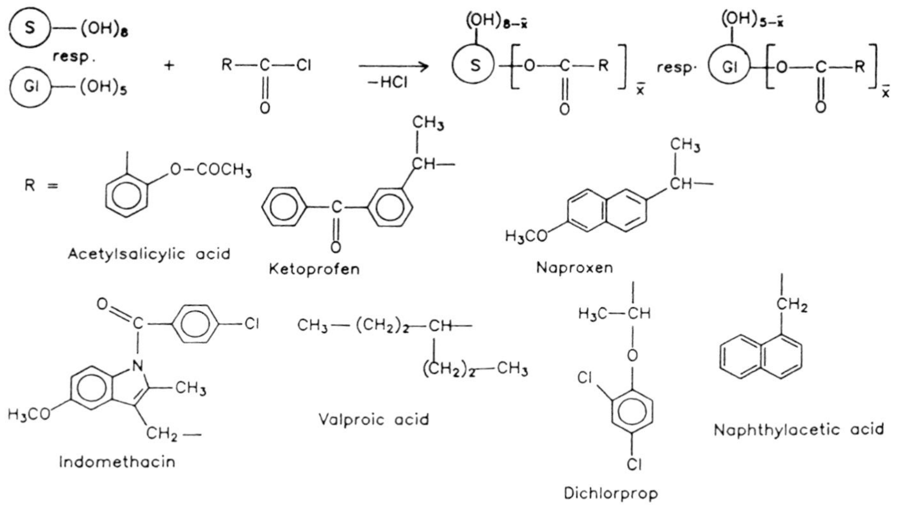

The carbohydrate esters of these drugs exhibit water solubilities 10 - 400 fold better than the basic substances provided that the DS does not exceed 2 for the sucrose ester and about 1 for the glucose esters because the solubility decreases rapidly with higher degree of substitution. Investigations on the activity of these esters are in progress.

References

1. H. Gruber: Hydrophile Polymergele mit reaktiven Gruppen, I. Herstellung und Polymerisation von Glucose- und Saccharosemethacrylaten. *Monatsh. f. Chem.* **112** (1981) 273-285.

2. H. Gruber: Hydrophile Polymergele mit reaktiven Gruppen, IV. Chelatharze mit schwefelhaltigen Ankergruppen auf Basis von Saccharosemethacrylaten. *Monatsh. f. Chem.* **112** (1981) 747-758.

3. H. Gruber, N. Gamsjäger: Hydrophile Polymergele mit reaktiven Gruppen, V. Chelatharze mit phosphorhaltigen Ankergruppen auf Basis von Saccharosemethacrylaten. *Monatsh. f. Chem.* **115** (1984) 1329-1333.

4.   H. Gruber: Hydrophile Polymergele mit reaktiven Gruppen, II. Saccharosemethacrylaten. *Monatsh. f. Chem.* **112** (1981) 445-457.

5.   H. Gruber: Hydrophile Polymergele mit reaktiven Gruppen, III. Chelatharze mit Formazan-Ankergruppen auf Basis von Saccharosemethacrylaten. *Monatsh. f. Chem.* **112** (1981) 587-593.

6.   H. Gruber, G. Greber: Phasentransferkatalysatoren auf Basis von Saccharose-Ethylenoxid-Addukten. *Monatsh. f. Chem.* **112** (1981) 1063-1076.

7.   W. Nußbauer, H. Gruber, G. Greber: Immobilisierte Tris(2,2'-bipyridin)ruthenium(II)-Komplexe als Photosensibilisatoren für die Wasserphotolyse. *Makromol. Chem.* **189** (1988) 1027-1033.

8.   A. Schernhammer: Immobilisierte Girard-Reagentien. *Doctoral Dissertation*, TU Wien, 1989.

9.   G. Greber, H. Gruber: Verfahren zur Herstellung von β-Amidoethylether-Derivaten von Kohlenhydraten. *Austrian Pat.* 369 383 (1982); *Chem. Abstr.* **98** (1983) 128080s.

10. G. Greber, H. Andres, W. Pichler: Verfahren zur Herstellung von neuen, vernetzten Harzen auf Basis von N-Methylol-Gruppen enthaltenden Kohlenhydrat-Derivaten. *Austrian Pat.* 359 287 (1980); *Chem. Abstr.* **94** (1981) 16573r.

11. W. Pichler: Herstellung und Anwendung neuer reaktiver Saccharosederivate. *Doctoral Dissertation*, TU Wien, 1981.

12. G. Greber, H. Andres, M. Weber: Zusatz zu durch UV-Licht vernetzbaren Polymeren. *Austrian Pat.* 381 324 (1986); *Chem. Abstr.* **104** (1986) 149346r.

13. G. Greber, H. Gruber: Verfahren zur Herstellung von neuen Kohlenhydrat-Derivaten mit etherartig gebundenen γ-Aminopropyl-Gruppen. *Austrian Pat.* 384 811 (1988); *Chem. Abstr.* **109** (1988) 6896n.

14. G. Greber, H. Gruber: Verfahren zur Herstellung von neuen Kohlenhydrat-Derivaten mit β-Nitroethyl- bzw. β-Aminoethyl-Gruppen. *Austrian Pat.* 381 498 (1986); *Chem. Abstr.* **104** (1986) 168767.

15. G. Greber, H. Gruber: Verfahren zur Herstellung neuer, grenzflächenaktiver Kohlenhydrat-Derivate. *Austrian Pat.* 382 381 (1987); *Chem. Abstr.* **108** (1988) 114719.

16. G. Greber, H. Gruber: Verfahren zur Herstellung neuer, grenzflächenaktiver Kohlenhydrat-Derivate. *Austrian Pat.* 386 413 (1988); *Chem. Abstr.* **110** (1989) 137536p.

17. G. Greber, H. Gruber: Verfahren zur Herstellung von neuen Kohlenhydrat-Estern der Acetylsalicylsäure. *Austrian Pat.* 382 624 (1987); *Chem. Abstr.* **110** (1989) 115263v.

5

Convenient Synthesis and Application
of Sucrose Sulfates

Wiesław Szeja

Silesian Technical University, 44-101 Gliwice, Poland

Summary. Salts of sugar sulfates play important roles in medicin as antiulcer and antipeptic agents. Their preparation and the selectivity in sulfation of sucrose is the topic of this paper. The influence of different solvents upon sulfation of sucrose is examined using different sulfur trioxide base complexes, $SO_3 \cdot DMF$ proving to be the reagent of choice. Special attention is paid to the economy of the preparation and isolation process. For this reason the use of a flow-reactor is presented. The suitability of the sucrose sulfates thus obtained for the preparation of surfactants, emulsifiers, and resins is discussed.

1. Introduction

The chemistry of sucrose has become subject of intensive investigations initiated mainly by the fact that the global production of this inexpensive and reproducible raw material exceeds the annual demand by several million tons. The development of improved synthetic methods and analytical techniques has led to the preparation and characterization of a large number of sucrose derivatives, that have been covered in a number of excellent reviews.[1-4] The actually achieved commercial success for sucrose-based products, however, has been rather insignificant so far. This might be commented upon according to Hickson:[1] "I believe it to be more the fault of the molecule than the chemist". It seems that the last word has not yet been said on this subject, and the wide application of sucrose as a raw material is a question of the nearest future. Its applicability will depend, to a large extent, on the price of other raw materials, particularly crude oil, as well as on the elaboration of economical methods for its derivatization. Although chemists dealing with synthetic sugar-chemistry exert little influence on the prices of the fundamental chemical raw materials, the latter aspect ought to be taken into consideration in an ever growing degree.

Much attention has been focused on the physiological properties of sulfated sugars, particularly on their enhanced antiulcerogenic activity;[5,6] the aluminium salt of sulfated sucrose ("Sucralfate"), for example, is a popular antiulcer drug with antipeptic activity.[7] In connection with this, many attempts have been made towards the synthesis of sucrose octasulfate.[8-13] The methods described apply the usual reagents, i.e. the complex formed from sulfur trioxide and pyridine.[14] Accordingly, the successful preparation of sucrose sulfates by means of this reagent in pyridine[8] or dimethylformamide[9] as a solvent has been reported, an essential difficulty being the capricious isolation of the product. The reaction mixture is usually neutralized with barium hydroxide, the barium sulfate is separated and the aqueous solution is concentrated, to yield the sulfated sucrose as the barium salt.

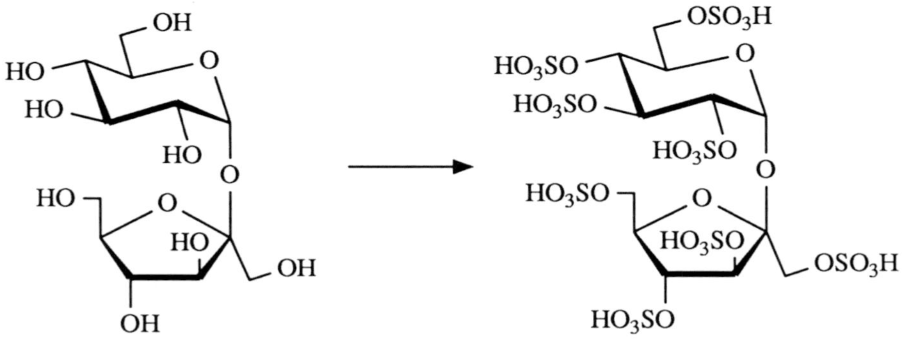

Sucrose   Sucrose octasulfate

2. The Sulfation Process

Analyzing the conventional methods of synthesizing sugar sulfates we arrived at the conclusion that the economy of the process is highly influenced by the cost of the solvents used, as well as by the laborious and energy-consuming technique of their purification. The necessity of disposing of the resulting wastes and sewages not only endangers the environment, but strongly affects the costs of obtaining sulfated sucrose. The purpose of taking up investigations in our laboratory was to eliminate or at least to reduce the obvious shortcomings of the sulfation process.

The simplest solution, conceivably, would be to find conditions by which the saccharide is sulfated directly with sulfur trioxide. Preliminary tests proved to be little encouraging since the high reactivity of this reagent merely generated mixtures of unidentified tarry products. Better results were achieved under heterogenous conditions, i.e. by treating a suspension of sucrose in dichloromethane with equivalent amounts of sulfur trioxide. This method was considerably simplified by sonification of the sugar suspension in dichloromethane for 30 minutes at 0 °C followed by dropwise addition of the solution of sulfur trioxide in the same solvent. Separation of the resulting sulfation mixture was achieved by absorbing the sulfates on Dowex-1 resin and their subsequent elution with an increasing concentration of sulfuric acid.[15] Nevertheless, the sulfur trioxide caused some degradation, the product formed mainly being a mixture of isomeric monosulfates with the glucose-6-sulfate (65 %)[16,17] as the major product. The disadvantage of sulfating with sulfur trioxide is its high reactivity, the strongly exothermic reaction resulting in partial decomposition of the sugar. This can be overcome by moderating the reactivity of SO_3 by using complexes of sulfur trioxide with organic compounds which differ in basicity. The bases employed were amines, amides, and ethers. The complexes were prepared at 0 °C by dropwise addition of equivalent amounts of sulfur trioxide to the solution of a base in dichloromethane. Sonically dispersed sucrose in a non-polar solvent was treated with a suspension of the sulfur trioxide complex. In order to estimate the progress of the reaction, the samples were neutralized with an aqueous barium hydroxide solution, and the amount of barium sulfate formed was determined. Separation and preliminary identification of sugar sulfates proved to be possible via paper chromatography.[18] The sulfation strongly depended on the structure of the complex. Thus, when 1 mole of sucrose was treated with 1 mole of adduct at -10 °C, the major product isolated was glucose-6-sulfate. Treatment of sucrose with the acetamide · SO_3 complex yielded sucrose mono- and disulfates in yields of 75 and 8 %, respectively. Under these conditions the complex SO_3 · pyridine was unreactive, and minor amounts of the sulfate (~5 %) were isolated.

More detailed investigations of the formation of sucrose sulfates were carried out applying complexes of SO_3 with amides differing in basicity. It has been found that within the temerature range of -20 to 0 °C the process proceeds so rapidly that the rate of sulfation could not be determined. Using sulfur trioxide · dimethyl-formamide, the most advantageous variant was to create conditions in which the product precipitates from the reaction mixture. For this purpose, tetrachloroethylene proved to be highly advantageous; it is mixable with SO_3 in any ratio, the solution may be stored for many hours at 0 °C, and the sulfur trioxide dissolved therein does not undergo polymerization to form a solid product – all facts that considerably simplify the generation of the $SO_3 \cdot$ dimethylformamide adduct. Accordingly, applying adequate proportions of reagents, the product of the sulfation of sucrose can be isolated in the form of an oil containing slight amounts of dimethylformamide.[19] These conditions could readily be applied to a flow-reactor, such that to the adduct, obtained by mixing the DMF solution in tetrachloroethylene with a solution of SO_3 in the same solvent, a supersaturated sucrose solution in DMF was introduced, followed by direction of the reaction mixture to the separator (see Fig. 1). The sulfated sucrose was isolated as an oil, whereas the solvent after addition of dimethylformamide is recycled.

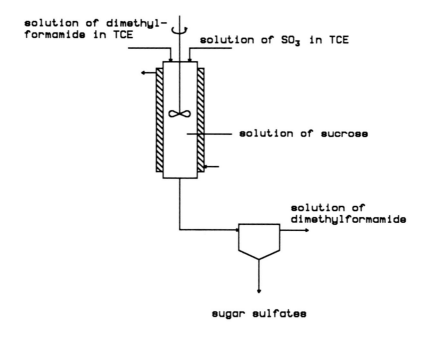

Fig. 1. Large-scale sulfation of sucrose in a flow reactor (TCE= tetrachlorethylene)

When the sucrose was treated with 8 moles of adduct, a mixture of hepta- and octasulfates was obtained in a yield of 92 %. After neutralization with sodium hydroxide solution, the product obtained was treated with basic aluminium chloride, $Al_2(OH)_5Cl$, to yield a product identical with the medicine known as Sucralfate, Antepsin, Iselpin, Ulcermin, Ulcogant, or Ulsanic.

3. On the Regioselectivity of Sulfation

The literature records numerous examples of reactions involving partial substitution of sucrose in which selectivity for certain positions, especially *O*-6, has been proposed.[1-4] In an attempt to examine the relative extent of sulfation of hydroxyl groups in the glucose and fructose portions of sucrose, model studies were untertaken on the partial sulfation of methyl α-D-glucopyranoside (**1**) and methyl β-D-fructofuranoside (**2**) in a two-phase solid-liquid system. Thus, an equimolar mixture of **1** and **2** in tetrachloroethylene was sonicated for 30 minutes and allowed to react with molar equivalents of the $SO_3 \cdot DMF$ complex at -10 °C for 5 minutes. The reaction mixture was neutralized with aqueous sodium carbonate, the solution of sugars was passed through the column, the sulfates were absorbed on the Dowex-1 anion exchange resin, the eluent was concentrated and the composition of glycosides was determined by ^1H NMR spectroscopy.

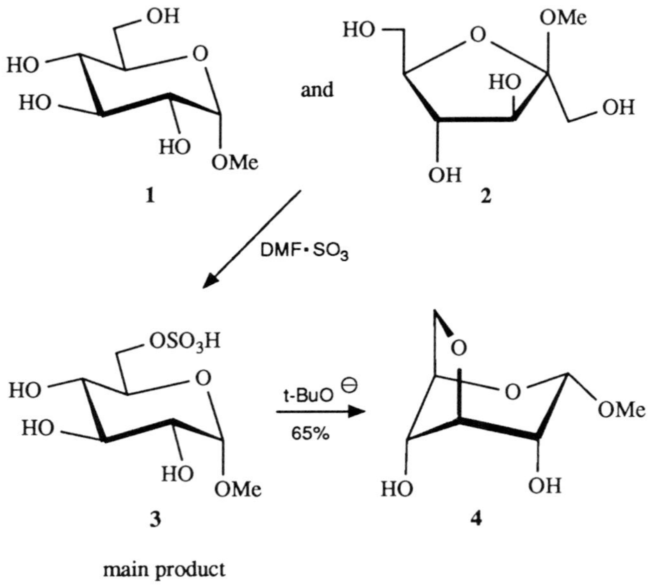

The results demonstrate that sulfation occurs to a considerably higher extent in the glucose moiety, the regioselectivity being around 8:1. When the resulting glucose sulfates containing **3** as the main component were treated with sodium *t*-butoxide, methyl 3,6-anhydro-α-D-glucopyranoside (**4**) was isolated in 65 % yield, clearly indicating that sulfation had preferentially occured at the primary hydroxyl group.[20]

The introduction of a sulfate group in the anomeric position was evaluated in a model monosaccharide system. It would appear that the C-1 hydroxyl is unique, either in not forming a sulfate or, if one is formed, this ester decomposes so readily that it cannot be isolated under the usual conditions.[21] Since direct sulfation of sugars leads to a mixture of products we have started with derivatives having the anomeric hydroxyl group free: treatment of 2,3:5,6-di-*O*-isopropylidene-D-mannofuranose (**5**) with the $SO_3 \cdot$ DMF adduct indeed yields a monosulfate, yet the product **6** was not isolated due to its high reactivity; however, adding thiophenoxide ion to the reaction mixture in the presence of potassium carbonate the thioglycoside **7** was formed in high yield (37 %). The α-configuration of glycoside was determined by means of ^{1}H NMR spectroscopy.[16]

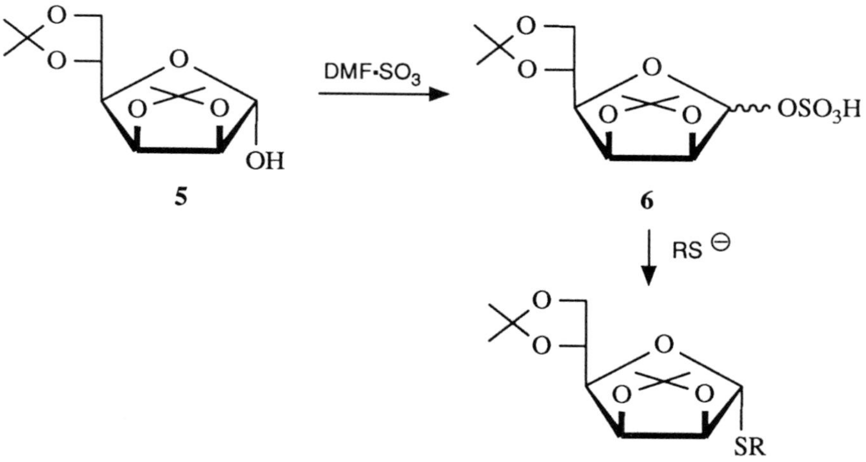

4. Applications of Sulfated Sucrose

This simple and efficient method of sulfating sucrose, described here, makes it possible to utilize the materials obtained in practice. Most attractively, they might be applied as surfactants, plastics, and pharmaceuticals. Long chain fatty acid esters of sucrose have indeed shown promise as surfactants and compare well in overall performance with other surface active compounds in their detergency and emulsification potential.[22] The commercial use of these esters has been limited though, partly because remnants of the solvent, usually dimethylformamide, remain rendering the product unsuitable as a food emulsifier. One can suppose that by sulfation of esters, this disadvantage can be omitted. The sulfation of sugar esters has been conducted as usual, by immersion of the reaction vessel in an ultrasonic cleaning bath. A solution of sucrose stearates in dichloromethane was treated with equimolar amounts of the $SO_3 \cdot DMF$ complex at 0 °C for 5 minutes, subsequent neutralization by solid sodium hydrogen carbonate, filtration of the sodium salts of the sucrose sulfate esters and washing with acetone. The product shows good emulsifying properties and stable emulsions of oil in water as well as water in oil are formed.[23]

$$
\text{1. MeO\overset{O}{\overset{\|}{C}}R} \quad \text{2. } SO_3 \cdot DMF
$$

R = $C_{17}H_{35}$

The other possible application of sulfated sucrose is in plastics industry, i.e., as commercial resins and polymers. A wide range of substances will react chemically with epoxy resins to form a modified polymer system.[24] One can expect that sulfated sucrose will perform several functions at the same time, since the sulfate group can act as cure accelerator, whereas the hydroxyl groups may alter the properties of the epoxy resins through crosslinking. Thus, copolymerization of mono- and disulfate fraction of sucrose (5 parts) with the epoxy resin Epidian 5 (100 parts) at 110 °C for 1 h gave a resin, of which the chemical resistance was determined by its percentage weight change after immersion in a reagent; on this basis it proved to be inferior to the resin prepared using triethylene tetramine as the curing agent. Good results were obtained when the epoxy resin was blended with other resins of the phenol-formaldehyde or formaldehyde-melamine type.[23] This system can be used in the paint industry, since such a mixture of resins and sucrose mono- and disulfates has a long life before breakdown under working conditions. The copolymerization takes place on heating to 140 - 160 °C. The chief properties of these epoxy resin coatings are their chemical resistence, excellent adhesion to metal, good electrical properties in terms of dielectric strength and resistance, thus making them highly useful as systems for the electric industry.

References

1. L. Hough: Sugar (J. Yudkin, J. Edelman, L. Hough, Eds.), Butterworth, London, 1972, pp. 49 ff.

2. R. Khan: The Chemistry of Sucrose. *Adv. Carbohydr. Chem. Biochem.* **33** (1976) 235-294. – Chemistry and New Uses of Sucrose: How Important ? *Pure Appl. Chem.* **56** (1984) 833-844.

3. C. E. James, L. Hough, R. Khan: Sucrose and its Derivatives. *Progr. Chem. Org. Natl. Products* **55** (1989) 117-184.

4. T. Vydra, K. Čapek, *Chemicke Listy*, **83** (1989) 686 ff.

5. S. Levey, S. Sheifield: Inhibition of the Proteolytic Action of Pepsin by Sulfate Containing Polysaccharides. *Gastroenterology*, **27** (1954) 625-628; *Chem. Abstr.* **50** (1956) 1097b.

6. W. Anderson, J. Watt: Carageenin Inhibition of Peptic Activity and Combination with Gastric Mucin. *J. Pharm. Pharmacol.* **11** (1959) 318; *Chem. Abstr.* **53** (1959) 16379c.

7. I. M. Samloff, J. Scand. *J. Gastroent.* **18** (1983) 7.

8. M. Namekata, T. Tanaka, N. Sakamoto, K. Moro: Oligosaccharide Sulfates and Monosaccharide Sulfates for Medical Purposes. Antiulcerogenic Properties of Sucrose Sulfate-Aluminium Complex. *Yakugaku Zasshi* **87** (1967) 889; *Chem. Abstr.* **68** (1968) 20831d.

9. K. Ochi, Y. Watanabe, K. Okui, M. Shindi: Crystalline Salts of Sucrose Octasulfate. *Chem. Pharm. Bull.* **28** (1980) 638-341.

10. Chugai Pharmaceutical Co.: Disaccharide Aluminium Polysulfates. *Jap. Pat. Appl.* (1965), *U.S. Pat.* 3,432,489, *Fr. Pat.* 1,500,571 (1967); *Chem. Abstr.* **69** (1968) 97100t.

11. T. E. Vasquez, R. L. Bridges, P. Braunstein (Univ. California, Berkeley): Diagnostic Procedures Using Radiolabeled Suralfate and Derivatives or its Precursors. *Eur. Pat.* 109,072 (1982); *Chem. Abstr.* **101** (1982) 60163u.

12. K. Steiner (Merck Patent GmbH): Use of Sucralfate in Treating Emesis and Diarrhoea in Veterinary Medicine. *Ger. Pat.* 3,322,078 (1983); *Chem. Abstr.* **102** (1985) 72871g.

13. K. Miura, H. Nagasaka, N. Takahashi, T. Ochiai, R. Takasaki (Mitsubishi Chem. Ind. Co.): Positive Photosensitive Compositions Useful as Photoresists. *Eur. Pat.* 136,110 (1983); *Chem. Abstr.* **103** (1985) 62600z.

14. R. B. Duff: Carbohydrates Sulphuric Esters. Demonstration of Walden Inversion on Hydrolysis of Barium 1,6-Anhydro-β-D-galactose-2-sulfate. *J. Chem. Soc.* **1949**, 1597-1600.

15. A. G. Lloyd: Fractionation of the Products of the Direct Sulphation of Monosaccharides on Anion-Exchange Resin. *Biochem. J.* **83** (1962) 455-460.

16. I. Fokt, W. Szeja, unpublished results.

17. R. L. Whistler, W. W. Spencer, J. N. BeMiller: D-Glucose-3-Sulfate and 6-Sulfate. *Methods Carbohydr. Chem.* **2** (1963) 298-303.

18. D. A. Rees: Paper Chromatography of Acidic Carbohydrates. *Nature* **185** (1960) 309-310.

19. W. Szeja, *Polish Pat.* (1989), in examination.

20. W. Szeja, unpublished results.

21. J. R. Turvey: Sulfates of the Simple Sugars. *Adv. Carbohydr. Chem.* **20** (1965) 183-219.

22. J. C. Colbert: Sugar Esters; Preparation and Application. *Noyes Data Corporation*, New Jersey, 1974.

23. T. Bieg, W. Szeja, *Polish Pat.* (1989), in examination.

24. W. G. Potter: Epoxide Resins. Butterworths, London, 1970.

6

Sucrose-based Hydrophilic Building Blocks as Intermediates for the Synthesis of Surfactants and Polymers

Markwart Kunz

Institut für landwirtschaftliche Technologie und Zuckerindustrie
an der Technischen Universität Braunschweig, D-3300 Braunschweig, Germany

Summary. Due to difficulties to secure pure mono-substituted derivatives of sucrose by direct etherification or esterification, new reaction pathways for its utilization as a basic chemical are presented.

In a two step sequence, comprising a biochemical transformation of sucrose into reducing disaccharides followed by chemical modification, specific mono-derivatized polyhydroxy compounds can be produced. Main topic of this account are investigations on the application of three types of reactions to the polyhydroxy compounds thus obtained, i.e. oxidative degradation, reductive amination (hydrogenation), and catalytic oxidation, as well as the utilization of the products for the preparation of novel surfactants and hydrophilic polymers.

1. Introduction

Surfactants and hydrophilic polymers are produced by the chemical industry in huge amounts. By far most of these products are based on petrochemicals. For the hydrophilicity being of major importance for the application-properties of surfactants and polymers, a large number of attempts have been reported to use carbohydrates, especially sucrose, as raw materials to make polymers and surfactants hydrophilic.[1-4]

The well known first approach has been the direct conversion of sucrose with reactive fatty or vinyl derivatives in suitable solvents like dimethylformamide or dimethyl sulfoxide :

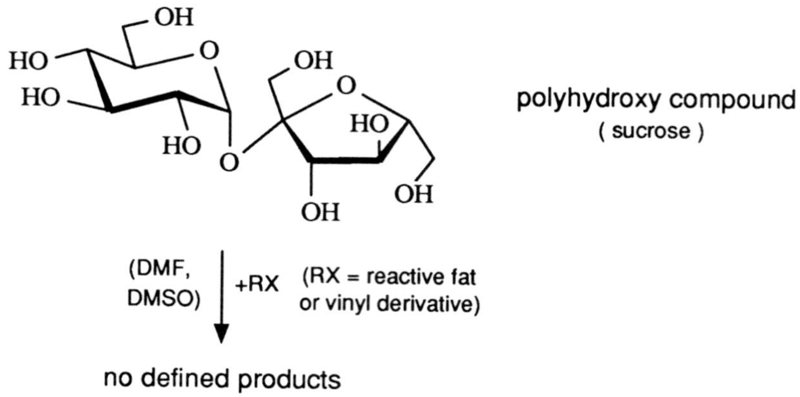

polyhydroxy compound
(sucrose)

(DMF, DMSO) | +RX   (RX = reactive fat or vinyl derivative)

no defined products

In spite of a large number of investigations, no specific, direct mono- or di-derivatization products of sucrose have become available with high yield. Nevertheless, some products made by direct conversion of sucrose are being used today, yet a further introduction into the market is impeded mainly by two disadvantages :

(i)   for low priced products yields are too poor or downstream processing is too expensive and

(ii) for high priced products the quality is too poor, because only mixtures of sucrose derivatives with varying degree of substitution are available.

About 20 years ago a second approach to use sucrose as a hydrophilic building block was attempted with the introduction of a carboxylic or an amino group as specific reacting group into the sucrose molecule. Hydroxy groups being poor leaving groups, they are not accessible for direct nucleophilic substitution. Hence, reaction sequences like the one outlined in Fig. 1 have been investigated.

Those sequences include the following reaction steps :

♦ Formation of sucrose esters with regioselective reacting compounds, e.g. special tosylates.

♦ Transformation of these products by replacing the tosyl group by a cyano or azide group and saponification resp. hydrogenation into monocarboxylic- or monoamino-polyhydroxy compounds, which should react selectively with fatty acid or vinyl reagents to give surfactants or polymerizable vinylsaccharides.

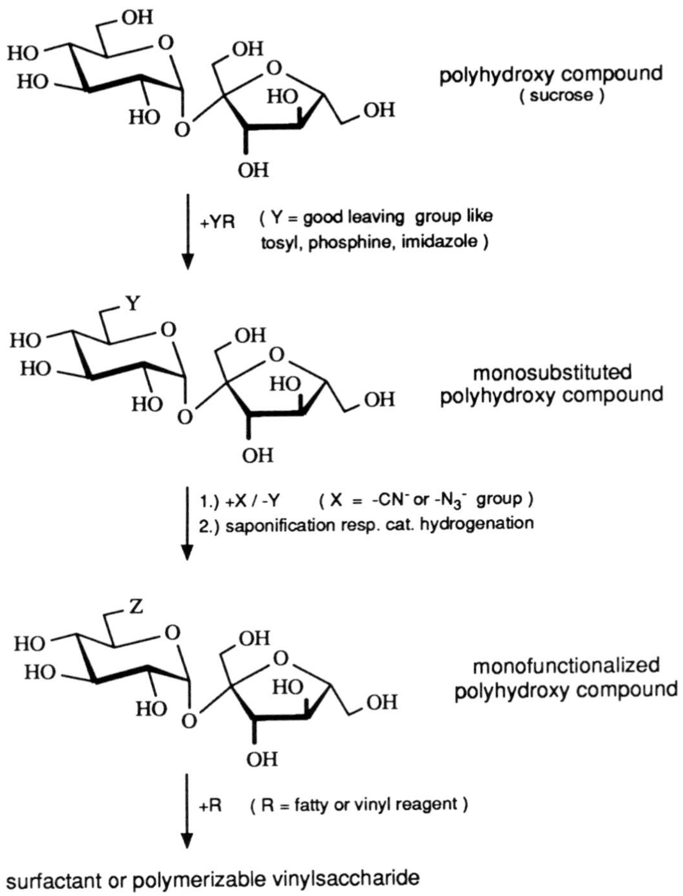

Fig. 1. Sugar as a basic chemical : reaction sequences are not selective enough, providing uniformly derivatized hydrophilic products only upon high purification efforts

There have been lots of investigations with different types of regioselective reagents (e.g. ref.[5,6,7]), but this reaction sequence was not successful, because it was too expensive and too complicated for products with industrial application profiles. This led to the search for a new approach.

Most products of the starch industry which find application in non-food manufacturing are made by a two-step reaction sequence, i.e. biochemical conversion of native starch, and chemical treatment of the degraded starch.

A similar approach for sucrose appears feasible comprising the biochemical transformation of the non-reducing sucrose into a reducing disaccharide, and the subsequent chemical transformation of the reducing disaccharide into specifically functionalized, stable polyhydroxy compounds, capable of reacting for instance with fatty acid or vinyl derivatives to give surfactants or vinylsaccharides, according to the following scheme :

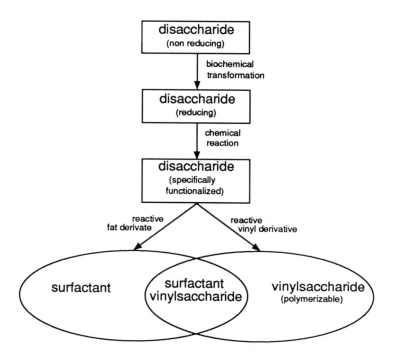

The aim of this paper is to give an impression of the scope of this new approach. A short description of the biochemical transformation of sucrose shall be followed by detailed information on the chemical reactions applied, including a technical approach for specific "carbohydrate-like" down stream processing. Finally, a short survey on surfactants and polymers prepared from the new intermediates is presented.

2. Biochemical Transformations

Two different biochemical methods are known to transform sucrose into reducing disaccharides :

♦ dehydrogenation, especially by *Agrobacterium tumefaciens*, which is under investigation at the Braunschweiger Zuckerinstitut by Buchholz.[8,9]

♦ transglucosylation of sucrose into glucosyl-fructoses. The *Protaminobacter rubrum*-catalyzed process leading to isomaltulose has been extensively investigated by Südzucker in Germany[10] and Mitsui in Japan.[11] The biochemical conversion of sucrose into leucrose by *Leuconostoc mesenteroides* is being applied industrially by Pfeifer & Langen in Germany.[12]

Correspondingly, this paper does not deal with special investigations concerning the biochemical processes. The reducing disaccharides accessible by the biochemical processes mentioned are palatinose (isomaltulose, **1**), isomaltose (**2**), trehalulose (**3**), 3-keto-sucrose (**4**), and leucrose (**5**) :

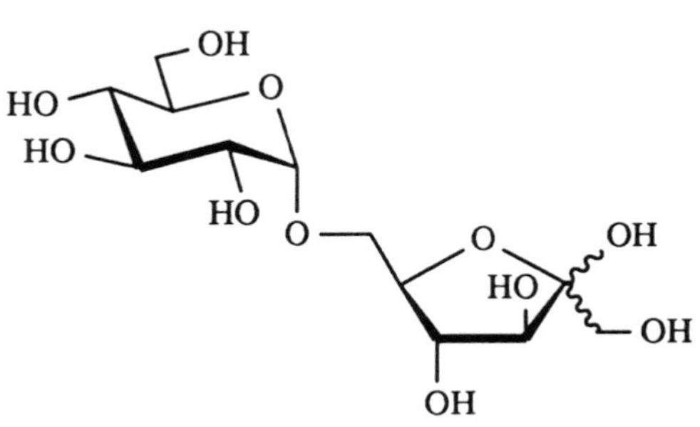

palatinose **(1)**

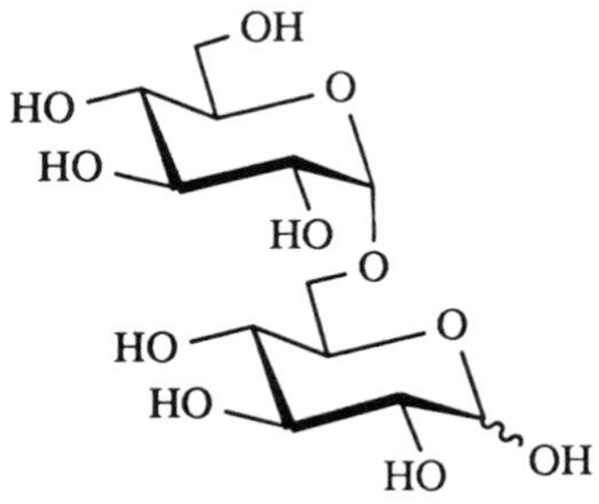

isomaltose **(2)**

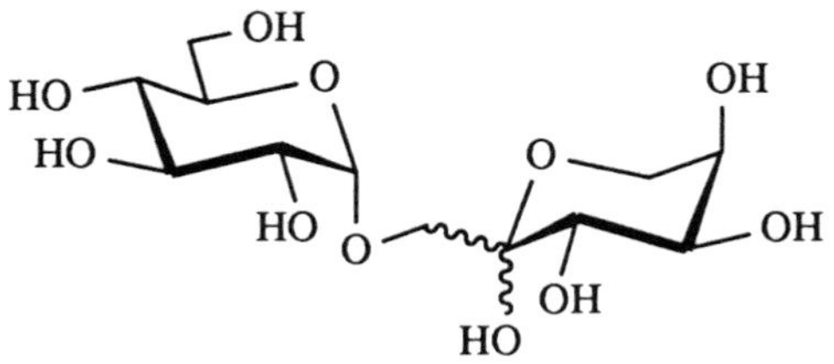

trehalulose **(3)**

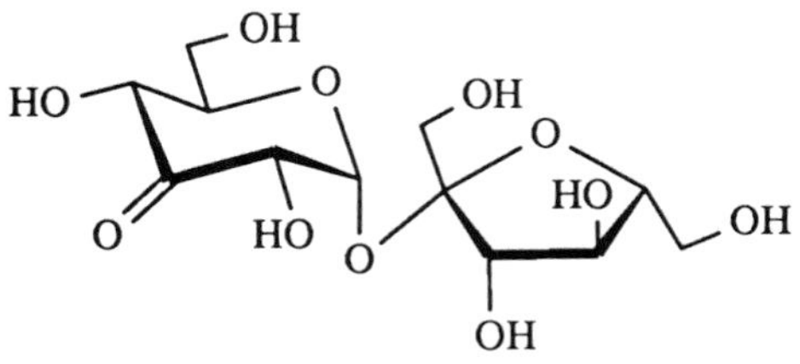

3 - keto - sucrose **(4)**

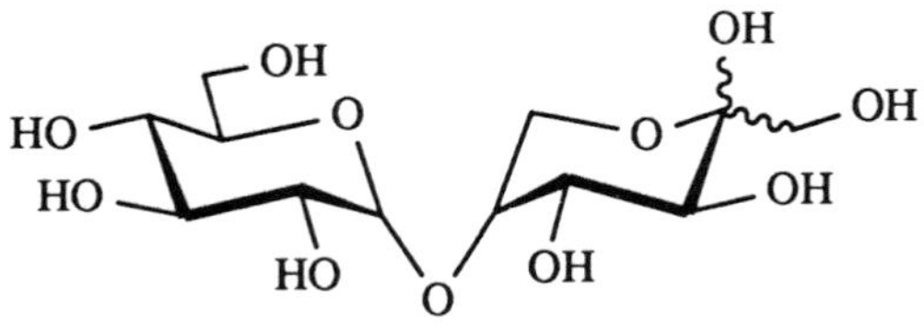

leucrose **(5)**

3. Chemical Reactions

The production of palatinose (**1**) is – as far as known – best introduced in industry, about 7000 tons having been produced in 1989. Therefore, this presentation will focus on this molecule, although the chemical reactions mentioned were applied to the other molecules, too.

From a technical point of view, the reactions investigated

♦ should be high in selectivity and reactivity,

♦ should use industrially established technical processes,

♦ should be using simple solvents like water or methanol, and

♦ should involve only standard industrial downstream processing, such as crystallization or ion-exchange chromatography.

To introduce specific reactive functional groups, the following reactions were investigated :

♦ oxidative cleavage according to Spengler and Pfannenstiel,[13]

♦ reductive amination with nickel-catalysts,

♦ catalytic oxidation with platinum and oxygen, and

♦ reduction with nickel-catalysts.

Oxidative Cleavage of Palatinose

D-Glucopyranosyl-$\alpha(1\rightarrow5)$-D-arabinonic acid, in the form of its potassium salt **8**, is readily prepared from palatinose in oxygen-saturated, strongly alkaline medium.[14,15] The reaction starts with the formation of an enediolate **6**, continues with oxygen addition to the 1,2-peroxide adduct **7** and is concluded by cleavage of the C-1–C-2 - bond with generation of formiate and the glucosyl-arabinonate **8** :

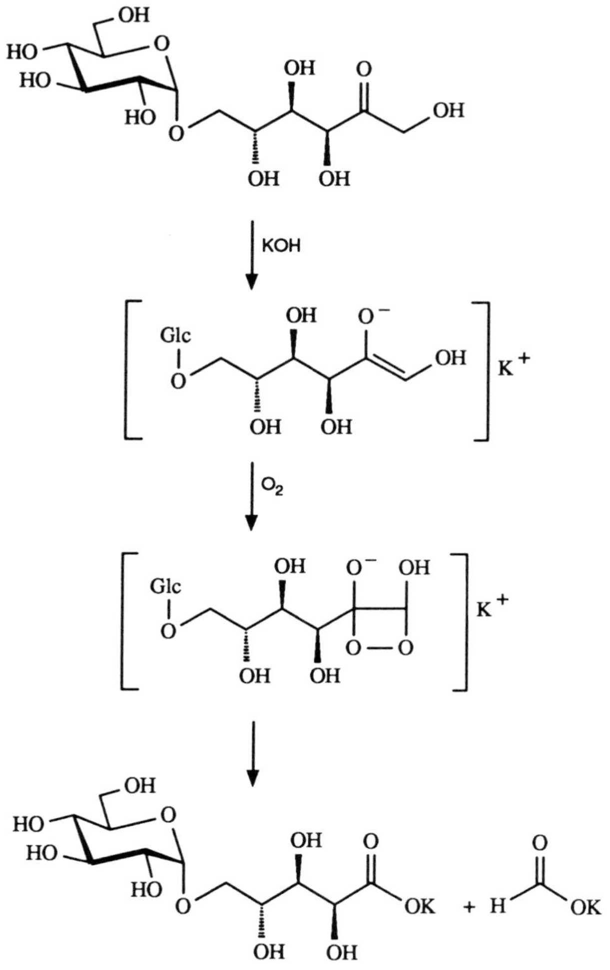

palatinose *
(1)

KOH

1,2 - enediolate
(6)

O₂

1,2 - peroxide
(7)

D-glucosyl -
α(1→5) -
D - arabinonate
(8)

The most suitable conditions found, were to perform the reaction in a vigorously stirred (4000 rpm), aqueous solution of palatinose (0.4 mol/l) and potassium hydroxide (1.4 mol/l) under oxygen saturation (corresponding to an O_2-concentration of about 40 mg/l). After about 4 h at 25 °C, the potassium salt of glucosyl-arabinonic acid (8) crystallized in a yield of 80 %. The yield found by g.l.c.-analysis exceeded 90 %. As far as known, this is one of the most specific reactions of oxidizing carbohydrates under alkaline conditions.

---

* Palatinose, adopting the β-D-furanoid form in the solid state[16] and, with high preference, also in solution,[17] is pictured here in its open-chain form to illustrate the ensuing products.

The energy of activation was determined to be about 67 kJ/mol. Applied to palatinose, this reaction has by far the highest oxidation rate, as compared to other simple mono- and disaccharides (Table 1) :

Table 1. Rate constants (20 °C) for the oxidative
cleavage of mono- and disaccharides

Educts		$k \cdot 10^3$
Ketoses:	isomaltulose	12.344
	fructose	4.231
	leucrose	3.002
	lactulose	2.497
Aldoses:	isomaltose	1.146
	melibiose	1.093
	lactose	0.563
	glucose	0.555

The slowest step in the reaction is the formation of the enediolate anion, i.e. **1 → 6**. Hence, the results shown reflect the fact, that the rate of enediolate formation depends on the tautomeric form present in aqueous solution. Palatinose, for example, has by far the highest oxidation rate and, de facto, is the only saccharide in Table 1, in which the reducing end, i.e. the fructose portion, exists exclusively in its two furanoid forms, with an α-*f* / β-*f*-ratio of about 1 : 4 at 20 °C.[17] Since all other sugars listed in Table 1 adopt pyranoid tautomeric forms either exclusively or with high predominance (D-fructose), and are oxidized distinctly faster, the conclusion is warranted that enediolate formation from a furanose form is considerably easier than from the respective pyranoid tautomer.

The oxidative cleavage of palatinose yields a specifically monofunctionalized polyol, which can easily be isolated. The reaction is high in selectivity (93 %) and very high in reactivity. But as a salt the product is not directly convertible into reactive compounds; it has first to be transferred into its lactone or into an ester. Besides that, only amides are formed as stable products. These facts limit the utilization of glucosyl-arabinonic acid **8** as reactive intermediate.

Reductive Amination

Applying the reductive amination to disaccharides, the main products should be aminoglycosyl-hexitols, in the case of palatinose (**1**) the D-glucopyranosyl-α(1→6)-2-amino-2-deoxy-D-mannitol (2-amino-GPM, **10**) and its D-sorbitol analog (2-amino-GPS, **11**). The equimolar mixture of **10** and **11**, accumulatingly is designated isomaltamine,[18] corresponding to the bulk sweetener isomalt or Palatinit®, which is the mixture of 1-*O*-(α-D-glucopyranosyl)-D-mannitol (GPM) and 6-*O*-(α-D-glucopyranosyl)-D-sorbitol (GPS).

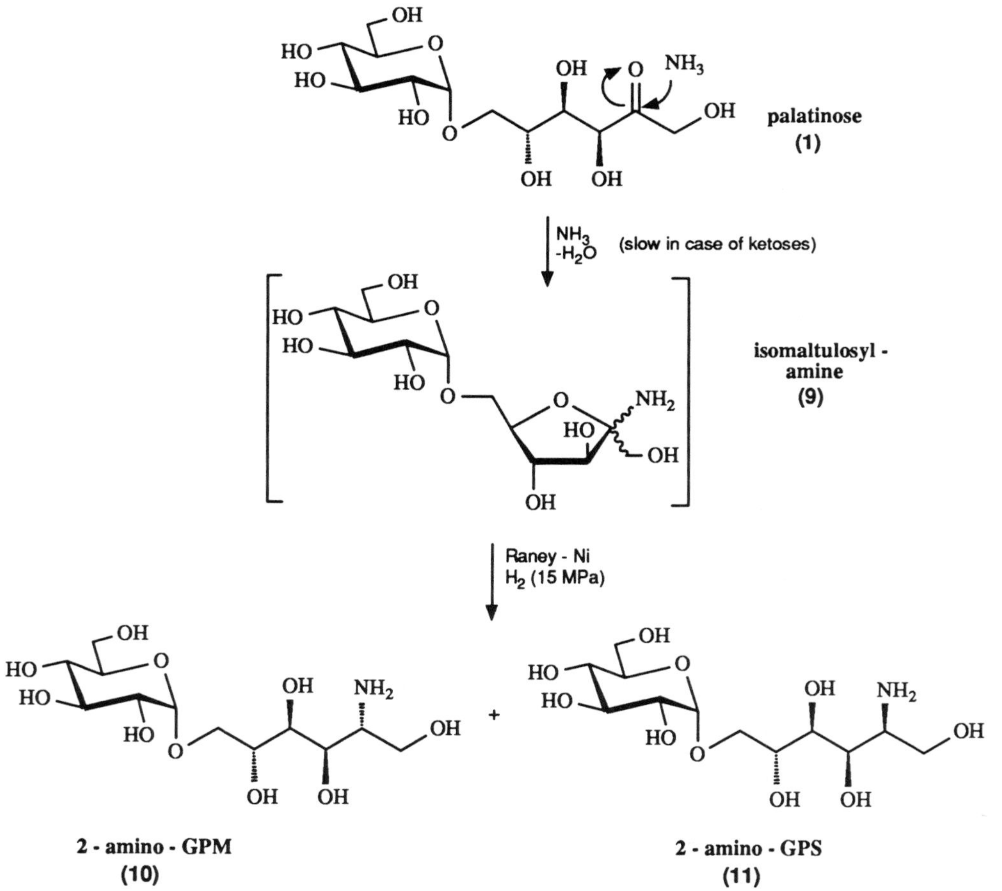

Reductive aminations of carbonyl compounds are well established in chemical industry, but rarely applied to carbohydrates due to the usual formation of complex product mixtures if ammonia is used as aminating reagent. As far as known, an effective reductive amination with ammonia in high yields has been reported only for a monosaccharide.[18] Preparatively, it is much easier to use hydrazine as the aminating reagent, a process first applied to carbohydrates by Lemieux;[19] however, hydrazine is too expensive for industrial applications.

After it had been established that the amination of palatinose with hydrazine yielded isomaltamine nearly quantitatively,[20] the reaction of palatinose with ammonia under reducing conditions was investigated in detail.

The first step, i.e. formation of the ketosylamine, being very slow, a pre-reaction time before hydrogenation of about 17 h at 20 °C is required. After this period the intermediate ketosylamines are easily hydrogenated, yet for minimization of parallel reactions such as isomerization, retroaldol-reaction, β-elimination, Heyns- and Amadori-rearrangement or Maillard-reaction, the reaction mixture must be well stirred. Tables 2 and 3 summarize the conditions.[20,21]

Table 2. Reductive amination of palatinose with hydrazine and ammonia as aminating reagents

| | Aminating reagent | |
	hydrazine	ammonia
Molar ratio : palatinose / amine	1 / 3.9	1 / 19.2
Pre-stirring time (h)	17	17
Yield (%)		
– neutral compounds	c. 2.0	9.7
– isomaltamine	96.6	78.7
– aminated byproducts	c. 1.0	c. 10.0

Table 3. Reductive amination of palatinose : hydrogenation conditions

Catalyst	Raney-nickel
Hydrogen pressure	15 Mpa
Temperature	50 °C
Stirrer Speed	1000 min^{-1}
Reaction time	24 h

Compared to the use of hydrazine, the yield of isomaltamine is lower and the formation of byproducts, mainly other mono-aminopolyols, is higher. But compared to other di- and oligosaccharides the yield of about 80 % for these aminoglycosyl-hexitols is high (cf. Table 4). Besides, the reaction rate of the reductive amination with ammonia is much higher than with hydrazine, as evidenced by the data in Fig. 2.

Table 4. Reductive amination of different disaccharides with ammonia (conditions : Table 2 and 3)

Disaccharide	Monoaminoglycosyl-hexitols (% yield)	Other amino-polyols
Palatinose	78.6	7
Leucrose	56.7	26
Trehalulose	59.1	29.2
Lactulose	< 30	50
Isomaltose	63.4	7.5
Maltose	16.9	42
Lactose	29	43.1

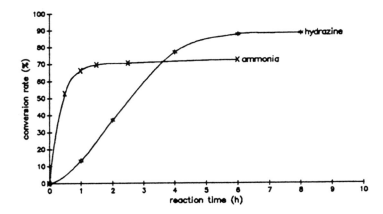

Fig. 2. Reductive amination of palatinose : comparative study of hydrazine and ammonia as aminating reagents

On the basis of these studies, isomaltamine – the equimolar mixture of **10** and **11**, which may also be termed "palatinamine" – has become an easily accessible product – two steps away from sucrose and prone to bulk scale generation, if required.

Of the two epimeric glucosyl-aminoalditols **10** and **11**, which can be separated by standard chromatography, one shows a high tendency for crystallization. It turned out to be the dihydrate of 2-amino-GPM (**10**·H$_2$O), as evidenced by an X-ray structural analysis[22] (cf. Fig. 3).

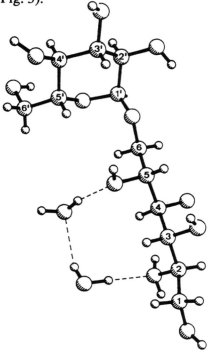

Fig. 3. Perspective view of the molecular structure of 1-*O*-(α-D-glucopyranosyl)-2-amino-2-deoxy-D-mannitol dihydrate (**10** · 2H$_2$O).[22] The two molecules of crystal water form a "double water bridge" between N-2 and O-5 of the aminomannitol portion. Carbon atoms have been numbered.

The molecular geometry of **10**-dihydrate shows the two molecules of crystal water fixed via hydrogen bonds to an aminomannitol chain in a nearly planar zigzag conformation, forming a "double water bridge" between N-2 and O-5, i.e. between two atoms five bonds apart.[22] This molecular structure, in fact, appears to be rather unique, the only other example with a nearly identical double water bridge being the "hydroxy analog" of **10**, i.e. glucosyl-α(1→1)-mannitol ("GPM"), which has been X-rayed already 10 years ago.[23] Correspondingly, the conformational features of 2-amino-GPS **11** may be surmised to closely resemble those observed[24] for the glucosyl-α(1→6)-sorbitol ("GPS", **12**). In **12**, the glucitol portion adopts a non-linear, bent-chain arrangement[24] due to unfavourable parallel 1,3-interactions of O-2 and O-4 in the extended-chain rotamer, which is avoided by rotation about the C-2 – C-3 bond – a situation that similarly holds for the 2-amino analog **11** :

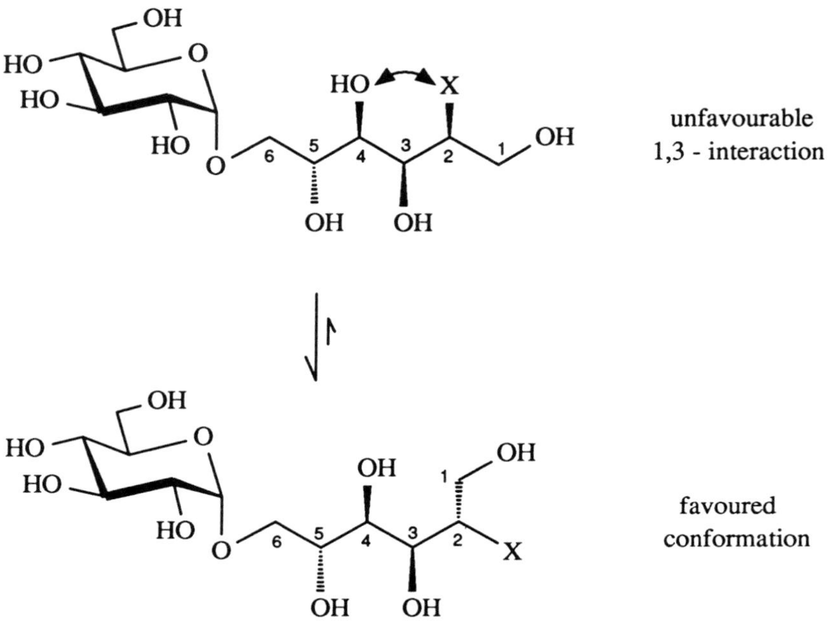

unfavourable
1,3 - interaction

favoured
conformation

11 : X = NH$_2$ (2-amino -GPS)
12 : X = OH   (GPS)

Catalytic Oxidation

Catalytic oxidations of carbohydrates – both aldoses and ketoses – are well established reactions. Especially Heyns[25,26] and, later, van Bekkum[27] did a lot of work in that field, resulting in the following selectivity rules :[26]

- ◆ carbonyl groups of aldoses are readily oxidized at room temperature,

- ◆ oxidation of hydroxyl groups requires stronger reaction conditions with primary hydroxyl groups reacting faster than secondary ones,

- ◆ in ketoses, the primary hydroxyl group adjacent to the anomeric centre is most readily oxidized, its reactivity being comparable to that of an aldose carbonyl group.

Correspondingly, one would expect a main product on catalytic oxidation of palatinose (**1**), namely, 6-*O*-(α-D-glucopyranosyl)-D-*arabino*-hexulosonic acid (**13**) :

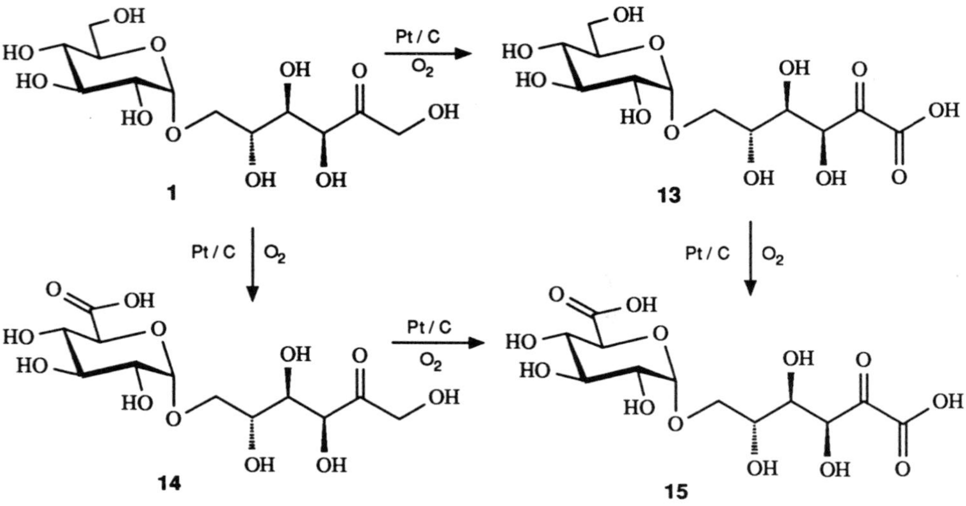

However, the 6'-hydroxyl group proved to be nearly as reactive as the 1-OH, thus two polyhydroxy-monocarboxylic acids, **13** and **14** respectively, and a polyhydroxy-dicarboxylic acid may be formed.

Under suitable conditions (20 °C), nearly only monooxidation takes place. At higher temperatures (~50 °C), the major product after 3 days of reaction is the one oxidized at both terminal hydroxyl groups, i.e. 6-*O*-(α-D-glucopyranuronyl)-D-*arabino*-hexulosonic acid (**15**). Reaction conditions are summarized in Table 5.[28] The isomeric monoacids are easily separated by anion-exchange chromatography.

Table 5. Catalytic oxidation of palatinose[28]

Concentrations applied:		
palatinose (mol/l H_2O)	0.1	
10 % Pt/C-catalyst (g/l H_2O)	5.0	

	Main products	
Reaction conditions:	mono-acids	di-acids
stirring speed (min-1)	1000	1000
temperature (°C)	20	50
pH	7.5	7.5
aeration (ml·min^{-1}·l^{-1})	250	250

Adams-catalyst is unsuited for this reaction due to its rapid inactivation by pure as well as atmospheric oxygen. Though the initial activity of platinum on carbon is lower as compared to Adams-catalyst, the over-all reaction rate is higher and little or no inactivation is observed when using atmospheric oxygen (cf. Fig. 4) :

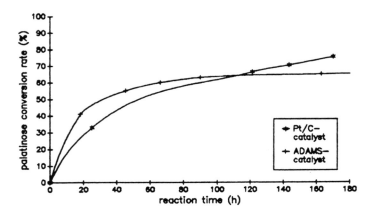

Fig. 4. Catalytic oxidation of palatinose (**1**) by atmospheric oxygen (0.014 g Pt/g carbohydrate)

Rate constants for the oxidation of palatinose with platinum on carbon were determined at different temperatures. Fig. 5 shows that at 35 °C as well as at 50 °C the oxidation at C-1 is faster than that at C-6'. In the second oxidation step, however, the ratio of the reaction rates is reversed at 50 °C, a phenomenon that, as of now, is not well understood.

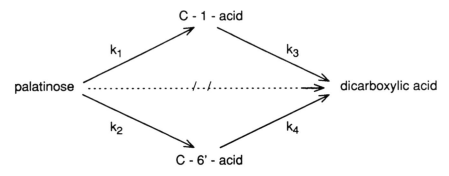

parallel consecutive reactions with common endproduct

| 35 °C | $k_1 > k_2$ | $k_3 < k_4$ |
| 50 °C | $k_1 > k_2$ | $k_3 > k_4$ |

Fig. 5. Scheme of the catalytic oxidation of palatinose (**1**)

The oxidation products themselves are unstable, decomposing readily under alkaline conditions. After hydrogenation, however, the resulting polyhydroxy mono- or di-carboxylic acids are stable.

All of the oxidation products can easily be transformed into polyhydroxy amino-carboxylic acids by the reactions described (vide supra) providing products of type **16 - 18**, i.e. most interesting intermediates for surfactants and polymers.

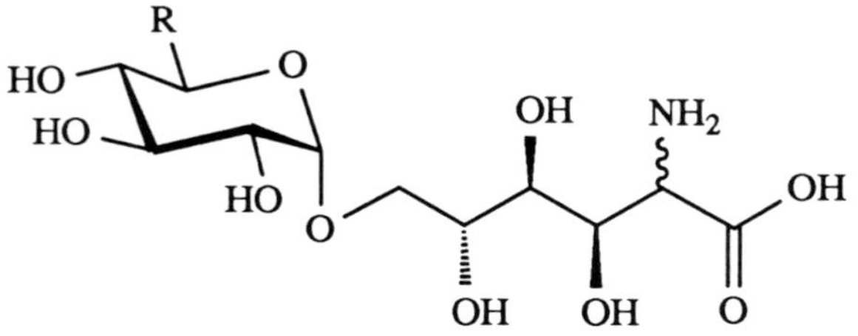

16   (R = CH$_2$OH)
17   (R = COOH)

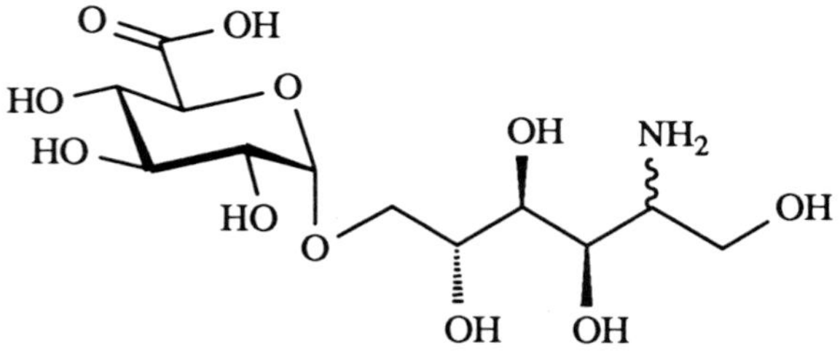

18

These data suffice to conclude, that the selective introduction of one or two carboxylic groups is possible by catalytic oxidation, with retention of the reducing property of the molecule. After reduction or reductive amination the resulting products are stable compounds. The amino acids produced are directly convertible into other products with high variability of possible reaction partners.

4. Ion-exchange Chromatography

Reductive amination and catalytic oxidation of palatinose yields isomeric compounds. For their separation, cheap and effective methods were necessary. Ion-exchange chromatography is – as far as known – the most widely applied chromatographic technique on an industrial scale. Separation of isomeric carbohydrates like glucose and fructose is carried out with such systems at a scale of several 100.000 tons per year. Besides, chromatography of molasses to enhance the yield of sucrose in its production is also well introduced. Since both, liquid fructose and sucrose cost less than 1 $ / lb, the ion-exchange chromatography applied on an industrial scale is an exceedingly inexpensive separation methodology.

Although the separation of amines, acids, and aminoacids by ion-exchange chromatography is well known for analytical purposes, the conditions used – special resins and buffer systems as eluents – are unsuited for industrial separation technologies.[29] After evaluating a series of different types of resins it was found that cation exchangers like Amberlite IR 120 in the ammonium form were appropriate for separations of aminopolyols with dilute aqueous ammonia as eluent. The main advantage of this chromatography is, that the separation of the compounds from the eluent is possible simply by evaporation of ammonia and water, and that the resulting condensates can be recycled.

Correspondingly, the polyhydroxy monocarboxylic acids could be separated using anion-exchange resins in the formate form and aqueous formic acid as eluent. Isolation of the separated polyhydroxy acid was also done by evaporation of the eluent, which subsequently, may also be recycled. Chromatographic conditions for both systems are given in Table 6.

Table 6. Ion-exchange chromatography

Separation of monoamino polyols	
column size:	$0.05 \cdot 1.00$ m
packing:	Amberlite IR 120 (NH_4^+-form)
eluent:	0.5 % NH_3 (aq. soln.)
throughput:	$6 \, g \cdot l^{-1} \cdot h^{-1}$
Separation of polyhydroxy monocarboxylic acids	
column size:	$0.05 \cdot 1.00$ m
packing:	Amberlite IRA 400 ($HCOO^-$-form)
eluent:	1.0 % $HCOOH$ (aq. soln.)
throughput:	$3 \, g \cdot l^{-1} \cdot h^{-1}$

Both chromatographic systems showed normal behaviour according to the van Deemter equation. However, the theoretical plate heights in the cation-exchange system were much smaller than those in the anion-exchange case, which is also reflected in the corresponding dependencies on the rate of elution (Fig. 6).

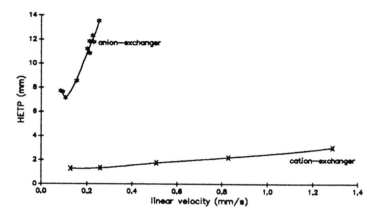

Fig. 6. Plot of HETP versus linear velocity – comparison of anion- and cation-exchanger systems

Isomeric polyhydroxy monoamines show very different pK_b-values. Amino-GPM (**10**) and amino-GPS (**11**) for example differ only in the configuration at C-2, but the difference of their pK_b-values is about 0.6. This, probably, is the main reason why they are easily separated by the ion-exchange chromatography. In addition, the shape of the molecules – expressed as their molecular weight – influences the chromatographic effect (cf. Fig. 7).

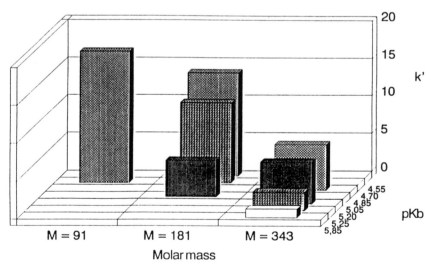

Fig. 7. Dependence of k'-values on molar mass and pK_b

5. Conclusions and Applications

The introduction of specific, reactive functional groups into sucrose-based polyols is possible using a two step biochemical / chemical sequence of industrial applicability.

The processes are simple, the resulting products are stable intermediates, and the yields are high. In terms of the potential application profiles of these sucrose-based hydrophilic disaccharide building blocks, their utilization for the preparation of new types of surfactants and polymers are being outlined in the sequel.

Surfactants

As reactive derivatives of fatty acids to be attached to the products described, fatty amines, fatty acid anhydrides and/or esters, as well as alkylisocyanates were used. Examples of the neutral surfactants synthesized are :

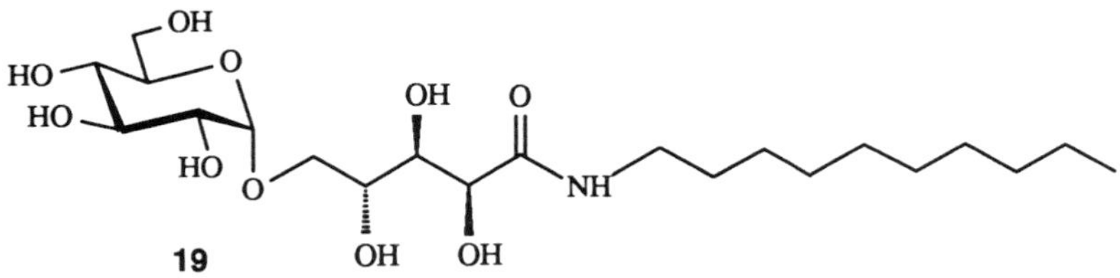

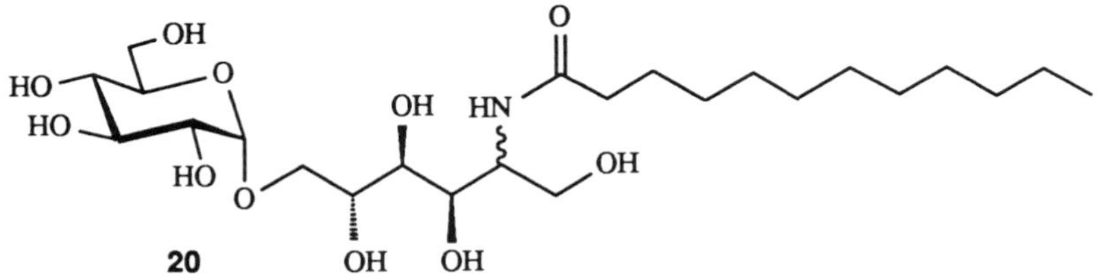

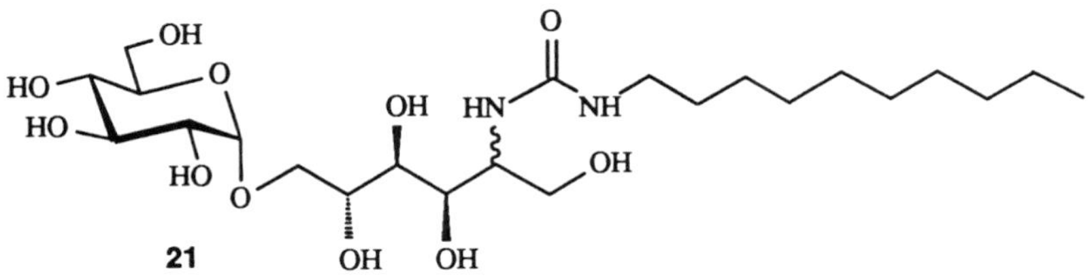

The first product, **19**, is made from the glucosyl-α(1→5)-arabinonic acid (**8**) with fatty amines.[30] Attempts to prepare esters from this acid with fatty alcohols were not successful, because these esters are unstable and are easily hydrolized. Product **20** is made from isomaltamine with fatty acid esters,[30] whereas **21** was obtained from isomaltamine by reaction with alkylisocyanates.[31] All products show typical behaviour of surfactants, but isomeric surfactants may differ in their properties as shown in Fig. 8.[32]

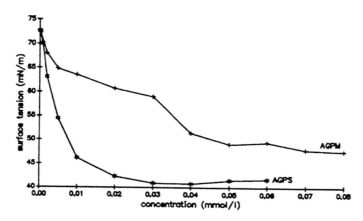

Fig. 8. Surface tension of aqueous solutions of the dodecylamides of GPM (AGPM) and GPS (AGPS) in relation to concentration

Some other properties of the surfactants **19 - 21** were also investigated.[33] One interesting fact is the reduction of the lachrymative property of dodecyl-sulfonates (rabbit test). Furthermore, the products showed low skin degreasing effects and very low skin irritating properties (zein test). In addition, some of them showed a good skin moisturizing effect. Other compounds formed stable gels at very low concentrations in aqueous solutions. Some products had very high foaming capacity. Because of these properties, we believe, some of these products may find interest in cosmetic applications.

Vinylsaccharides

Klein[34-37] proposed new types of semi-synthetic polymers based on well defined mono(vinyl)-saccharides :

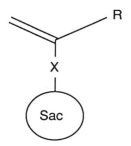

General structure of vinylsaccharides
R= H, CH_3, COOH, CH_2-COOH
X= ester, ether, amide, urea
Sac= saccharide side group

To synthesize such products, a first reaction step may be the introduction of an amino group into a saccharide. The resulting aminopolyols can react with suitable vinyl compounds to give polymerizable vinyl monomers. In case of polyhydroxy aminoacids, the resulting products are anionic vinyl monomers. When alkylamines are used as aminating reagents, surfactant monomers are the result :

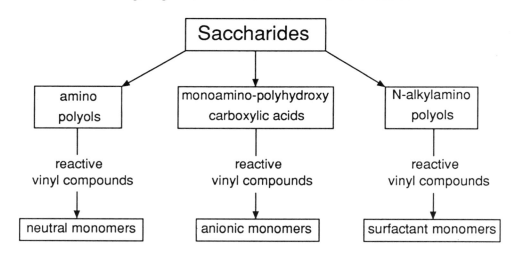

Different types of reactive vinyl derivatives under investigation are :

acrylic and methacrylic acid derivatives (halides, anhydrides, esters),

vinylisocyanato derivatives (IEM),

glycidylic esters of acrylic and methacrylic acid,

glycidylallylethers and allylamines.

As far as this has been investigated, it is possible to make well defined mono(vinylsaccharide)s on the basis of the hydrophilic intermediates, described above, which are readily polymerized. The resulting linear homopolymers have been characterized.[20,38,39]

One main macroscopic effect of water-soluble polymers is the viscosity enhancement in aqueous solutions. It is interesting to note, that the Staudinger-indices of polymers made from isomeric vinylsaccharides, e.g. those based on maltose, lactose, and palatinose are different (Fig. 9). Introduction of a hydrophobic alkyl group reduces the Staudinger-index whereas the introduction of a carboxyl group enhances it even in saline solution.

Especially aqueous solutions of poly(vinylsaccharide)s with carboxylic groups showed interesting behaviour. The values obtained for one of these polymers (Pol. II) – oxidized at C-1 in the carbohydrate moiety – showed an exponent "a" of about 1 according to Mark-Houwink-equation, while normal polyacrylates (Pol. I) show an exponent of about 0.6 - 0.7.[40] Up till now values of 1 or higher are known only for biopolymers of the xanthan-type (Pol. III, Fig. 10).[41]

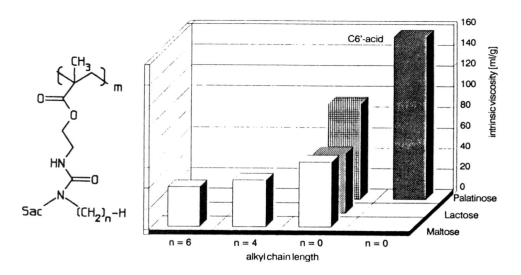

Fig. 9. Intrinsic viscosity of urea type polymers normalized on M_w= 2.6 · 10^6 g mol^{-1} in 0.1 M Na$_2$SO$_4$ at 25 °C

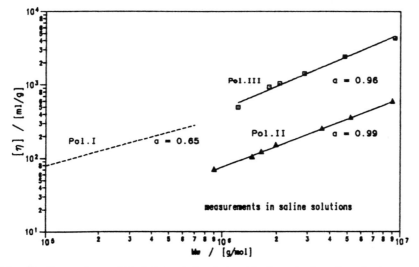

Fig. 10. Comparison of Mark-Houwink-relationship for anionic and nonionic poly(vinylsaccharide)s

Besides that, the anionic poly(vinylsaccharide)s showed different precipitation behaviour towards mono- or bivalent cations. The poly(vinylsaccharide) based on palatinose oxidized at the C-1-OH, i.e. **1 → 13** (cf. above) didn't precipitate with any metal cation investigated including calcium. This offers interesting future application possibilities.

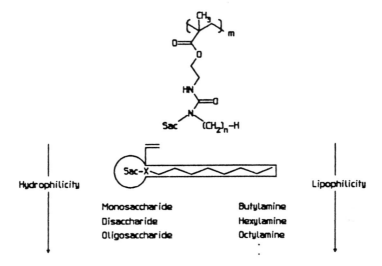

Fig. 11. Surfactant polymers : variation of hydrophilic and lipophilic properties by specific monomer syntheses

Another interesting group of new poly(vinylsaccharide)s are surfactant polymers (Fig. 11).

With the methods developed, it is possible to produce monomers with a wide range of hydrophilic-lipophilic balance. The longer the carbohydrate chain, the more hydrophilic character the monomer will have, whereas its lipophilic character increases with the length of the alkyl chain.

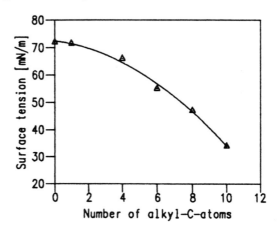

Fig. 12. Surface tension of 0.1 % polymer solutions as a function of alkyl chain length

Polymers based on disaccharides with an alkyl chain length of about 8 to 10 carbon atoms showed surface tensions of about 30 to 40 mN $\cdot$ m^{-1} in 0.1 % aqueous solution (Fig. 12).

Table 7. Surface tension of solutions of monomers and polymers rsp. starting materials at 25 °C

Compound	Surface tension of 0.1 % solution (mM$\cdot$m^{-1})
water	72
maltose	72
1-amino-maltitol	72
N-octyl-maltitol	50
IEM/N-octyl-monomer (mono)	46
IEM/N-octyl-polymer	47
PAAm	72
acrylamide/ (mono)-copolymer (16 : 1 molar ratio)	53

Surprisingly, the vinylsaccharide monomers and their corresponding polymers had nearly the same effect on surface tension. Also a copolymer made with acrylamide in a molar ratio of 18 : 1 exhibited nearly the same effect on surface tension as the surfactant homopolymer (Table 7).

Acknowledgements. The author wants to thank Prof. Dr. E. REINEFELD, former head of the Braunschweiger Zuckerinstitut, and Prof. Dr. K. BUCHHOLZ, now in charge of this institute, for the opportunity to work there. Thanks are also due to Prof. Dr. J. KLEIN (GBF Braunschweig) and Prof. Dr. F. W. LICHTENTHALER (TH Darmstadt) for their kind cooperation during recent years. The helpful assistance of H. RÖGER, W. BEHRENS, A. HAJI BEGLI, S. ENGELKE, S. RIEGER, J. KOWALCZYK, and H. PUKE is gratefully acknowledged. Last but not least, the author wants to thank the *German Ministry for Research and Technology* for supporting some of this work.

References

1. J. L. Hickson (Ed.): Sucrochemistry. *ACS Symposium Series No. 41*, Amer. Chem. Soc., Washington, 1977.

2. J. C. Colbert: Sugar Esters – Preparation and Applications. *Chemical Technology Review No. 32*, Noyes Data Corporation, New Jersey, London, 1974.

3. E. Reinefeld: Use of Sucrose in the Chemical Industry. *Zuckerind.* **112** (1987) 1049-1056.

4. C. E. James, L. Hough, R. Khan: Sucrose and its Derivatives. *Progr. Chem. Org. Natl. Products* **55** (1989) 117-184.

5. L. Hough, S. P. Phadnis, E.Tarelli: The Direct Preparation of 1',6,6'-Trideoxy-sucrose from Sucrose. *Carbohydr. Res.* **44** (1975) C 12.

6. G. Descotes, J. Mentech: Deoxygenation Regiospecifique en 6,6'-du Saccharose. *Third European Symposium on Carbohydrates*, Grenoble, Sept. 1985, Abstract D.3-4P.

7. O. Mitsunobu: The Use of Diethyl Azodicarboxylate and Triphenylphosphine in Synthesis and Transformation of Natural Products. *Synthesis* **1981**, 1-18.

8. K. Matalla, E. Stoppok, K. Buchholz: Specific Modification of Sucrose to Building Blocks for Synthesis. *Dechema Biotechnology Conference* (D. Behrens, A. J. Driesel, Eds.), Vol. 3, Part A, VCH Weinheim **1989**, 117-121.

9. K. Buchholz, M. Kunz: Biotechnologische und chemische Wege zu Synthesebausteinen aus Saccharose. *Zuckerind.* **115** (1990) 20-24.

10. H. Schiweck: Isomalt (und Isomaltulose). *Ullmanns Enzyklopädie der Technischen Chemie. 4. Aufl.*, VCH Publishers, Weinheim, Bd.24. (1983) 780-781.

11. J. Shimizu, K. Suzuki, Y. Nakajima, (Mitsui Sugar Co.): Palatinose Production Using Immobilized α-Glucosyltransferase. *Jap. Pat. Appl.* 80/113.982 (1980); *Neth. Pat. Appl.* NL 81-03,911 (1980); *Chem. Abstr.* **97** (1982) 53994m.

12. D. Schwengers, H. Benecke (Pfeifer & Langen KG): Sweetener and its Use. *DE-PS* 34 46 380 (1984); *Eur. Pat.* 185 302 (1985); *Chem. Abstr.* **105** (1986) 77815p.

13. O. Spengler, A. Pfannenstiel: Über die Oxydation reduzierender Zucker durch Sauerstoff. *Z. Wirtschaftsgr. Zuckerind.* **85** (1935) 546-552.

14. F. W. Lichtenthaler, R. G. Klimesch (Süddeutsche Zucker AG): Derivate und Reduktionsprodukte der D-Glucopyranosyl-α(1→5)-D-arabonsäure. *DE-OS* 3,248,404 (1982); *EP* 114 954 (1983); *Chem. Abstr.* **102** (1985) 7034x.

15. H. Röger, H. Puke, M. Kunz: Untersuchungen zur oxidativen Spaltung von reduzierenden Disacchariden. *Zuckerind.* **115** (1990) 174-181.

16. W. Dreissig, P. Luger: Die Struktubestimmung der Isomaltulose. *Acta Crystallogr., Sect. B* **29** (1973) 514-521.

17. F. W. Lichtenthaler, S. Rönninger: α-D-Glucosyl-D-fructoses: Distribution of Furanoid and Pyranoid Tautomers in Water, Dimethyl Sulphoxide, and Pyridine. *J. Chem. Soc., Perkin Trans.2,* **1990**, 1489-1497.

18. H. Kelkenberg: Sugar-based Detergents. New Components for Cleaning Compositions and Cosmetics. *Tensides - Surfactants -Detergents* **25** (1988) 8-13; *Chem. Abstr.* **108** (1988) 223509x.

19. R. U. Lemieux (National Research Council, Ottawa): Amino Alcohols. *US Pat.* 2,830,983 (1958); *Chem. Abstr.* **52** (1958) 14668c.

20. J. Klein, W. Behrens, M. Kunz (Süddeutsche Zucker AG): Isomaltamines and their N-Acyl Derivatives. *Ger. Offen.* 3,625,931 (1986); *Eur. Pat.* 255 033 (1987); *Chem. Abstr.* **110** (1989) 95711j.

21. S. Rieger, M. Kunz: unpublished results.

22. D. Schomburg, M. Kunz: unpublished results.

23. H. J. Lindner, F. W. Lichtenthaler: Extended Zigzag Conformation of 1-*O*-α-D-Glucopyranosyl-D-mannitol. *Carbohydr. Res.* **93** (1981) 135-140. – For a graphical presentation of the structure, see p. 25 of this monograph.

24. F. W. Lichtenthaler, H. J. Lindner: The Preferred Conformations of Glycosyl-alditols. *Liebigs Ann. Chem.* **1981**, 2372-2383.

25. K. Heyns, H. Paulsen: Selektive katalytische Oxidationen mit Edelmetallkatalysatoren. *Angew. Chem.* **69** (1957) 600-608.

26. K. Heyns, H. Paulsen, G. Rüdiger, J. Weyer: Configuration and Conformation Selectivity in Catalytic Oxidation with Oxygen on Platinum Catalysts. *Fortschr. Chem. Forsch.* **11** (1969) 285-374.

27. H. E. van Dam, A. P. G. Kieboom, H. van Bekkum: Glucose-1-phoshate Oxidation on Platinum-on-carbon Catalysts: Side-reactions and Effects of Catalyst Structure on Selectivity. *Recl. Trav. Chim. Pays-Bas* **108** (1989) 404-407, and references cited therein.

28. H. Puke, M. Kunz: unpublished results.

29. D. T. Gjerde, J. S. Fritz: Ion Chromatography, 2nd Ed., Huethig, Heidelberg, 1987.

30. M. Kunz: Saccharosederivate – Zucker als hydrophiler Baustein. *Zuckerind.* **113** (1988) 273-278.

31. B. Schneider, M. Kunz (Südzucker AG): Patent applied.

32. H. K. Cammenga, D. Hamann, M. Kunz: unpublished results.

33. N. B. Desai: Kosmetische Spezialprodukte aus nachwachsenden Rohstoffen. *Parfümerie und Kosmetik* **70** (1989) 332-338, 340-341.

34. J. Klein, D. Herzog, A. Haji Begli: Emulsion Polymerization of Polymethacryloylglucose. *Makromol. Chem. Rap. Comm.* **6** (1985) 675-678.

35. J. Klein, D. Herzog: Synthesis of some Poly(vinylsaccharides) of the Amide Type and Investigation of their Solution Properties. *Macromol. Chem.* **188** (1987) 1217-1232.

36. J. Klein, K. Blumenberg: Synthesis and Cationic Polymerization of 6-*O*-Vinyl-1,2:3,4-di-*O*-isopropylidene-D-galactopyranose. *Makromol. Chem. Rap. Comm.* **7** (1986) 621-625.

37. J. Klein, K. Blumenberg: Synthesis and Polymerization of 6-*O*-Methylallyl-galactose Derivatives. *Makromol. Chem.* **189** (1988) 805-813.

38. J. Klein, M. Kunz, J. Kowalczyk: New Surfactant Polymers Based on Carbohydrates. *Makromol. Chem.* **191** (1990) 517-528.

39. J. Klein, J. Kowalczyk, S. Engelke, M. Kunz, H. Puke, *Makromol. Chem. Rap. Comm.* **11** (1990), in press.

40. R. Arnold, S. R. Caplan: Solutions of Polymethacrylic Acid. *Trans. Faraday Soc.* **51** (1955) 857-863.

41. A. Ach, *Doctoral Dissertation*, Techn. Universität Braunschweig, 1988.

7

Enzymatic Sucrose Modification and Sac

Klaus Buchholz, Eberhard Stoppok, Klaus Matalla, l
and Hans-Joachim Jördening

Institut für landwirtschaftliche Technologie und Zuc
an der Technischen Universität Braunschweig, D-3300 Brau.   ⸺g, Germany

Summary. Biotechnological methods might contribute significantly to the new dynamics of carbohydrate research since biocatalysts offer the potential for high selectivity in synthesis and derivatization steps under technically favourable conditions.

Two examples are being described in this account, i.e. the oxidation of sucrose and isomaltulose to the 3-keto-derivatives by *Agrobacterium tumefaciens*, and the synthesis of oligosaccharides by the enzyme dextran sucrase, which required a detailed study of reaction kinetics in order to obtain high yields.

Introduction

Disaccharides, although available in ton scale with very high purity and reasonable prices, are not yet used on a large scale as chemicals. One classical problem is the high functionality which makes selective reaction routes difficult to design. Biotechnology offers, by means of selective enzymatic transformations, the solution to this problem; the combination with technically established chemical reactions provides promising prospects for the design of carbohydrate building blocks and products with potential application profiles.

For the approaches undertaken, the following conditions were selected in order to provide favorable perspectives for scale up and transfer of the reactions to the technical scale: conversion in aqueous systems, no protecting groups, few reaction steps and simple, straightforward product isolation and purification.

Main aims of the investigations were: both high yield and concentration of products as well as low by-product concentration in order to avoid purification steps for sequential chemical reaction steps. Therefore reaction parameters and kinetics

...gated in more detail. Experimental and analytical methods will be ...ed elsewhere.

Two reaction pathways were investigated: The first is the specific oxidation of sucrose at position 3 of the glucose portion by *Agrobacterium tumefaciens*. It was originally described and investigated by De Ley et al..[1,2] The second are side-reactions of the synthesis of dextran by dextran sucrase, which yield oligosaccharides by glucosylation of saccharides as acceptors and sucrose as the glucosyl donor. Robyt and Eklund[3,4] have shown the rather wide potential of these reactions. Kinetic analysis should provide the basis for maximizing the oligosaccharide yield and minimize dextran formation.

Parameters of Sucrose Oxidation by *Agrobacterium tumefaciens*

A specific dehydrogenase of *Agrobacterium tumefaciens* catalyzes the oxidation of a range of saccharides, preferentially disaccharides, to the 3-keto-derivatives,[1,2] as, for example, the conversion of sucrose to D-*ribo*-hexopyranos-3-ulosyl-$\alpha(1{\rightarrow}2)\beta$-D-fructofuranoside (3-keto-sucrose, cf. Fig. 1). The products are excreted into the culture suspension by the bacterium. However, since they cannot be easily separated from this medium, a two stage procedure with separate fermentation and subsequent disaccharide oxidation proved more favorable. Both steps therefore could be optimized separately.

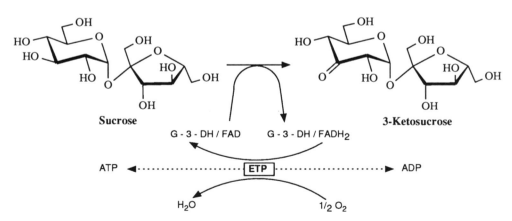

Fig. 1. Oxidation of sucrose by *Agrobacterium tumefaciens* dehydrogenase (G-3-DH) with FAD as coenzyme; electron transfer enzymes (ETP) are also indicated

It was necessary to indentify fermentation conditions which gave a reasonable cell yield with good activity of glycoside-3-dehydrogenase first. Literature data concerning this problem turned out to be not reproducible.[5] However, a standard fermentation procedure yielding a reasonable amount of active cell mass has been developed.[6]

The kinetics and yield of disaccharide oxidation depend on catalyst (biomass) and substrate concentrations (sugar, oxygen) as well as on the pH and the temperature. Experimental details of investigations concerning these parameters will be published elsewhere.[7] At high catalyst concentration the specific oxidation rate decreases due to limitation by the oxygen transfer rate. The influence of the sucrose concentration approximately follows typical enzyme kinetics, as seen in Fig. 2 representing a Haldane plot for conditions where oxygen is not rate limiting. Constants for simple Michaelis-kinetics with

$$v = V_{max} * S * (K_S + S)^{-1} \; ; \; V_{max} * S * v^{-1} = K_{M+S} \text{ (Haldane plot)}$$

are: V_{max}= 1.03 g/g * d (g product per g bacterial dry weight and day); K_S= 3.3 m mol/l; v: reaction rate at 25 °C and pH 7.

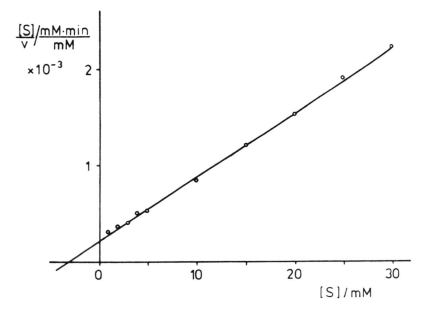

Fig. 2. Haldane plot of the oxidation rate as a function of sucrose concentration (oxygen concentration constant, $1 \leq O_2 \leq 50$ % of saturation).

The influence of the oxygen concentration suggests complex reaction pathways. Experiments with high catalyst concentration indicate low specific reaction rates at limiting oxygen concentration, which, however, are below the sensitivity of the oxygen electrode.[6] A quasi-saturation on the range of low (about 1 %) up to about 50 % oxygen saturation in solution (by air) is followed by further increasing reaction rates at higher oxygen concentration (Fig. 3). These unconventional but reproducible results suggest at least two different types of enzymes for electron transfer to oxygen, one with high and another with low affinity for oxygen.[7]

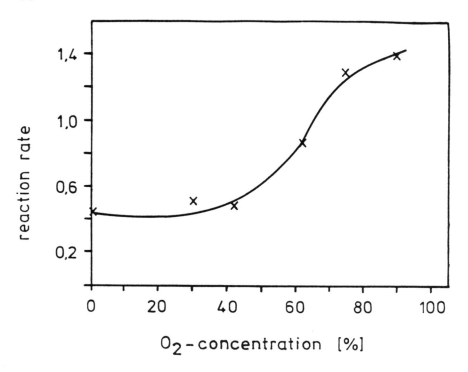

Fig. 3. Reaction rate (g ketosucrose x 10^4 / g bact. dry mass/min) as function of oxygen concentration (% of saturation at a constant sucrose concentration of 25 g/l).

The maximum yield of 3-keto-sucrose is about 60 % at low (5 g/l), and 40 % at high (100 g/l) sucrose concentration (Fig. 4). The overall yield for both the fermentation and oxidation steps is about 25 %. Notably, the fermentation requires further investigations for improving yields.

The oxidation of palatinose [isomaltulose, D-glucopyranosyl-$\alpha(1{\rightarrow}6)$-fructose], and leucrose, the $\alpha(1{\rightarrow}5)$-linked isomer, – both are industrial products[8,9] – gives distinctly higher yields, as is shown in Fig. 4 for 3-ketopalatinose. The reason could be that the products are converted (or metabolized) much slower by the bacterium in subsequent metabolic steps. Interestingly, the chemical stability shows similar trends.

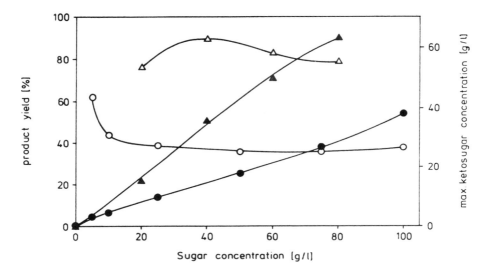

Fig. 4. Product yield and maximum product concentration at different initial substrate concentrations: yield of sucrose (O), palatinose (Δ), 3-ketosucrose (●), 3-ketopalatinose (▲).

Stability of, and Derivatives from 3-Keto-disaccharides

The stability depending on pH and temperature deserves major attention since purification and further derivatization depend much on these characteristics. First the chemical structures, molecular weight, and the position of the new carbonyl function, have been established by mass spectrometry and NMR, confirming the oxidation in position 3 of the glucosyl moiety.[10,11]

3-Keto-sucrose is hydrolyzed at increasing rates with decreasing pH, whereas it is rather stable at neutral and slightly alkaline pH[12] (cf. Fig. 5). Fructose is a reaction product, whereas another reaction intermediate originating from the 3-keto-glucose moiety is rather unstable. The reaction sequence involved with the latter is not yet established, an analogous mechanism to that of 3-keto-methylglucose might be discussed.[13] It has already been mentioned, that the 3-keto-products from palatinose and leucrose are much more stable.

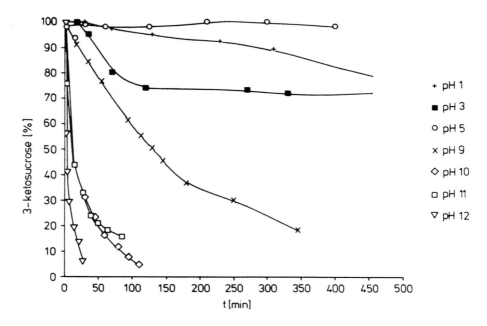

Fig. 5. Rates of decomposition of 3-ketosucrose at various pH-values (20 °C)

A few simple reactions to obtain derivatives from the 3-keto-disaccharides have been performed. One obvious procedure to apply is the reductive amination which gives monoamino-disaccharides from sucrose as the original starting material, or diamino-sugars from palatinose or leucrose, respectively.[14]

These might be valuable intermediates, e.g., for acylating reactions to give surfactants with technically favorable properties.[15] Another substitution product results from addition of cyanide to give the cyanohydrin of 3-keto-sucrose.[16] Thus the oxidation reaction by *A. tumefaciens* opens a straightforward way to a range of specifically functionalized disaccharides which might be of industrial interest.

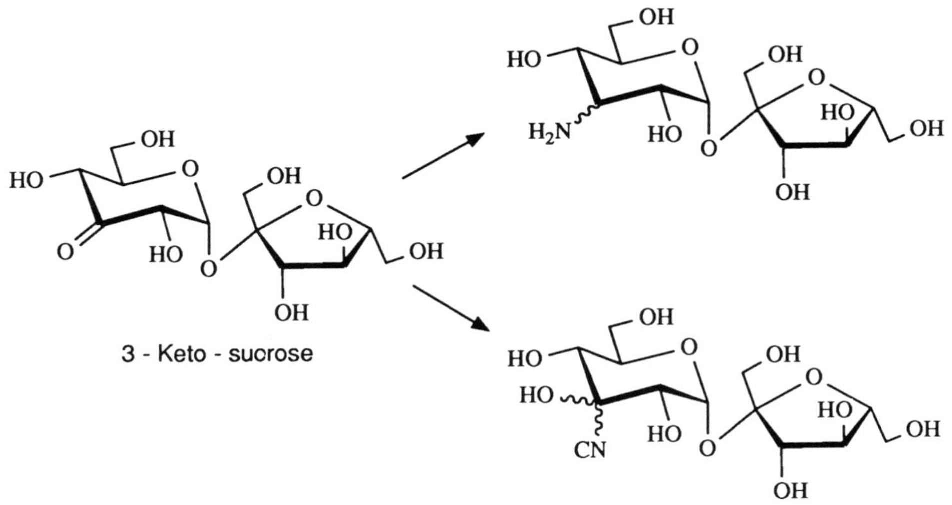

3 - Keto - sucrose

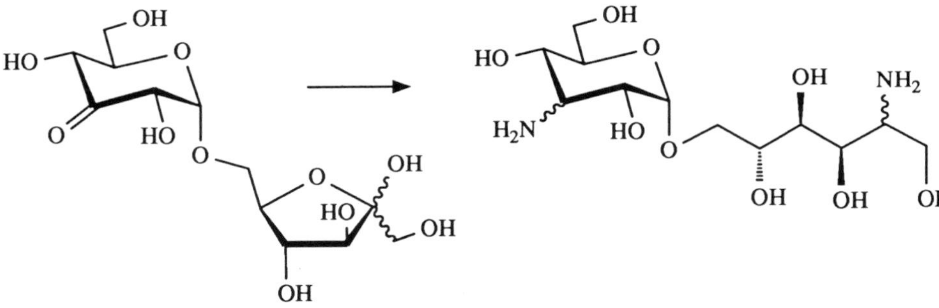

3 - Keto - palatinose

Oligosaccharide Synthesis by Dextran Sucrase

Transglucosylation, a side reaction of dextran synthesis by dextran sucrase, offers promising potential for oligosaccharide synthesis.[3,4] Under appropriate conditions high yields of one product can be obtained, as has been shown on a technical scale for leucrose, a disaccharide applied as a sweetener.[9] The enzyme specifically catalyzes the transfer of a glucosyl unit from sucrose to an acceptor, which preferentially can be a mono-, di- or trisaccharide, or also a derivative such as D-glucosyl-α(1→5)-arabonic acid, as has been shown recently.[17]

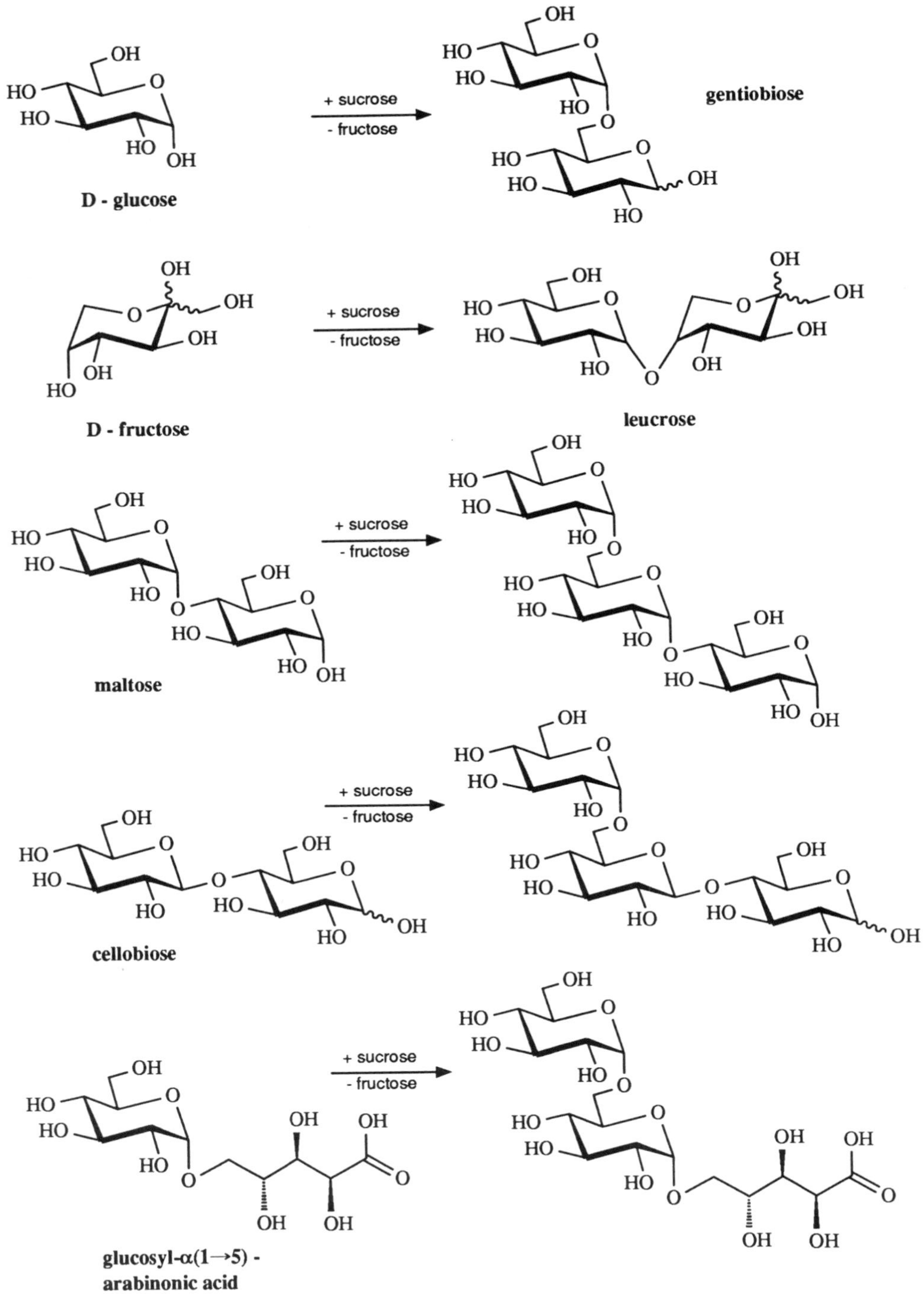

D - glucose   + sucrose / - fructose →   **gentiobiose**

D - fructose   + sucrose / - fructose →   **leucrose**

maltose   + sucrose / - fructose →

cellobiose   + sucrose / - fructose →

glucosyl-α(1→5) - arabinonic acid   + sucrose / - fructose →

Fructose is obtained as a by-product. Several parameters and notably the concentrations of enzyme, substrate, and acceptor, control the product yield and catalyst productivity. Therefore, the reaction kinetics are essential for the establishment of optimal reaction conditions. The kinetics of dextran synthesis have been investigated in detail by Ebert and Schenk,[18] and Ludwig.[19]

An extended scheme for the kinetics comprising acceptor reactions and inhibition has been proposed recently.[20] It is based on a rather complete scheme of the pathways included, as far as they are known from the literature (Fig. 6).

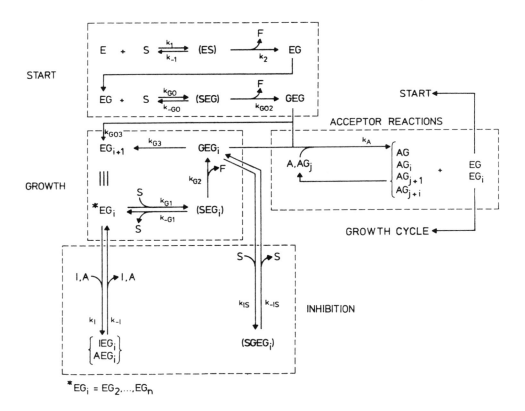

Fig. 6. Pathways of dextran (AGi) and oligosaccharide (AGj) synthesis from sucrose (S) and acceptors (A, AGj) by dextransucrase (E) (G: glucosyl unit, I: inhibitors, F: fructose; kinetic constants k_i also given).

A set of kinetic measurements with maltose as an acceptor shows that the yield of the trisaccharide panose increases from about 40 % with equal concentrations of substrate and acceptor ($\propto$ 10 K_M) to over 70 % when that ratio is 1 : 10 (Fig. 7).

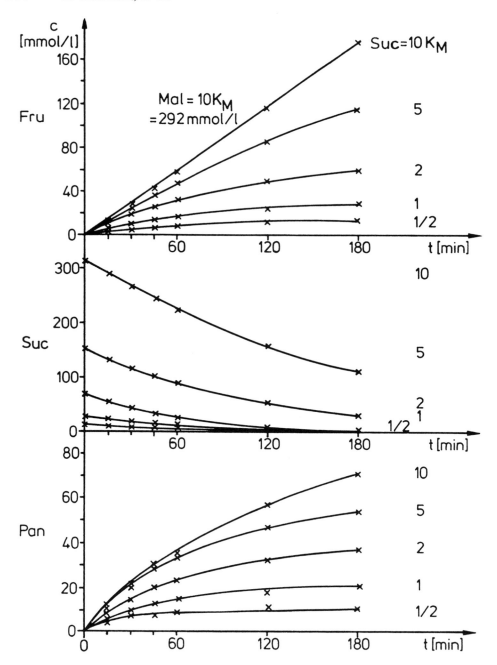

Fig. 7. Kinetics of panose (Pan) formation from sucrose (Suc) and maltose (Mal), with fructose (Fru) as a side product, catalyzed by dextran sucrase (25 °C, pH 5.4).

Since the trisaccharide is also an acceptor and since inhibition phenomena play a role (Fig. 6), the optimization requires further investigations, notably in the range of concentrated solutions. High product yields at high concentrations can thus be obtained, which is important for technical purposes.

Immobilization of Dextran Sucrase

For industrial application continuous processing and / or the re-use of the catalyst are essential. Enzyme immobilization offers the most straight-forward approach for that problem.[21] However the investigations concerning dextran sucrase immobilization reported in the literature were unsuccessful with respect to yield and stability. One problem involved might be testing with sucrose as a substrate, which yields dextran and will lead to occlusion of the porous support. This situation can be overcome by using an appropriate ratio of maltose and sucrose, thus minimizing dextran formation in the test.

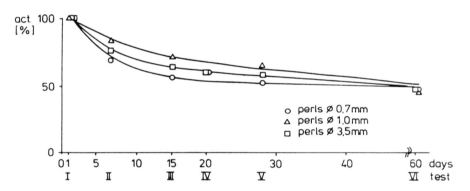

Fig. 8. Operational stability of dextran sucrase immobilized in alginate with different particle size (∅); reactions and tests with 20 g/l sucrose and 100 g/l maltose, 25 °C, pH 5.4.

A rather unusual method for enzyme immobilization, i.e. entrapment in alginate, has been used by Schwengers[9] and was further investigated concerning activity, yield, and stability under operating conditions.[22,23] Both activity and stability depend much on reaction conditions (concentrations), acceptor type and particle diameter. Under appropriate conditions, the yield of activity immobilized is more than 75 %, and with maltose as an acceptor, the stability is rather high under operating conditions with a half life of more than 50 days (Fig. 8).[22] It might be assumed that this rather uncommon method for enzyme immobilization works successfully because dextran is a by-product with rather low yield preventing leakage of the enzyme from the alginate matrix.

Conclusions

Although a range of disaccharides and, notably, sucrose are available on an industrial scale at high purity and at low or moderate price, their use in synthesis is limited, despite their advantageous properties in several respects. The difficulties of specific derivatization and synthesis can be overcome by enzymatic routes. Several requirements of technical importance, such as reaction in water without protecting groups, one or few reaction steps, acceptable or high yields at high concentration in solution can be met by reaction engineering approaches.

It has been shown that with *Agrobacterium tumefaciens*, which oxidizes several disaccharides regioselectively, the 3-keto-derivatives can be made available at the conditions mentioned.

The specific synthesis of a rather broad range of oligosaccharides by glucosylation with sucrose as glucosyl donor can also be achieved with high yield with immobilized dextran sucrase which is appropriate for continuous processing.

Acknowledgement. Funds from the Federal Ministry for Research and Technology are gratefully acknowledged.

References

1. M. J. Bernaerts, J. de Ley: Microbiological Formation and Preparation of 3-Ketoglycosides from Disaccharides. *J. Gen. Microbiol.* **22** (1960) 129-136.

2. J. van Beumen, J. de Ley: Hexopyranoside: Cytochrome C Oxidoreductase from *Agrobacterium tumefaciens*. *European J. Biochem.* **6** (1968) 331-343.

3. J. F. Robyt, S. H. Eklund: Relative Quantitative Effects of Acceptors in the Reaction of *Leuconostoc mesenteroides* B-512 F Dextransucrase. *Carbohydr. Res.* **121** (1983) 279-286.

4. J. F. Robyt, S. H. Eklund: Stereochemistry Involved in the Mechanism of Action of Dextransucrase in the Synthesis of Dextran and Formation of Acceptor Products. *Bioorganic Chem.* **11** (1982) 115-132.

5. W. M. Kurowski, A. H. Fenson, S. J. Pirt: Factors influencing the Formation and Stability of D-Glucoside-3-dehydrogenase Activity in Cultures of *Agrobacterium tumefaciens*. *J. Gen. Microbiol.* **90** (1975) 191-202.

6. K. Matalla, E. Stoppok, K. Buchholz: Specific Modification of Sucrose to Building Blocks for Synthesis. *Dechema Biotechnol. Conf.*, Frankfurt, Vol. 3 (1989) 117-123.

7. K. Matalla: Mikrobielle Oxidation von Disacchariden. *Doctoral Dissertation*, Techn. Universität Braunschweig, 1990.

8. R. Weidenhagen, S. Lorenz (Süddeutsche Zucker AG): Verfahren zur Herstellung von Palatinose (6-α-Glucosido-fructofuranose). *Ger. Pat. Appl.* 10 49 800 (1957); *Chem. Ztrbl.* **1959**, 13962.

9. D. Schwengers, H. Benecke (Pfeifer & Langen): Süßungsmittel, Verfahren zu Herstellung und Verwendung derselben. *Ger. Pat.* 34 46 380 (1986); *Sugar Ind. Abstr.* **49** (1987) 101; *Chem. Abstr.* **105** (1986) 77815p.

10. M. Noll: Oxidation von Palatinose mittels *Agrobacterium tumefaciens*. VCI-BMFT-Symposium Saccharidchemie, Hamburg 1990, and personal communication.

11. J. Walter: Untersuchungen zur mikrobiellen Leucrose-Oxidation durch *Agrobacterium tumefaciens*. *Diploma Thesis*, Techn. Universität Braunschweig, 1990.

12. M. Walter: Personal communication.

13. O. Theander: The Oxidation of Glycosides. Degradation of Methyl-α-D-3-keto-glucopyranoside, Methyl-β-D-3-ketoglucopyranoside and Methyl-β-D-2-ketogluco-pyranoside in Lime Water. *Acta Chem. Scand.* **12** (1958) 1887-1896.

14. M. Kunz et al. (Verein der Zuckerindustrie): Diamine. *Eur. Pat. Appl.* 90 109 667.7 (1990).

15. M. Kunz: Saccharosederivate – Zucker als hydrophiler Baustein. *Zuckerind.* **113** (1988) 273-278.

16. D. Miehe: Personal communication.

17. D. Prinz: Personal communication.

18. K. H. Ebert, G. Schenk: Mechanisms of Biopolymer Growth: The Formation of Dextran and Levan. *Adv. Enzymol.* **30** (1968) 179-221.

19. H.-P. Kindler, M. Ludwig: Untersuchungen zum Mechanismus der enzymatischen Dextran-Synthese mit einer Strömungsapparatur. *Chem.-Ing.-Tech.* **24** (1975) 1035.

20. K.-D. Reh, K.-J. Jördening, K. Buchholz: Kinetics of Oligosaccharide Synthesis by Dextran Sucrase. *10th Enzyme Engineering Conference*, Kashikojima/Japan 1989.

21. K. Buchholz: Immobilisierte Enzyme – Kinetik, Wirkungsgrad und Anwendung. *Chem.-Ing.-Tech.* **61** (1989) 611-622.

168   *K. Buchholz, et al.*

22. I. Knop: Immobilisierung von Dextransucrase für die Oligosaccharidsynthese. *Diploma Thesis*, Fachhochschule Emden 1990.

23. K.-D. Reh: Personal communication.

8

Polyfructose :
a New Microbial Polysaccharide

Margaret A. Clarke, August V. Bailey, Earl J. Roberts, and Wing S. Tsang

Sugar Processing Research, Inc., New Orleans, Louisiana, USA 70124

Summary. A microbial polysaccharide that is a polymer of fructose has been produced, in good yield and high purity, by a strain of *Bacillus polymyxa* from sucrose. The compound, which has been given the trivial name polyfructose, consists entirely of fructose, with one glucose unit in the initial sucrose group.

The structure has been shown, by NMR and methylation analysis, to be a β-(2→6)-linked backbone, with up to 12 % branching through β-(1→2)-linkage. X-ray crystallography has indicated that the compound is amorphous.

Polyfructose can be produced from pure sucrose, or from sucrose in sugarcane or sugarbeet molasses or syrups, or from sugarcane or sugarbeet juice. The product made from molasses or juices requires some clean-up, for example, with DEAE cellulose, to attain the white color of the polyfructose made from pure sucrose. Gel permeation chromatography has shown polyfructose to have a narrow molecular weight range centered on 2×10^6 daltons.

The product is soluble in water, and is readily hydrolyzed at high temperatures and acid pH to fructose. Hydrolysis rates, using several acids at various temperatures, and the products formed, are presented. Polyfructose is not hygroscopic and can be stored at atmospheric conditions for several months. The compound is therefore an easily stored material that is a source of fructose, and forms fructose syrup upon acidification.

The products of enzyme hydrolysis of polyfructose are presented: the compound is resistant to most enzymes other than fructanases. Potential applications of polyfructose, including applications as a sweetener source, as an encapsulation material, and as a sweetness potentiator are considered.

Introduction

As part of an ongoing study to develop new products from the agricultural resources provided by sugar-producing crops, a search was initiated for microorganisms to produce polymeric compounds for industrial use. Polysaccharides were the first group of polymers considered. Dextrans, polymers of glucose synthesized from sucrose, are important industrial polysaccharides.[1]

Fructans are natural polymers of fructose. Depending on the linkage types, fructans are classified into two groups: the levans, with mostly β-(2→6)-linkages and the inulins with β-(2→1)-linkages:

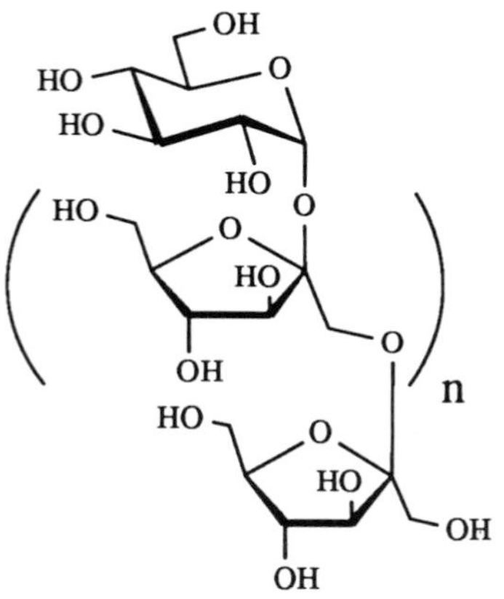

inulin
β (1 → 2)

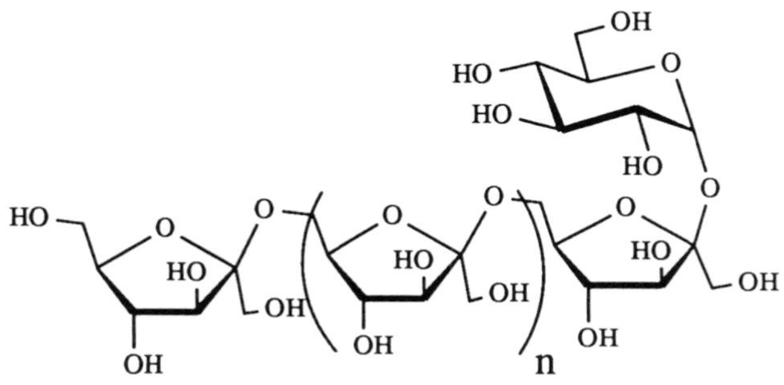

levan
β (2 → 6)

Many fructans of both types have branched chains. Levans and inulins of low molecular weight are abundantly found in plants, while high molecular weight fructans are produced by many microorganisms.[2-4] A variety of them produce extracellular polysaccharides in the form of capsules attached to the cell wall, or as slime secreted into the growth medium. These materials are used in the organism's defense mechanism, or as a food reservoir. Some bacteria produce fructan, among which *Bacillus* spp. predominate. Oral bacteria such as *Rothis dentocariosa*, *Streptococcus salivarius* and *Odontomyces viscosus* accumulate fructan in human dental plaque.[5-9]

Most research on the biosynthesis of fructan has been conducted using *Bacillus subtilis*, *Aerobacter levanicum*, and *Streptococcus salivarius*.[10-19]

Recently, fructans produced by *Zymomonas mobilis*[20-22] have also been investigated.

Microbial fructans or levans, like dextran, were first found in sugar factories.[2,23,24] These polysaccharides caused difficulties in the beet sugar manufacturing process by increasing the viscosity of process juice and syrups. Since their discovery in 1891, fructans have received little attention and have never been exploited for industrial applications.

Isolation of a Fructan-producing Bacterium

Fig. 1 shows the production scheme using a levan producing bacterium. About 1 g of rotting sugarcane stalks and the adhering soil particles were added to 100 ml of basal medium and incubated at 30 °C with constant shaking. The isolation medium consisted of sucrose 80 g to 150 g; peptone 2 g; yeast extract 2 g; K_2HPO_4, 2 g; $(NH_4)_2SO_4$, 0.3 g; in a liter of water. The growth culture was then transferred to fresh media every 7 - 10 days.

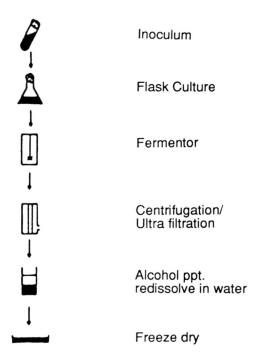

Fig. 1. Scheme for production of levan by *B. polymyxa*

A detailed isolation procedure was reported elsewhere.[25] The organisms have been registered at USDA, Northern Regional Research Center, Peoria, Illinois, and identified as NRRL B-18475 and B-18476. These, plus several other strains, are discussed in this paper.

Production of Polyfructose

The *B. polymyxa* (NRRL B-18475) produces a large quantity of fructan when grown on 4 - 16 % sucrose solution. The organism converted the fructose moiety of sucrose to fructan; of the remaining glucoses, most were used as the carbon source for microbial growth and a small amount accumulated in the growth medium. No fructan was produced when the organism was grown on glucose or fructose.

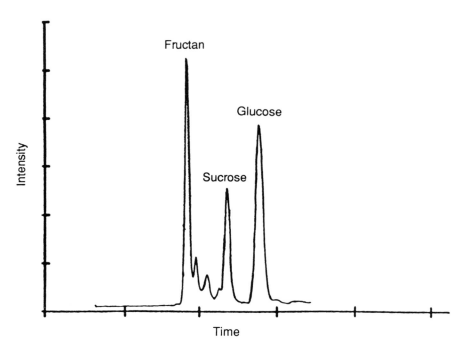

Fig. 2. HPLC profile of microbial levan products at Day 2

The composition of the products was monitored by HPLC (Sugar Analyzer, Waters Associates; HPX-87C column, BioRad Corp. with deionized water, 40 ppm as mobile phase). During fermentation, the sucrose levels dropped and fructan started to appear in 2 days; thereafter, sucrose level gradually decreased as fructan increased. Glucose was the major byproduct. The pH of the growth medium fell from 7.0 to 4.7 indicating acid production. In reports of other fructan production, maintaining pH above 5.5 was important because the optimum pH for fructansucrase is between 5.5 - 7.0 and fructan may be hydrolyzed at a lower pH.[2] Optimum temperature for growth and fructan production was around 30 °C.

Composition of a typical fermentation mixture is shown in Fig. 2, and progress of fructan formation over an 8 day period is shown in Fig. 3. Optimum period for production of high molecular weight fructan appears to be 3 days. Fermentation times were extended to 16 days, with little improvement in yield. The detailed production is reported elsewhere.[26] conditions of ionic strength and pH are under investigation.

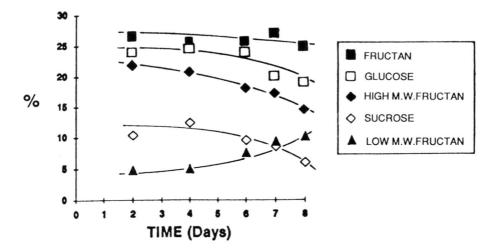

Fig. 3. Composition of products from *B. polymyxa*

Polyfructose can be produced from sugar juices (sugarbeet juice and sugarcane juice) and molasses; typical yields are shown in Table 1. These are production yields on juice and molasses added instead of sucrose to the growth medium. Beet and cane molasses were not treated or cleaned up (e.g. desludged) in any way in this set of trials, but the crude product reported here from molasses was yellowish in color, and required a clean-up procedure, such as filtration over DEAE-cellulose, to produce a white product in slightly lower yield.

Table 1. Yields of polyfructose from different sucrose sources

Source of sucrose[a]	% yield of levan[b]	% of original sucrose unreacted	glucose formed, % of original sucrose
Beet Molasses	37	21	–
Beet Juice	28	23	8.5 %
Cane Molasses	22	45	7
Cane Molasses	57	38	20

a) The medium is made up to contain 8 % of sucrose irrespective of source; data
   are for 2 d incubation period
b) based on available fructose

Structure and Properties

Experimental methods and materials used in determining the structure of this fructan, or levan, which has been given the trivial name of polyfructose, are reported in detail elsewhere.[27]

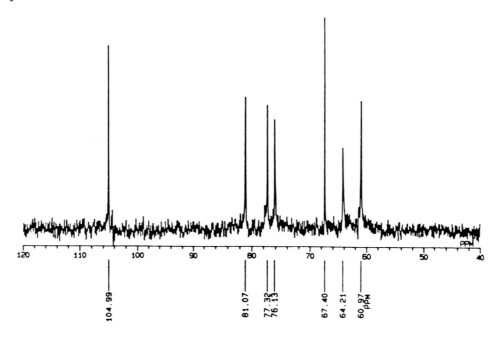

Fig. 4. ^{13}C-NMR spectrum of polyfructose (100 MHz, dioxane as internal standard at δ= 67.40)

Carbon-13 NMR-spectra (as shown in Fig. 4) indicate that all fructose molecules were in the same conformation.[27] Proton NMR have indicated that only fructose is present in the molecule. Because sucrose is the initial molecule in the chain, there must be some terminal glucose residues. However, since the molecular weight of polyfructose has been shown to be about 2×10^6 daltons, the ratio of fructose to glucose is about 12,000 : 1, and so glucose present would not be observed on NMR spectra or by HPLC analysis. Comparison with literature assignments of NMR peaks indicated that polyfructose is of the β-(1→2)-linked type.[20] Methylation analysis with GLC and MS detection indicated that 71 % of the fructose moieties form a β-(2→6)-linked backbone, with 12 % branch points of β-(2→1)-linkage, and 13 % terminal groups.[27,28]

Polyfructose is not hygroscopic; and lyophilized material has been stored under ambient conditions (25 - 30 °C; 70 - 90 % relative humidity) for several months. It is soluble in water, although some turbidity or opalescence is always present. This opalescence is apparently characteristic of fructans in solution. Addition of a few drops of acid has been observed to remove the opalescence while not causing any hydrolysis or breakdown of the fructan. It can readily be hydrolyzed in acid below pH 3.5 to form fructose. The addition of heat (standard or microwave) greatly increases the rate of hydrolysis, as shown in Table 2:

Table 2. Acid hydrolysis to fructose.

conditions	time	temp.	% fructose
0.5 % citric acid	48 h	r.t.	
	2 min	microwave	33
	5 min	microwave	76
	7 min	microwave	100
0.5 % ascorbic acid	15 min	100 °C	100

Enzyme Hydrolysis

It was of interest to conduct enzyme hydrolysis of polyfructose to gain more information on the structure. The polysaccharide showed no reaction with amylase, dextranase or other glucanases (exception below).

No fructofuranosyl fructosidase, or fructanase, could be obtained, so various available crude enzyme preparations were presented to the polysaccharide. Two of these commercially available enzymes, Gamanase (Novo Biochemicals, Inc.) a "debranching enzyme", or crude pullulanase, had similar effects on polyfructose. Both hydrolyzed the polymer to a smaller polysaccharide of about 20,000 daltons. Gel permeation chromatograms for molecular weight determination (Sephacryl S-500; water) are shown in Fig. 5. It should be noted that HPLC[27] of the fermentation mixture showed two peaks (or a split peak) for fructan; similar HPLC analysis of the smaller (now referred to as low molecular weight fructan or levan) showed a single peak. Apparently, a fructanase exists as a contaminant in the commercial enzyme preparations used, and is similar to a fructanase present in the strains NRRL-B-18495 and NRRL-B-18476. Carbon-13 NMR-spectra of the enzyme-hydrolyzed products were similar (cf. Fig. 6A). Comparison of NMR-spectra as shown in Fig. 6B indicates, by the increase in terminal groups, that the enzymes have hydrolyzed β-(2→6)-linkages in the polyfructose backbone to form smaller backbone segments that maintain their branch structures.

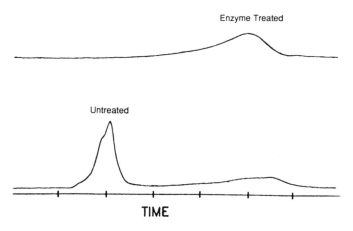

Fig. 5. GPC profile of microbial levan

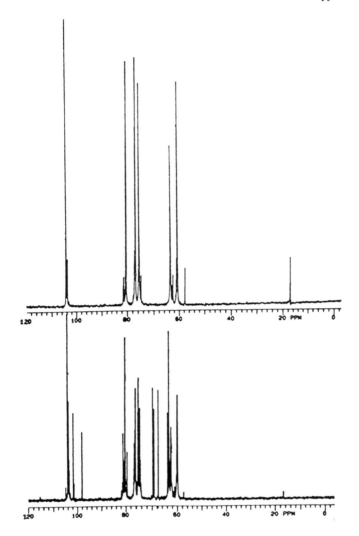

Fig. 6. ^{13}C NMR Spectra of untreated (A, top) and enzyme treated levan (B, below)

The lower molecular weight levan is much more soluble than the high molecular weight, giving a clear solution. Differences in rate of acid hydrolysis of the high and low molecular weight products are shown in Table 3, and differences in viscosity of solutions of the two in Fig. 7. Angular rotations and melting points are listed in Table 4.

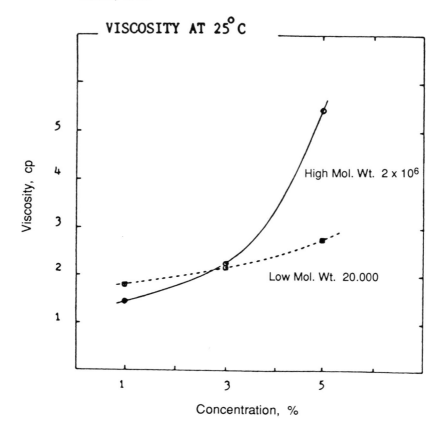

Fig. 7. Viscosity profiles of high and low molecular weight polyfructose

Table 3. Hydrolysis of levans at 65 °C in 0.5 % oxalic acid

Time (min)	% fructose	
	Low Mol. Wt. (20,000)	High Mol. Wt. 2 x 10^6
0	0	0
10	29	24
20	68	46
30	96	68
40	100	88
60		92
80		100

Table 4. Physical properties of polyfructose

	$[\alpha]_D^{24}$	m.p. (°C)
High Mol. Wt. (2 x 10^6 daltons)	-42.3°	> 200
Low Mol. Wt. (20,000 daltons)	-62.2	92

Applications of Polyfructose

Polyfructose being a stable, non-hygroscopic, storable, low-cost source of fructose, many applications in the food and beverage industry are possible. Hydrolyzed solutions of polyfructose are, in effect, fructose syrups. The question of the use of this compound as a liquid sweetener source is an economic one. Polyfructose in its lower molecular weight form offers the possibility of sweetness potentiation.

The viscosity profiles of polyfructose indicate that it can be used as a thickener, perhaps combining added sweetness and enhanced mouthfeel in a single product.

Polyfructose can be used to encapsulate flavoring and coloring agents. As one form of encapsulation, it can be made into tablets containing flavor, color or other additives which will store well under atmospheric conditions because polyfructose does not absorb water readily. Polyfructose is a stable, non-toxic, non-hygroscopic compound that is expected to be digested in the stomach, contributing calories equivalent to its hydrolysis product, fructose.

References

1. G. P. Meade, J. P. C. Chen: Cane Sugar Handbook (11th Edition). Wiley / Interscience, New York, 1985.

2. G. Avigad, D. S. Feingold: Fructosides formed from Sucrose by a Coryne Bacterium. *Arch. Biochem. Biophys.* **70** (1957) 178-184.

3. E. J. Vandamme, D. G. Derycke: Microbial Inulinases: Fermentation Process, Properties, and Applications. *Adv. Appl. Microbiol.* **29** (1983) 139-176.

4. H. G. Pontis, E. Del Campillo: Biochemistry of Storage Carbohydrates in Green Plants (P. M. Dey, R. A. Dixon, Eds.). Academic Press, New York 1985, Chapter 5, pp. 205-227.

5. R. S. Manly, D. T. Richardson: Metabolism of Levan by Oral Samples (Saliva and Plaque). *J. Dent. Res.* **47** (1968) 1080-1089; *Chem. Abstr.* **70** (1969) 65949f.

6. E. Newbrun, S. Baker: Physicochemical Characteristics of the Levan Produced by *Streptococcus salivarus. Carbohydr. Res.* **6** (1968) 165-170.

7. M. Higuchi, Y. Iwami, T. Yamada, S. Araya: Levan Synthesis and Accumulation by Human Dental Plaque. *Arch. Oral Biol.* **15** (1970) 563-567; *Chem. Abstr.* **73** (1970) 33187b.

180   *M. A. Clarke, et al.*

8. J. R. Loewenberg, E. T. Reese: Microbial Fructosanes and Fructosananes. *Can. J. Microbiol.* **3** (1957) 643-650; *Chem. Abstr.* **51** (1957) 13068b.

9. A. Fuchs, J. M. DeBruijn, C. L. Niedveld: Bacteria and Yeast as Possible Candidates for the Production of Inulinases and Levanases. *Antonie Van Leewenhoek* **51** (1985) 333-343; *Chem. Abstr.* **104** (1986) 107853y.

10. S. Hestrin, D. Avineri-Shapiro, M. Aschner: The Enzymic Production of Levan. *Biochem. J.* **37** (1943) 450-456.

11. T. H. Evans, H. Hibbert: Bacterial Polysaccharides. *Adv. Carbohydr. Chem.* **2** (1946) 204-233.

12. D. S. Feingold, M. Gehatia: The Structure and Properties of Levan, a Polymer of D-Fructose produced by Cultures and Cell-free Extracts of *Aerobacter levanicum*. *J. Polymer Sci.* **23** (1957) 783-790; *Chem. Abstr.* **51** (1957) 9797a.

13. R. Dedonder: Levansucrase from *Bacillus subtilis*. *Methods Enzymol.* **8** (1966) 500-505.

14. M. Takeshita: Translucent Colony Form of the Gram-negative, Levan-producing *Aerobacter levanicum*. *J. Bacteriol.* **116** (1973) 503-506.

15. T. Tanaka, S. Yamamoto, S. Oi: Structures of Heterooligosaccharides synthesized by Levansucrase. *J. Biochem.* **90** (1981) 521-526.

16. P. Mantsala, M. Puntala: Comparison of Levansucrase from *Bacillus subtilis* and from *Bacillus amyloliquefaciens*. *FEMS Microbial Lett.* **13** (1982) 395-399; *Chem. Abstr.* **96** (1982) 213043b.

17. E. W. Lyness, H. W. Doelle: Levansucrase from *Zymomonas mobilis*. *Biotechnol. Lett.* **5** (1983) 345-350; *Chem. Abstr.* **99** (1983) 2215s.

18. P. Perlot, P. Monsan: Production, Purification, and Immobilization of *Bacillus subtilis* Levan Sucrase. *Ann. New York Acad. Sci.* **434** (1984) 468-471; *Chem. Abstr.* **102** (1985) 130391p.

19. S. Yamamoto, M. Iizuka, T. Tanaka, T. Yamamoto: The Mode of Synthesis of Levan by *Bacillus subtilis* Levan Sucrase. *Agric. Biol. Chem.* **49** (1985) 343-349.

20. K. D. Barrow, J. G. Collins, P. L. Rogers, G. M. C. Smith: The Structure of a novel Polysaccharide isolated from *Zymomonas mobilis* determined by Nuclear Magnetic Resonance Spectroscopy. *Eur. J. Biochem.* **145** (1984) 173-179.

21. T. D. Mays, E. L. Dally (IGI Biotechnology, Inc.): Microbiological Production of Polyfructose. *US Pat.* 689238 (1985); *Chem. Abstr.* **106** (1987) 48658e.

22. J. F. Kennedy, D. L. Stevenson, C. A. White, L. Viikari: The Chromatographic Behavior of a Series of Fructooligosaccharides derived from Levan produced by the Fermentation of Sucrose by *Zymomonas mobilis*. *Carbohydr. Polym.* **10** (1989) 103-113; *Chem. Abstr.* **111** (1989) 78521a.

23. E. O. Lippman: Über das Lävulan, eine neue in der Melasse der Rübenzucker-fabriken vorkommende Gummiart. *Ber. Dtsch. Chem. Ges.* **14** (1881) 1509.

24. F. Schneider, H. P. Hoffman-Walbeck, M. A. F. Abdou: Polysaccharide Producing Microorganisms in Sugar Factories. *Pseudomonas fluorescens*, isolated from Sugar Beet Tissue. *Zucker* **22** (1969) 465-473.

25. Y. S. Han, M. A. Clarke, *J. Agric. Food Chem.* **38** (1990) 393-396.

26. Y. S. Han, M. A. Clarke: Agricultural and Synthetic Polymers: Utilization and Biodegradability. *ACS Adv. in Chem. Series* **1990**, 210-219.

27. M. A. Clarke, E. J. Roberts, W. S. Tsang, M. A. Godshall, Y. W. Han, L. Kenne, B. Lindberg: Structural Studies on a Fructan from Sugar Beet and Sugar Cane Juice. *Proc. Conf. Sugar Proc. Res.* **1990**, 139-146.

28. B. Lindberg, J. Lönngren, J. L. Thompson: Methylation Studies on Levans. *Acta Chem. Scand.* **27** (1973) 1819.

9

Leucrose, a Ketodisaccharide of Industrial Design

Dieter Schwengers

Pharma-Division, Pfeifer & Langen, D-4047 Dormagen, Germany

Summary. Although the keto-disaccharide leucrose has already been identified in 1952 as a natural by-product during formation of dextran by fermentation from sucrose with *Leuconostoc mesenteroides* bacteria, still little was known about its properties so far. Only after development of an enzymatic production process in 1986 leucrose has become available in any quantity required. In the course of its higher availability the interest in this disaccharide increased considerably as a sweetener in the food sector and as a building block for synthesis in the chemical industry. A survey is given on the applications of leucrose in both of these areas, with special emphasis on its physiological properties and on newer syntheses of glycosides, esters, oxidation products and acid amides.

Introduction

Pfeifer & Langen is one of the pioneers of research into and production of dextrans and dextran derivatives, which began more than 40 years ago. As competition increased, the manufacture of dextran with the help of bacteria such as *Leuconostoc mesenteroides* or *Streptococcus bovis* had to be made more efficient. As evidenced by the data collected in Table 1, the yield of high-molecular dextran is strongly dependant on the sucrose concentration of the fermentation solution in which the bacteria grow. If the fermentation is to be economic the concentration must be 10 - 12 %.

Secondary products of the manufacturing process are monosaccharides, primarily fructose, and a mixture of disaccharides and low-molecular dextrans that cannot be precipitated by 75 % alcohol.

Table 1. Yields of carbohydrate fractions from dextran-sucrase at varying sucrose concentrations (g per 100 g sucrose)

Sucrose (%, w/V)	Mono-saccharides	Di-saccharides	Total Dextran	HM WT Dextran	LM WT Dextran
(Theory)	52.6	0	47.4	47.4	0
2	52.2	1.9	45.9	45.9	0
4	50.0	4.4	45.6	45.6	0
5	51.3	4.3	44.4	44.4	0
10	51.2	6.8	42.0	39.0	3.0
15	55.5	9.4	35.3	25.3	10.0
20	56.9	11.2	31.9	17.9	14.0

In 1952 Stodola et al.[1] proved that the previously unknown disaccharide leucrose is present in this mixture of disaccharides. In 1955 the same authors[2] succeeded in synthesizing leucrose as a result of the action of the enzyme dextransucrase, which they isolated from a bacterial culture of *Leuconostoc mesenteroides* on a 55 % aqueous sucrose solution; the yield was 7.9 %. They were able to identify leucrose as 5-*O*-(α-D-glucopyranosyl)-β-D-fructopyranose, i.e. a ketodisaccharide, that like sucrose, consists of one molecule of glucose and one molecule of fructose:

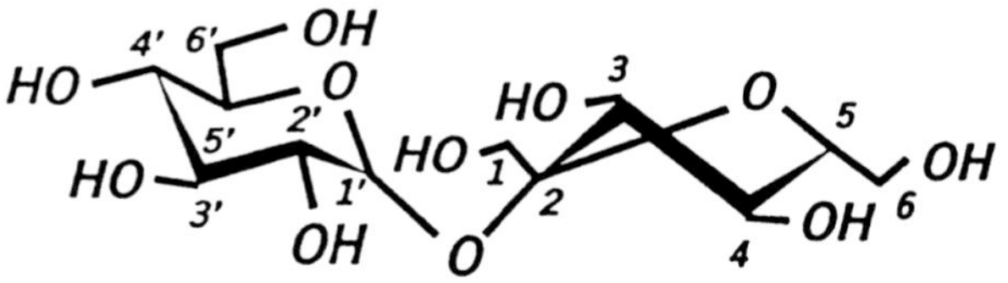

Sucrose
(α (1→ 2) linkage)

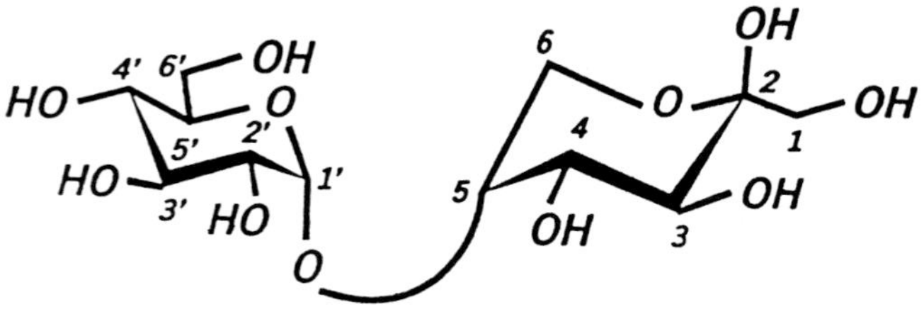

Leucrose
(α (1→ 5) linkage)

Leucrose, together with other disaccharides, including isomaltose and palatinose, the α(1→6)-linked isomer of sucrose, seems to be a general secondary product of dextran-producing strains of bacteria. An example of this is provided by studies of Siebert,[3] which indicated that leucrose is formed from sucrose in the oral cavity during the formation of dental plaque polysaccharides. It has also been detected alongside other disaccharides as an ingredient of honey.[4-6]

Because leucrose is formed in very small quantities only during the cultivation of microorganisms [2,7,8] and since its isolation from the complex product mixture is exceedingly difficult, for many years very little was known of its properties.

Manufacture of Leucrose

In 1986 Pfeifer & Langen was granted a patent[9] for a biotechnical manufacturing process that permits leucrose to be produced in any quantity desired. The process is based on a suggestion by Stodola that the enzyme dextransucrase, an $\alpha(1\rightarrow6)$-glucosyltransferase, should be used in the manufacture of leuctrose, instead of the bacterial culture that produces this extracellular enzyme. The decisive step beyond Stodola's suggestion was the result of systematic studies of the formation of low-molecular dextrans, based on our current understanding of the reaction sequence involved in the formation of dextran.

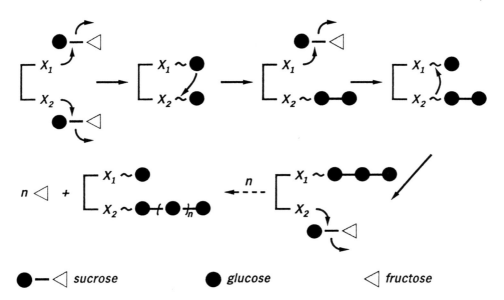

●—◁ *sucrose*   ● *glucose*   ◁ *fructose*

Fig. 1. Schematic presentation of the mechanisms proposed for the action of dextransucrase[10-12]

Robyt et al.[10-12] (Fig. 1) assume that dextran-sucrase has two equally active centres X_1 and X_2, and that one end of the growing dextran chain alternates between these two centres, picking up one additional glucose molecule with each transfer from one centre to the other.

According to this model, interruption of the formation of high molecular dextran by acceptors outside the sphere of the enzyme is not favoured, and this is supported by the finding that low molecular intermediate stages cannot be detected during dextran formation. The following model, presented in Fig. 2, permits a better explanation of our experience with various acceptors that can be used to control or even totally suppress dextran formation:

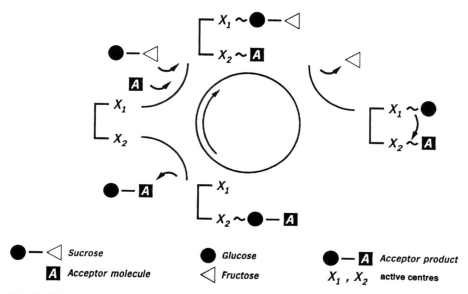

Fig. 2. Schematic presentation of the mechanism proposed for the action of dextransucrase in acceptor-reactions

Sucrose (glucosyl donor) is hydrolysed at a site X_1 on the dextransucrase, forming an activated glucosyl-enzyme complex and simultaneously releasing fructose. An acceptor is bound at another site X_2. In the next step the activated glucose at X_1 is transferred to the acceptor at X_2, forming an $\alpha(1\rightarrow6)$-bond. The acceptor product thus formed can again bind to X_2, where it is available for the transfer of another glucose molecule.

With glucose as an acceptor, we have exploited this transfer reaction in the industrial manufacture of an isomaltose / oligosaccharide mixture with an average molecular weight of 1000 Dalton (Table 2)

Table 2. Acceptor reactions of dextransucrase (DS)

Substrate		Acceptor	Catalyst	Product
Glu – Fru	+	Glu	DS	Glu – Glu + Fru
Glu – Fru	+	Glu – Glu	DS	Glu – Glu – Glu + Fru
n Glu – Fru +		Glu – Glu	DS	Glu – $(Glu)_n$ – Glu +
nFru				
Glu – Fru_1	+	Fru_0	DS	Glu – Fru_0 + Fru_1
Glu – Fru_2	+	Fru_1	DS	Glu – Fru_1 + Fru_2

In mixtures with sucrose, 6-*O*-methylglucose completely suppresses dextran formation and brings about slow but complete hydrolysis of the sucrose into glucose and fructose. This means that 6-*O*-methylglucose blocks the active centre on the dextransucrase where the transfer reaction occurs, while the centre responsible for the formation of the glucosyl-enzyme complex remains active.[13]

Under certain conditions, dextran formation can also be suppressed by fructose. Unlike 6-*O*-methylglucose, fructose acts as an acceptor; the enzyme separates the glucose from the fructose portion of sucrose and transfers it to a fructose molecule (cf. Table 2). The newly formed bond between the glucose and the fructose molecules differs from that in sucrose, thus forming the leucrose molecule.

The fructose molecules released in the reaction mixture are reused until, in the ideal case, all of the sucrose has been transformed into leucrose. So far we have obtained transformation rates of approximately 90 %. In a secondary reaction the dextransucrase molecule transfers glucose molecules to water, which acts as an acceptor, forming glucose and fructose. For this reason, yields of 100 % cannot be achieved.

The manufacture of leucrose (Fig. 3) starts with the extraction of the enzyme dextransucrase. For this purpose, *Leuconostoc mesenteroides* bacteria are cultivated in a 2 % sucrose solution, where they secrete the enzyme into the culture solution within 15 - 20 hours. After the bacteria have been removed with a separator, the solution containing the enzyme is adjusted to a specific enzyme activity by subjecting it to ultrafiltration to remove water, and is then added to a 65 % aqueous solution of 1/3 sucrose and 2/3 fructose at 25 °C. The duration of the reaction depends on the amount of enzyme used. When the reaction is complete, the leucrose that has been formed is separated from the fructoses in chromatography columns containing a weakly cross-linked polystyrene cation exchange resin carrying calcium ions.

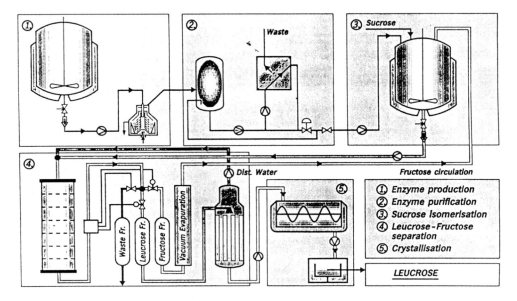

Fig. 3. Flow sheet for the production of leucrose

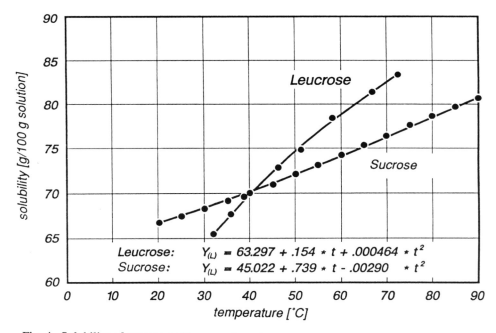

Leucrose: $Y_{(L)} = 63.297 + .154 * t + .000464 * t^2$

Sucrose: $Y_{(L)} = 45.022 + .739 * t - .00290 * t^2$

Fig. 4. Solubility of sucrose and leucrose in water

Properties of Leucrose

Leucrose crystallizes as a monohydrate with a melting point of 156 - 158 °C. The purity of the crystals, on an anhydrous basis, is at least 99 %. ^{1}H- and ^{13}C-N.M.R. spectroscopy[14,15] and an X-ray crystal structure analysis[16] support its structure as a 5-*O*-(α-D-glucopyranosyl)-β-D-fructopyranose. When leucrose is dissolved in water, there is only a slight mutarotation from $[\alpha]_D^{20}$= -8.7 to -7.6° within 7 minutes, since the β-pyranoid form is present to 98 %. In pyridine, dimethylsulphoxide at room temperature, and water at 60 °C, the proportion of the α-pyranoid form almost reaches 10 %.[17]

Sensory tests showed that the sweetness of leucrose is temperature- and concentration-dependant. It is about 50 % less sweet than sucrose.[18]

The outstanding physiological property of leucrose is that bacteria in the mouth do not ferment it to dentally harmful organic acids.[19-21]

Table 3. Basic data of leucrose

Purity		99.6 - 99.8
Water content		app. 4.6 %
spec. rotation		minus 7.5
Melting point		156 - 158 °C
Concentration at saturation	50 °C	72 %
(g / 100 g of solution)	70 °C	83 %
	80 °C	88 %
Cariogenity		none
Effect of laxation		none
Relative sweetening power		40 - 50 %

It would appear that there is no enzyme system in the human mouth capable of breaking the α(1→5)-intersaccharidic linkage. Furthermore, leucrose is neither an acceptor for, nor a donator of glucose, and therefore does not participate in the bacterial formation of dextran, which leads to the build up of plaque in the oral cavity. Leucrose therefore does not cause caries.

This remains true even after leucrose has been incorporated in food, because its glucosidic bond exhibits enormous stability against acid attack. In a comparative study, acid hydrolysis was measured at 71 °C in 0.3 % hydrochloric acid for leucrose, sucrose, maltose and palatinose[22] (cf. Fig. 5). Whereas sucrose underwent 50 % hydrolysis within 4 minutes, 110 hours passed before leucrose is hydrolysed to the same extent. Hence, there is no danger that the cariogenic monosaccharides glucose and fructose will be formed under the influence of acids and heat during processing in foods.

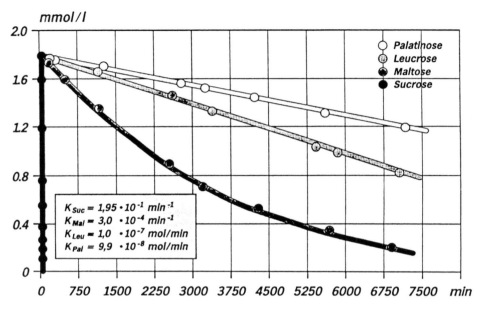

Fig. 5. Hydrolysis (71 °C, 0.08 N HCl) of sucrose, maltose, leucrose and palatinose

On the other hand, leucrose resembles sucrose in that it is broken down by the saccharase / isomaltase and glucoamylase / maltase enzyme complexes in the human small intestine, and is resorbed as glucose and fructose.[23] In the case of leucrose, the breakdown of the intersaccharidic bond proceeds approximately 54 % slower than in sucrose, but this is still fast enough to ensure that no leucrose enters the large intestine, where it would give rise to problems of incompatibility. In several studies on volunteers,[24,25] no gastrointestinal reactions such as flatulence or diarrhoea were observed; such reactions are known to be caused by all sugar alcohols. The intake of leucrose by diabetics[25] leads to a more gradual increase in the blood fructose level than with sucrose.

When foods are subjected to heat, chemical reactions between reducing sugars and free amino groups from amino acids and proteins play a role in the formation of substances that give the food its brown appearance and its aroma; these substances are largely responsible for the appearance and the taste of the processed food. The Maillard reaction is especially important in this context. In contrast to fructose, leucrose first undergoes this reaction after longer reaction times or at higher temperatures. Initially, a ketosylamine is formed as was shown in a model reaction with ammonia[26,27] which led to the isolation of a 1:1 anomeric mixture of the peracylated leucrosyl-amines **1** and **2** in 29 % yield.

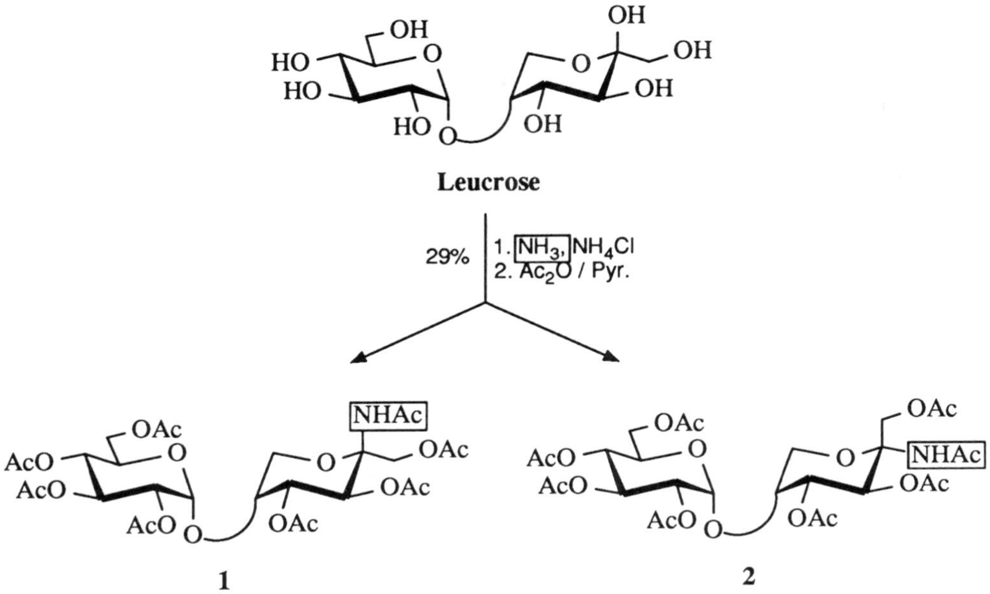

Leucrose

29% | 1. $\boxed{NH_3}$ NH$_4$Cl
2. Ac$_2$O / Pyr.

1   **2**

Possible Applications for Leucrose in the Food Sector

Because its price is higher than that of sucrose and it is less sweet, leucrose will not be able to compete with sucrose in the low price segment of the market. It can however be used for the manufacture of confectionary, especiallly products for children, where any risk to dental health should be excluded. As the pleasant, neutral sweet taste of leucrose enhances the intrinsic aroma of fruit preparations, it will also find customers in this area.

Chemical Reactions with Leucrose

Carbohydrates have been a focus of interest for many years as hydrophylic building blocks for surfactants and emulsifiers. As a result, fatty acid esters of disaccharides, especially of sucrose, are already manufactured on an industrial scale. However, they do not exhibit any tailor-made functional properties, because, due to the presence of three primary hydroxyl groups of nearly identical reactivity, product mixtures of various compositions are obtained. The fact that the glucosidic bond can be hydrolysed in the weakly acidic pH range is also a disadvantage. Leucrose, with its stable $\alpha(1{\rightarrow}5)$ linkage and only two primary hydroxyl groups, should therefore be a more suitable building block for syntheses than sucrose.

The easy synthesis of alkyl glycosides is also in favour of leucrose. Almost quantitative yields of the methyl leucroside (3)[13,22] can be obtained in the presence of an acid catalyst. Glycosides with longer alkyl chains,[12] e.g. 4 with butyl or octyl residues, can be preferentially obtained by transacetylation of the methyl glycoside:

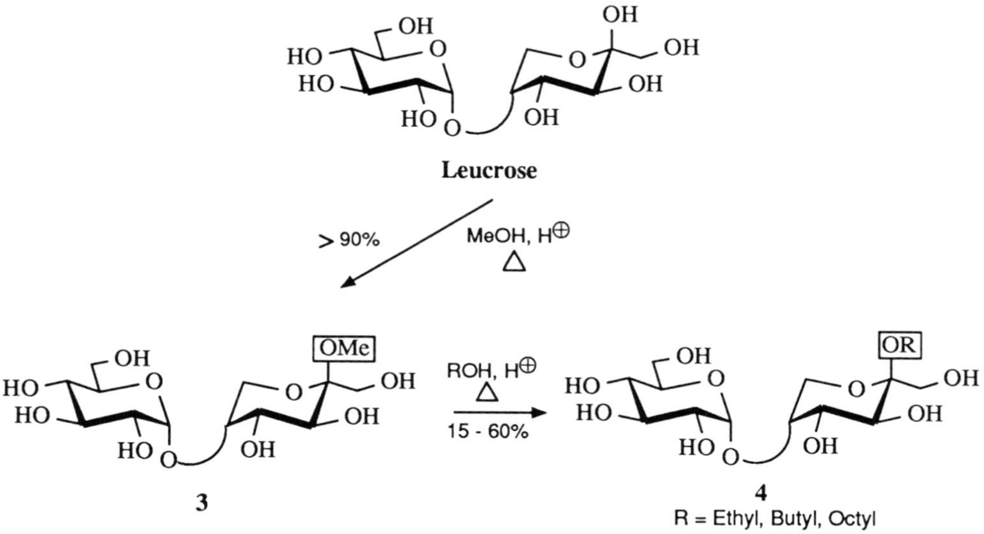

Leucrose

> 90%   MeOH, H$^{\oplus}$
$\triangle$

ROH, H$^{\oplus}$
$\triangle$
15 - 60%

3

4
R = Ethyl, Butyl, Octyl

The introduction of a long alkyl chain into the leucrose molecule imparts some lipophilic properties, so that it is possible to obtain esters by transesterification with methyl esters of carboxylic acids in the absence of a solvent. The main product is a mixture of esters, of which the major component apparently is a diester,[11] formed by transesterification of both primary hydroxyl groups; Thiem and Kleeberg[27] have isolated such a diester in pyridine after reacting it with methane sulfonyl chloride.

Another strategy for generating building blocks for syntheses for washing-active substances is based on studies by Lichtenthaler[28] with palatinose; it involves defined oxidative breakdown of leucrose with atmospheric oxygen in an alkaline medium. After optimization work on a pilot-plant scale, a 95 % yield of the potassium salt of the 4-*O*-(α-D-glucopyranosy)-D-arabonic acid (5) was obtained. It is practically impossible to prepare the free acid, because it forms the 1,5-lactone 6 most readily. Good yields of the acyclic acid amides, e.g. **7**, can be obtained by reaction of **6** with primary amines in methanol.[29]

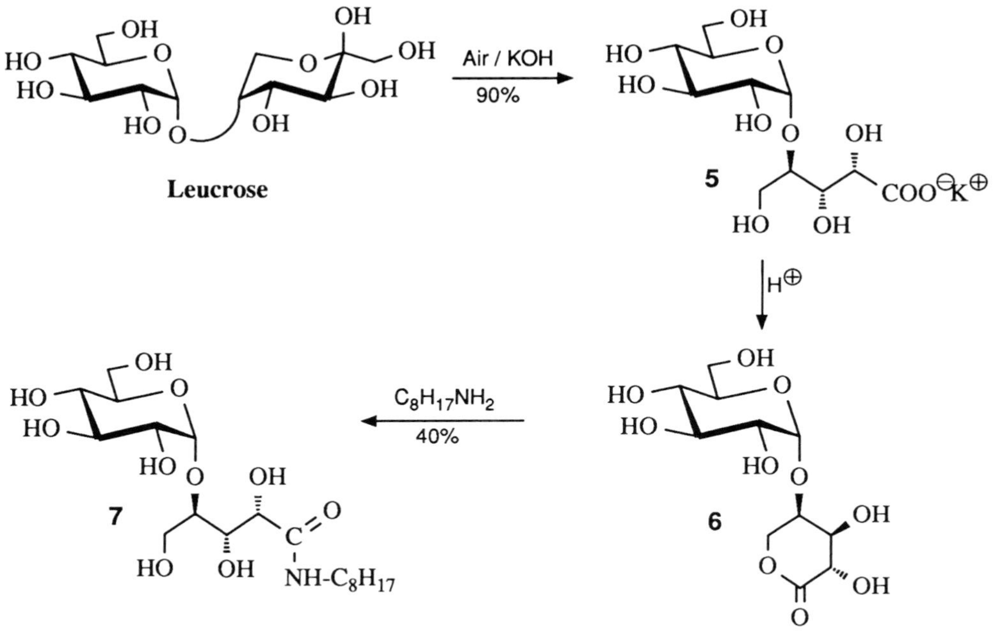

The above mentioned reactions of leucrose offer a good basis for the creation of industrially interesting products. The potential syntheses to be performed with leucrose are by no means exhausted.[27,29]

References

1. F. H. Stodola, H. I. Koepsell, E. S. Sharpe: A New Disaccharide Produced by *Leuconostoc mesenteroides. J. Am. Chem. Soc.* **74** (1952) 3202-3203.

2. F. H. Stodola, E. S. Sharpe, H. J. Koepsell: Preparation, Properties, and Structure of Leucrose. *J. Am. Chem. Soc.* **78** (1956) 2514-2518.

3. G. Siebert, D. Schwengers, unpublished results.

4. T. Watanabe, K. Aso: Isolation of Kojibiose, Nigerose, Maltose and Isomaltulose from Honey. *Tohoku J. Agric. Res.* **11** (1960) 109-117; *Chem. Abstr.* **54** (1960) 23111h.

5. Y. Motomura, K. Aso: On the Sugar Composition in the Pollen of *Typha latifolia. Tohoku J. Agric. Res.* **12** (1961) 173-178; *Chem. Abstr.* **56** (1962) 5228e.

194   *D. Schwengers*

6. I. R. Siddiqui, B. Furgula: Isolation and Characterization of Disaccharides from Royal Jelley. *J. Apicult. Res.* **4** (1965) 89-94; *Chem. Abstr.* **63** (1965) 18537a.

7. E. J. Bourne, I. H. Hutson, H. Weigel: Oligosaccharides in Dextran-producing Cultures of *Streptococcus Govis. Biochem. J.* **79** (1961) 549-553.

8. H. Ruttlow, R. Friese, K. Täufel: Saccharides of Dextran By-products and their Conversion in the Animal Intestinal Tract, with Special Reference to Leucrose. *Die Nahrung* **8** (1964) 523-531; *Chem. Abstr.* **62** (1965) 3143e.

9. D. Schwengers, H. Benecke (Pfeifer & Langen): Sweetener and its Use. *DBP-DE* 34 46 380 (1984); *Eur. Pat.* 185 302 (1985); *Chem. Abstr.* **105** (1986) 77815p.

10. J. F. Robyt, T. F. Walseth: Production, Purification, and Properties of Dextransucrase from *Leuconostoc mesenteroides. Carbohydr. Res.* **68** (1979) 95-111.

11. J. F. Robyt, T. F. Walseth: The Mechanism of Acceptor Reactions of *Leuconostoc mesenteroides* Dextransucrase. *Carbohydr. Res.* **61** (1978) 433-445.

12. J. F. Robyt, A. J. Corrigan: The Mechanism of Dextransucrase Action. Activation of Dextransucrase from *Streptococcus mutans* by Dextran and Modified Dextran and the Nonexistence of the Primer Requirement for the Synthesis of Dextran. *Arch. Biochem. Biophys.* **183** (1977) 726-731.

13. E. Kirchner, D. Schwengers, unpublished results.

14. J. P. Kamerling, M. J. A. DeBie, J. F. G. Vliegenthart: A PMR Study of the Anomeric Protons in Permethylsilyl Oligosaccharides, a Determination of the Configuration of the Glycosidic bond. *Tetrahedron* **28** (1972) 3037-3047.

15. A. De Bruyn, J. Van Beeumen, M. Anteunis, G. Verhegge: Proton NMR-Study of some D-Aldohexopyranosyl-D-fructos(id)es in Water. *Bull. Soc. Chim. Belg.* **84** (1975) 799-811.

16. J. Thiem, M. Kleeberg, K. H. Klaska: Neue Synthese und Kristallstruktur der Leucrose. *Carbohydr. Res.* **189** (1989) 65-77.

17. F. W. Lichtenthaler, S. Rönninger: α-D-Glucopyranosyl-D-fructoses – Distribution of Furanoid and Pyranoid Tautomers in Water, Dimethyl Sulphoxide, and Pyridine. *J. Chem. Soc., Perkin Trans. 2*, **1990**, 1489-1497.

18. Leucrose-Basisdaten. Anwendungstechnische Informationen, Pfeifer & Langen, Köln.

19. F. Forsthuber, G. Siebert: Stepwise Procedure for Sugar Substitutes – Preliminary Study with Enzymes. 4. Glycosyltransferase from *Streptococcus mutans. Z. Ernährungswiss.* **27** (1988) 48-56; *Chem. Abstr.* **109** (1988) 108961x.

20. H.-J. Gülzow, E. Polihronu-Panagiotu. *Dtsch. Zahnärztl. Z.* **45** (1990) 280-282.

21. S. C. Ziesenitz, G. Siebert, T. Imfeld: Cariological Assessment of Leucrose as a Sugar Substitute. *Caries Research* **23** (1989) 351-357.

22. M. Kleeberg, *Doctoral Dissertation*, Universität Münster 1988.

23. H. Heymann, *Doctoral Dissertation* in progress, Med. Hochschule Hannover.

24. B. Ückert, M. Seiberling, D. Schwengers, E. Fritschi, H. Maier-Lenz: 29th Spring Meeting, Deutsche Gesellschaft für Pharmakologie und Toxikologie, Springer International, March 1988.

25. Biodesign, Freiburg, unpublished results.

26. J. Thiem, M. Kleeberg, D. Schwengers: Leucrose - ein Ketodisaccharid von industriellem Zuschnitt. *Alimenta* **28** (1989) 20-30.

27. J. Thiem, M. Kleeberg: Synthesis and Reactions of Leucrose and its Exocyclic Glycal. *Carbohydr. Res.* **205** (1990) 333-345.

28. F. W. Lichtenthaler, R. G. Klimesch: Derivate und Reduktionsprodukte der D-Glucopyranosyl-α(1→5)-D-arabonsäure. *Ger. Offen.* DE 3,248,404 (1982); *U.S. Pat.* 4,618,675 (1983); *Chem. Abstr.* **102** (1985) 7034x.

29. J. Thiem, M. Kleeberg: Oxidation und Reduktion des Ketodisaccharides Leucrose. *Chem. Ztg.* **113** (1989) 239-242.

10

Prospects for Industrial Utilization of Sugar Beet Carbohydrates

James N. BeMiller

Whistler Center for Carbohydrate Research, Smith Hall, Purdue University, West Lafayette, Indiana, USA

Summary. Use of sucrose as a practical organic intermediate requires consideration of various factors and criteria; some reactions that meet these criteria are presented. With regards to beet sugar, utilization of by-products may also be required; potential uses for the pectin and arabinan are discussed. One class of carbohydrate-based high-value products, α-substituted cyclic ethers, are described with representative compounds and synthetic approaches.

For at least 40 years, it has been recognized that sucrose is a pure, plentiful, replenishable, relatively inexpensive chemical commodity, but research with the objective of using it as an industrial raw material has not met with great success. Sucrose can be the chemical intermediate for a number of products:

Table 1. Products from sucrose

D-Glucose + D-Fructose	Derivatives:
D-Sorbitol (D-Glucitol) + D-Mannitol	Carboxylic acid esters
Products of fermentation or enzyme action:	Sulfonic acid esters
Acids (gluconic, itaconic, levulinic, citric, etc)	Ethers
Ethanol	Silyl ethers.
Antibiotics	Cyclic acetals
Flavor enhancers	Polyurethans
Gums (dextrans, fructans, xanthan)	Deoxychloro derivatives
Polyhydroxybutyrate and related polymers	Hydroxymethylfurfural
Oligosaccharides (neosugar, isomaltulose, etc.)	
Sucrose tricarboxylic acid	
Sucrose dialdehyde	
Sucrose anhydrides	

Not all of these products are sucrose-specific. In addition, sucrose as a commodity material that can compete with other chemical intermediates sold in bulk at a low price should be distinguished from sucrose as a precursor of fine chemicals – value-added products that will, without doubt, be most profitable to the company that does the conversion.

When considering what new industrial uses for sucrose can be developed and commercialized, attention is often focused on replacement of plastics, primarily because of the millions of tons of plastics that are produced each year. The interest is often justified in terms of the potential for enhanced degradability, because that might have to be the justification for a more expensive material. But there are many other important commercial chemicals that could potentially be made from sucrose. Examples are agricultural pesticides, food ingredients, industrial coatings, industrial and institutional cleaners, printing ink compounds, oil well drilling chemicals, paper additives, water-treatment compounds, ore-refining chemicals, etc. And for several of these, sucrose derived products could have the advantage of being more environmentally friendly.

My philosophy is to use the unique properties of sucrose, i.e., to use it as a unique carbohydrate. But first, more basic research is needed to determine what uniquely can be done with sucrose. Then, we must determine what economically viable applications there might be for the products, taking into account the following criteria for the use of sucrose:

Criteria for Use of Sucrose as a Commodity Chemical

- Reactions, whether chemical or biochemical, and / or products should be unique to sucrose.

- Reactions should involve a minimum of chemistry and inexpensive reagents, give a high yield of specific products, and be amenable to scale-up.

- Reactions must be done either in aqueous solution or without solvent.

- Only bases can be used as catalysts. Use of acids in any step must be avoided.

- Product isolation and purification should be simple, inexpensive, and amenable to large-scale production.

One approach was reported by Richards and Shafizadeh.[1] They found that thermal degradation (pyrolysis) of sucrose produced the D-fructofuranosyl carboxonium (oxacarbenium) ion which reacts with sugar hydroxyl groups to produce oligosaccharides (41 %) and polysaccharides (46 %) or with added alcohols to produce glycosides. The reaction might be limited only by one's imagination of the nature of ROH.

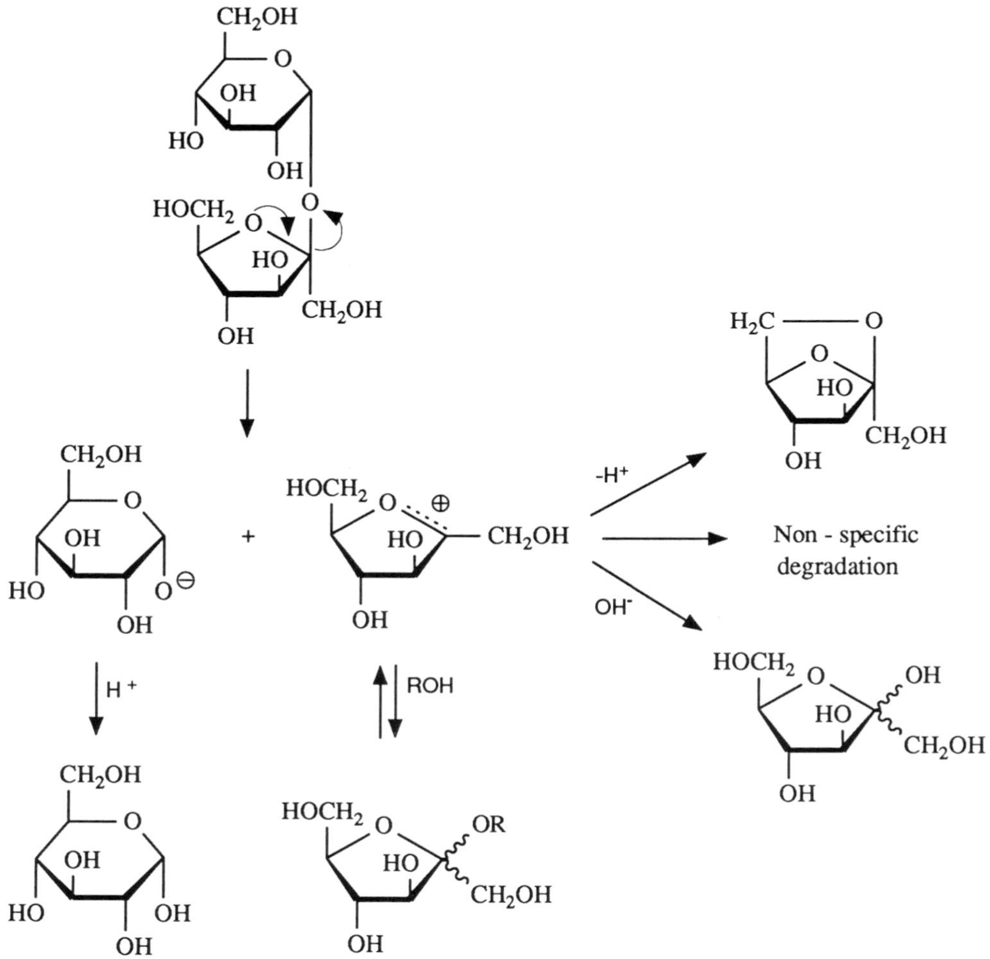

Any consideration of the production of polymers by chemical reaction of sucrose should include two kinds:

(1) Polymers with weak linkages, i.e., polymers with dilute acid lability. (I do not know what would be a use for such a polymer; but if there is one, sucrose could be a good building block for this material). However, it should be kept in mind that

the desired polymers for most applications are linear polymers, and I assume that it will be difficult to make linear, i.e., non-cross-linked polymers, from sucrose without the use of blocking groups.

(2) Water-absorbing polymers. There are applications for water-absorbing, as opposed to water-soluble or water-insoluble, polymers; and polymers containing hydrophilic blocks seem to be a logical approach. What cannot be predicted is whether sucrose might be the best polyol to use in the preparation of such polymers. An activated diacid (carboxylic or sulfonic) can be used to make a polyester or a diisocyanate can be used to make a polycarbamate, but again branching and crosslinking will be a problem.

In summary, in considering what can be done with sucrose as a commodity chemical, it can be looked upon as nothing more than a polyol; but it is a unique polyhydroxy compound and, if it is going to have a chemical (as opposed to an economic) advantage over other polyols, the advantage must lie in its unique nature. One is the very acid-labile glycosidic linkage. Another is the fact that it contains three or four reactive hydroxyl groups. Both could be considered as disadvantages. How can we make either or both advantages in producing materials for specific applications? What products require one or both of these special chemical properties?

With regards to high-value products, compounds with pharmacological activity can be made from sucrose, e.g., the anti-ulcerative drug Sucralfate (aluminum sucrose octasulfate) which reduces stomach acidity. Fermentation products such as antibiotics and the less complicated citric acid are listed in Table 1. When sucrose is used in the production of any of these, most of the profit goes to the company that makes the value-added product. And in the case of fermentation products, sucrose is seldom a unique source of carbon and energy, the production of dextrans and levans (fructans) being exceptions.

Another potential approach to value-added products is the preparation from carbohydrates of cyclic ethers, sometimes called *C*-glycosides, or more properly, *C*-glycosyl compounds. *C*-Glycosyl compounds, as for example the β-D-glucopyranosyl compound **1** are acid stable as opposed to β-D-glucopyranoside **2**, which as an acetal is hydrolyzed by acid.

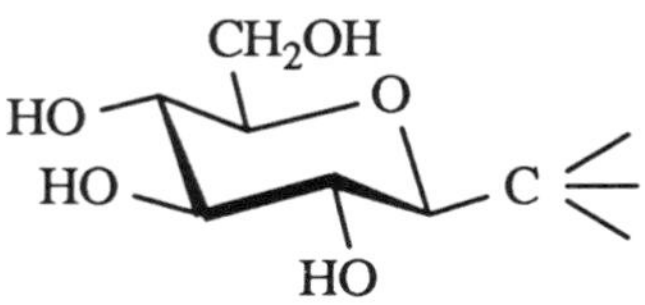

1

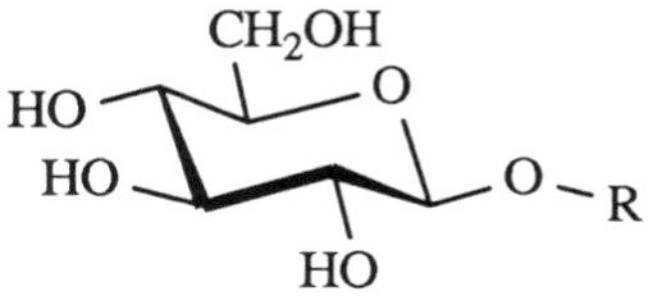

2

The compounds on which we have concentrated are **4** (deacetylated) and **5**, which can be made by a straightforward route[2] from β-D-galactopyranosyl cyanide **3**.[3-5] Identical reaction sequences can be used to make similar compounds with the α- and β-D-glucopyranosyl and α- and β-D-mannopyranosyl configurations.[6-8]

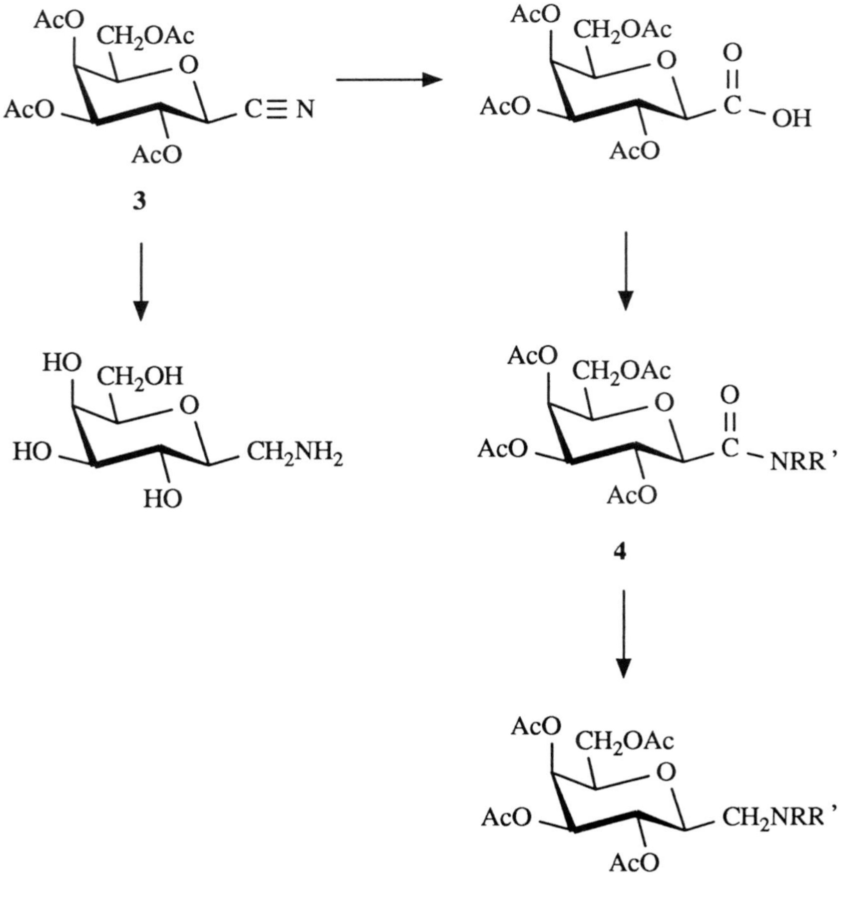

3

4

5

Natural products that can be made from sugars in this way are many and include the antimicrobial antibiotics, the pseudomonic acids (**6**), and the antifungal antibiotic, ambruticin (**7**); it is largely goals of synthesis of such compounds that has driven the interest in *C*-glycosyl compounds.

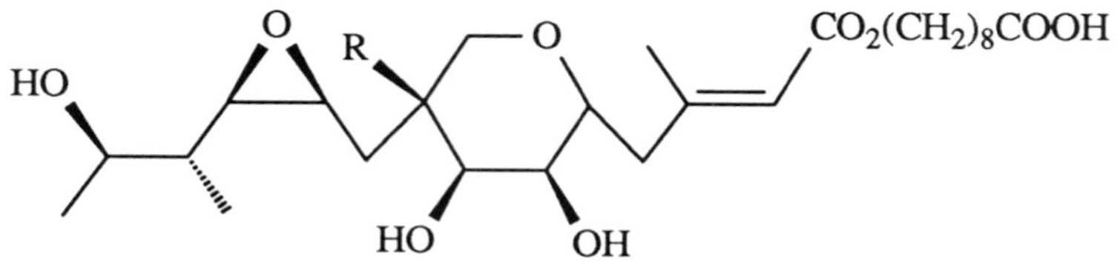

Pseudomonic Acids   (6)

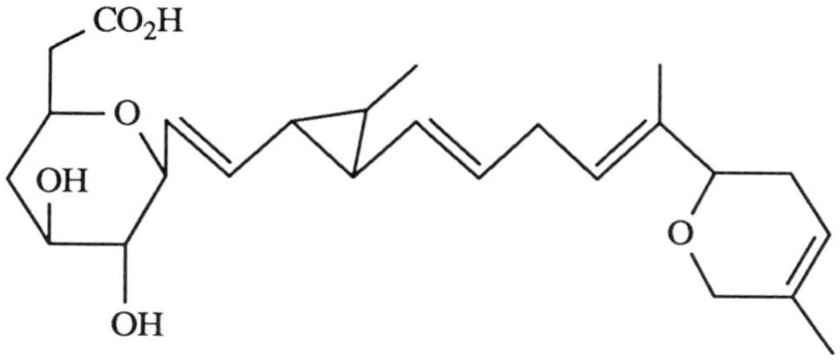

Ambruticin   (7)

Approaches to the synthesis of *C*-glycosyl compounds are outlined in Table 2. Which method is used has been determined largely by the desired end-product.

Now, I turn my attention to other possible sources of chemical intermediates, the byproducts of beet processing – pulp or fiber → pectin, arabinan, and cellulose – the use of some of which may be required to make use of sucrose as a chemical intermediate economical.

Sugar beet pectin has been most extensively studied by Thibault at Nantes.[9-15] Sugar beet pectin has unique properties, one of which is the inability to gel; so it should not be considered as a competitive product to apple or citrus pectin for making normal or dietetic jams, jellies, marmalades, or preserves, although it can be cross-linked oxidatively.[12,13,15,16] It can be considered to be a source of D-galacturonic acid and unsaturated mono- and oligosaccharides, all of whose chemistry could be explored.

Table 2.   Some approaches to the synthesis of *C*-glycosyl compounds (α-branched cyclic ethers)

I.	Replacement of Anomeric Constituents (via a carboxonium ion)	1.	via reactions of glycosyl halides
		2.	via replacement of an hemiacetal ester group
II.	Transformations of Unsaturated Monosaccharides	1.	via 1,2-additions to glycals
		2.	via Claisen rearrangement of glycal derivatives
		3.	via Lewis acid-catalyzed reactions of glycals with *C*-nucleophiles
		4.	via palladium-catalyzed reactions of unsaturated sugars
III.	Cyclization of acyclic sugar derivatives	1.	via intramolecular nucleophilic displacements
		2.	via intramolecular Michael addition reactions
		3.	via cyclodehydrations of alditols
		4.	via deamination
		5.	via cyclization of unsaturated alcohols
IV.	Other Syntheses	1.	from substrates with an unprotected anomeric hydroxyl group via reduction
		2.	via cycloadditions and cyclocondensation reactions

D-Galacturonic acid can be produced by acid- or enzyme-catalyzed hydrolysis of pectin. Degradation products from the oxidative and non-oxidative treatment of D-galacturonic acid with alkali have been determined.[17] Thirteen hydroxy monocarboxylic acids and 26 hydroxy dicarboxylic acids were indentified. In the absence of oxygen, the yield of dicarboxylic acids was about 53 %, the principal ones being 3-deoxy-*lyxo*-hexaric (~21 %), malic (~8), and 3-deoxy-*xylo*-hexaric (~7 %) acids; and the yield of monocarboxylic acids about 39 %, the principal ones being lactic (~24 %), and 3-deoxytetronic (~8 %) acids. In the presence of oxygen, the yield of dicarboxylic acids was about 82 %, the principal ones being arabinic (~22 %), threaric (~18 %), tartaric (~14 %), and malic (~7 %) acids; and the yield of monocarboxylic acids about 18 %, the principle one being glycolic acid (~13 %). Minor amounts of other mono- and dicarboxylic acids are also formed. Yields of the major products are respectable considering the source; but mixtures of nonvolatile acids pose a major separation and purification problem, making this an unlikely source of these hydroxylated acids.

2-Hydroxy-2-cyclopenten-1-ones[18] and mono-, di-, tri-, and tetramethyl benzoquinones are formed in only very low yields by treatment of pectic acid[18,19] and sucrose[20] with strong alkali at high temperatures under nitrogen.

The unsaturated monomer and oligomers can be made from pectin by treatment with either an enzyme or base. Subsequent hydrolysis would, of course, produce an α-keto acid (with a distal aldehyde group). The question arises, whether this is a useful intermediate, and in how far the dehydration to furan derivatives can be prevented.

Arabinan is easily recoverable from sugar beet pulp, the only known source of a relatively pure arabinan, via extraction with hot lime water (yield ca. 1 %).[21] Upon hydrolysis, it would yield L-arabinose. The only current demand for L-arabinose that I know of is for the preparation of arabinosyl cytosine (AraC), and for that a different source of the pentose is used, in the U.S. at least. Perhaps, the chemistry and production of L-arabinose also needs to be investigated.

The residue would be largely cellulose (beet pulp contains little lignin). At this time, we do not know how unique is this cellulose. Can it be pulped and used for newsprint? Does it have special absorbent properties? Is it a suitable dietary fiber? Only research will answer these and other questions.

References

1. G. N. Richards, F. Shafizadeh: Mechanism of Thermal Degradation of Sucrose. A Preliminary Study. *Aust. J. Chem.* **31** (1978) 1825-1832; *Chem. Abstr.* **89** (1978) 163877t.

2. J. N. BeMiller, M. R. Yadav, V. N. Kalabokis, R. W. Myers: N-Substituted (β-D-Galactopyranosylmethyl)amines and *C*-β-Galactopyranosylformamides. *Carbohydr. Res.* **200** (1990) 111-126.

3. B. Helferich, K. L. Bettin: Notiz zur Synthese nichtreduzierender Disaccharide. *Chem. Ber.* **94** (1961) 1159-1160.

4. B. Coxon, H. G. Fletcher, Jr.: The Structure of 2,3,4,6-Tetra-*O*-acetyl-β-D-galactopyranosyl Cyanide and some Derivatives therefrom. Synthesis of 1-Deoxy-D-*galacto*-heptulose. *J. Am. Chem. Soc.* **86** (1964) 922-926.

5. R. W. Myers, Y. C. Lee: Synthesis and Characterization of some Anomeric Pairs of Per-*O*-acetylated Aldohexopyranosyl Cyanides (Per-*O*-acetylated 2,6-Anhydro-heptononitriles). On the Reaction of Per-*O*-Acetylaldohexopyranosyl Bromides with Mercuric Cyanide in Nitromethane. *Carbohydr. Res.* **132** (1984) 61-82.

6. M. Chmielewski, J. N. BeMiller, D. P. Cerretti: Derivatives of *S*-α- and β-D-Hexopyranosyl Thiophosphates. *Carbohydr. Res.* **96** (1981) 73-78.

7. M. Chmielewski, J. N. BeMiller, D. P. Cerretti: Reverse Anomeric Effect of the Carbamoyl Group of 2,6-Anhydroheptonamides. *J. Org. Chem.* **46** (1981) 3903-3908.

8. G. Grynkiewicz, J. N. BeMiller: Synthesis of Aldopyranosyl Cyanides. *Carbohydr. Res.* **112** (1983) 324-327.

9. F. Michel, J.-F. Thibault, C. Mercier, F. Heitz, F. Pauillaude: Extraction and Characterization of Pectins from Sugar-Beet Pulp. *J. Food Sci.* **50** (1985) 1499-1500; *Chem. Abstr.* **103** (1985) 177137w.

10. F. M. Rombouts, J.-F. Thibault: Feruloylated Pectic Substances from Sugar-Beet Pulp. *Carbohydr. Res.* **154** (1986) 177-187.

11. J. M. Rombouts, J.-F. Thibault: Enzymic and Chemical Degradation and the Fine Structure of Pectins from Sugar-Beet Pulp. *Carbohydr. Res.* **154** (1986) 189-203.

12. J.-F. Thibault, J. M. Rombouts: Effects of some Oxidizing Agents, Especially Ammonium Peroxosulfate, on Sugar-Beet Pectins. *Carbohydr. Res.* **154** (1986) 205-215.

13. F. Guillon, J.-F. Thibault: Characterization and Oxidative Crosslinking of Sugar Beet Pectins after mild Acid Hydrolysis and Arabanases and Galactanases Degradation. *Food Hydrocolloids* **1** (1987) 547-549; *Chem. Abstr.* **108** (1988) 169538r.

14. F. Guillon, J.-F.Thibault: Further Characterization of Acid- and Alkali-Soluble Pectins from Sugar Beet Pulp. *Lebensm. Wiss. Technol.* **21** (1988) 198-205; *Chem. Abstr.* **110** (1989) 37921s.

15. J.-F. Thibault: Characterization and Oxidative Crosslinking of Sugar-Beet Pectins extracted from Cassettes and Pulps under different Conditions. *Carbohydr. Polymers* **8** (1988) 209-223; *Chem. Abstr.* **109** (1988) 131144s.

16. J.-F. Thibault, C. Garreau, D. Durand: Kinetics and Mechanism of the Reaction of Ammonium Persulfate with Ferulic Acid and Sugar-Beet Pectins. *Carbohydr. Res.* **163** (1987) 15-27.

17. K. Niemelä, E. Sjöström: Non-Oxidative and Oxidative Degradation of D-Galacturonic Acid with Alkali. *Carbohydr. Res.* **144** (1985) 93-99.

18. K. Niemelä: Alkyl-substituted Benzoquinones formed on the Degradation of Pectic Acid with Alkali. *Carbohydr. Res.* **195** (1989) 131-133.

19. K. Niemelä: The Formation of 2-Hydroxy-2-cyclopenten-1-ones from Poly-saccharides during Kraft Pulping of Pine Wood. *Carbohydr. Res.* **184** (1988) 131-137.

20. H. Kato, M. Mizushima, T. Kurata, M. Fujimaki: Formation of Alkyl-*p*-benzoquinones and Catechols through Base-catalyzed Degradation of Sucrose. *Agric. Biol. Chem.* **37** (1973) 2677-2678; *Chem. Abstr.* **80** (1974) 83454b.

21. J. K. N. Jones, Y. Tanaka: *Methods Carbohydr. Chem.* **5** (1965) 74.

11

Practical Routes from Mono- and Disaccharides to Building Blocks with Industrial Application Profiles

Frieder W. Lichtenthaler, Eckehard Cuny, Dierk Martin,
Stephan Rönninger, and Thomas Weber

Institut für Organische Chemie, Technische Hochschule Darmstadt
D-6100 Darmstadt, Germany

Summary. A variety of "reaction channels" are presented leading from inexpensive, bulk-scale accessible mono- and disaccharides to building blocks with potential industrial application profiles. Thereby, aspects of practibility are emphasized, such as the use of simple reactions and reagents, of simple protecting groups, the generation of stable, crystalline, readily purifiable products in a reasonable number of steps as to allow overall yields around 50 %, and large-scale adaptability without major surgery in the reaction scheme. An overview is given on the use of the building blocks, mostly of the dihydro-pyranone, 2-oxo- and 2-oximinoglycosyl, and glycosyl-HMF type for the practical acquisition of a variety of biologically interesting natural products is given, as well as on the industrial potential of glucosyloxymethyl-furfural ("GMF"), a unique hybrid between sugar- and petrochemistry.

1. Introduction

The attractiveness of low molecular weight carbohydrates as an organic raw material arises from the fact that they are available in bulk quantities, are exceedingly cheap and, on top of this, are enantiomerically pure. This propitious situation becomes particularly evident when comparing the bulk quantity prices of the 10 least expensive sugars – all with kg prices well below DM 100.--, and, hence, ideal starting materials for organic synthesis – with those of other chiral compounds and basic organic chemicals (Table 1): the seven cheapest sugars are not only less expensive than any of the other enantiomerically pure products, such as hydroxy or amino acids, but they are even cheaper than basic organic bulk chemicals as, for example, benzaldehyde or aniline. Actually, they are in the price range of the standard solvents in which organic reactions are usually performed.

Table 1. Cost of nature-derived low molecular weight raw materials as compared to basic organic chemicals and solvents

		Price per kg (in DM)[*]
Sugars	Sucrose	0.75
	D-Glucose	1.15
	D-Lactose	1.20
	D-Fructose	2.50
	D-Maltose	5.--
	D-Isomaltulose	5.--
	D-Leucrose	5.--
	D-Xylose	12.--
	L-Sorbose	35.--
	D-Galactose	85.--
Hydroxy Acids	L-Tartaric Acid	10.--
	L-Ascorbic Acid	24.--
Amino Acids	L-Glutamic Acid	15.--
	L-Aspartic Acid	15.--
	L-Lysine	20.--
	L-Leucine	40.--
Achiral Chemicals	Benzaldehyde	12.--
	Aniline	12.--
	Acetaldehyde	15.--
Solvents	Methanol	0.90
	Acetone	1.10
	Toluene	1.40
	Dichloromethane	1.50

[*] World-market prices on bulk delivery

The uniqueness of this situation becomes even more imposing when looking at these sugars in greater detail (Table 2). Sucrose, with a world production of over a 100 mill. tons per year,[1] is the most abundantly produced organic bulk chemical of low molecular weight, followed by its component sugars, D-glucose and D-fructose.

D-Lactose and D-maltose, readily available in large quantities from whey[3] and starch,[4] respectively, have some applications in the form of their reduction products maltitol and lactitol that are sweetening agents, and of lactulose,[3] a lactose isomerization product; however, despite of their basic chemistry being fairly well developed[5,6] there, at present, does not seem to be an qualified chemical use on an industrial scale for these disaccharides.

Two isomers of sucrose, the $\alpha(1\rightarrow6)$- and $\alpha(1\rightarrow5)$-linked glucosyl-fructoses isomaltulose and leucrose, although known since the mid-fifties,[7,8] have only recently become accessible on an industrial scale, the respective biotechnological processes being based on *Protaminobacter rubrum*[9-13] and *Leuconostoc mesenteroides*-induced[14] transglucosylations. Both, isomaltulose and leucrose wait for a broad, thorough exploitation of their chemistry.

Table 2. Accessibility and world market prices of low molecular weight carbohydrates

	world production (tons/year)[*]	Price (DM/kg)[**]	Source (Supplier)
D-Sucrose	106 500 000	0.75	World Market
D-Glucose	5 000 000	1.15	Cerestar
D-Sorbitol	650 000	1.20	Cerestar
D-Lactose	180 000	1.20	Meggle
D-Fructose	50 000	2.50	Südzucker
D-Mannitol	8 000	5.--	Cerestar
D-Maltose	3 000	5.--	Cerestar
D-Isomaltulose	7 000	5.--	Südzucker
D-Leucrose	10	5.--	Pfeifer & Langen
D-Gluconic acid[†]	40 000	7.--	Fluka
D-Xylose	1 000	12.--	Cerestar
Dianhydrosorbitol	1 000	12.--	Cerestar
Xylitol	3 000	14.--	Cerestar
L-Sorbose	25 000	35.--	Fluka
D-Galactose	?	85.--	Fluka

[*] Exact data are available for the 1989 world production of sucrose;[1] all other data given are estimations based on information from producers and/or suppliers.

[**] Exept for the Fluka-applied products, which refer to 100 kg batches, prices are those attainable on the world market for bulk delivery (ton range) or, in the EEC, after allowing for EEC refunds according to regulation No. 1010/86 for industrial utilization of sucrose.

[†] In the form of its sodium salt.

D-Xylose is the cheapest pentose, readily accessible from wood- or straw-derived-xylans, L-sorbose is the most readily, large-scale available L-sugar due to its technical production from D-sorbitol in the Vitamin C fabrication process.[15] The sugar alcohols D-xylitol[16] and D-sorbitol,[17] both of comparatively high yearly production, are mostly used as food ingredients due to their sweetening properties, yet their utilization as a raw material for broad-scale organic preparative purposes is essentially non-existent. The same holds for D-gluconic acid,[18] the sizable annual production going into its use as a chelating agent and a catalyst for textile printing.

In view of the large-scale accessibility of these mono- and disaccharides it must appear surprising that sugars are not utilized on a much larger scale as a raw material for chemical industry in general, and for the construction of enantiomerically pure non-carbohydrate products, in particular. There are reasons for this, of course. Sucrose, lactose, and maltose, for example, provide an interesting chemistry,[2,5,6] yet are unsuited for many synthetic transformations due to their acid-sensitive intersaccharide linkage. Their component monosaccharides are devoid of this deficiency, yet *direct* utilization of their vast synthetic potential is impeded by a number of obstacles: they are overfunctionalized with hydroxyl groups of similar or identical reactivities, they have considerably more chiral centers along the six-carbon chain than required for non-sugar target molecules, and, they lack suitable functional groups such as olefinic or carbonyl unsaturation to which modern preparative organic methodology can directly be applied.

MONOSACCHARIDES AS CHIRAL EDUCTS

D-GLUCOSE

SHORTCOMINGS : overfunctionalized with hydroxyl groups
too many chirality centres
lack of C=C and C=O functionalities

These adverse conditions have elicited considerable efforts to reduce the number of chiral centers as well as hydroxyl groups with the simultaneous introduction of useful functionality. One approach involves the shortening of the aldose carbon chain, or, more simply, its bisection, as exemplified by the use of D-mannitol-derived 2,3-*O*-isopropyliden-D-glyceraldehyde.[19] Whilst this product and its L-ascorbic acid-derived enantiomer have developed into the presently most popular enantiomerically pure three-carbon synthons, it must be objected that Nature graciously provides us – via photosynthesis – with six-carbon compounds, and if their synthetic potential is to be used efficiently, elaboration of building blocks from sugars should retain the carbon chain.

Indeed, the most frequently used alternative is the gradual step-by-step carving out of a target molecule segment from a monosaccharide whereby usually a reaction sequence evolves, that is specifically tailored for the synthetic target. The number of complex, non-carbohydrate natural products synthesized via this approach is enormous,[20] yet the fact cannot be concealed that the large majority of these total syntheses are exceptionally long, and that their transferability to large scale is essentially impossible with respect to the reagents used, the number of steps required, the expenditure of work involved, and the overall yields attainable.

By consequence, it appears to be an urgent necessity to develop *new*, economically sound industrial application profiles, which, are these eventually to be found, necessitates the adoption of *practical criteria* and their careful consideration in synthetic planning:

Criteria of Practicality for Building Block Aquisition

- ◆ retention of the carbon chain of the sugar
- ◆ selection of reactions that allow for simple reagents and uncomplicated, non-chromatographic workup
- ◆ use of simple protecting groups
- ◆ aiming for stable, crystalline, readily purifiable products
- ◆ reasonably high overall yields (75 % per step on the average)
- ◆ overall reaction sequences that have the potential of being transposable to the hectogram scale

Realization of most or all of these criteria calls for the conversion of the monosaccharide into a useful, versatile six-carbon synthon, preferably a stable building block with one or two chiral centers and with synthetically flexible functional groups. Since the efficiency of this conversion largely determines the practical value of the overall synthesis, restriction is necessary to what is preparatively "makeable" in 4 - 5 steps and with overall yields of 40 - 50 %. This, in turn, reduces the number of methodical entries sugar ⟹ six-carbon building block to comparatively few "reaction channels" which, nolens volens, are different for each sugar, since an optimal compliance with their individual stereochemical intricacies is imparative for achieving adequate preparative results.

The initial stage of any "reaction channel"

sugar ⟹ enantiomerically pure building block

invariably involves fixation of the sugar in the respective tautomeric form. For D-glucose, the few preparatively useful "entry reactions" have been elaborated before the turn of the century already: mercaptalization to the *acyclic* dithio acetals, isopropylidenation to *furanoid* systems, or the generation of *pyranoid* structures, such as glucosides, glucals, and hydroxyglucalesters.[21]

SIMPLE TAUTOMERICALLY FIXED DERIVATIVES

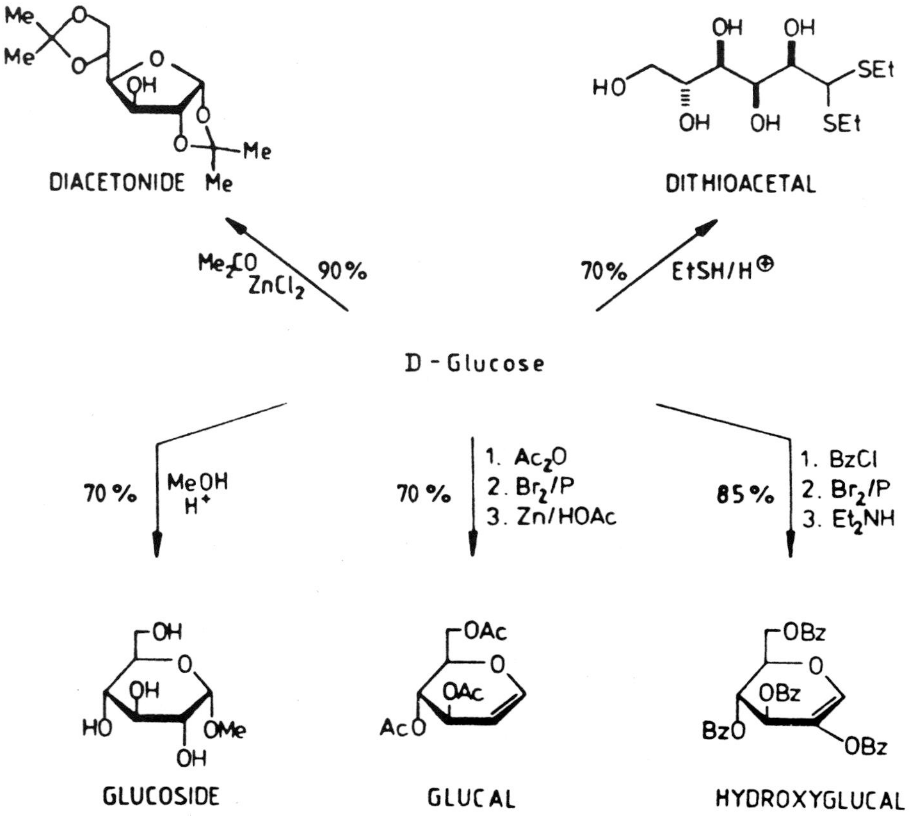

DIACETONIDE

DITHIOACETAL

D - Glucose

GLUCOSIDE

GLUCAL

HYDROXYGLUCAL

Notable exceptions are the acquisition of pyranoid derivatives such as the spirocyclic diacetone-fructose **1**[23,24] and the tetrabenzoyl-β-D-fructopyranosyl bromide **2**.[25,26] The most readily accessible open-chain derivative is not the dithioketal as in the case of hexoses, but the 6-chloro-tetraacetate **3**, smoothly generated on acetylation and PCl_5 treatment.[27] Of derivatives fixed in the furanoid form, the 1,3,4,6-tetrabenzoate **4** is surprisingly well accessible on high-temperature (60 °C) benzoylation of fructose in pyridine.[28]

D - Fructose : simple , large scale - accessible derivates

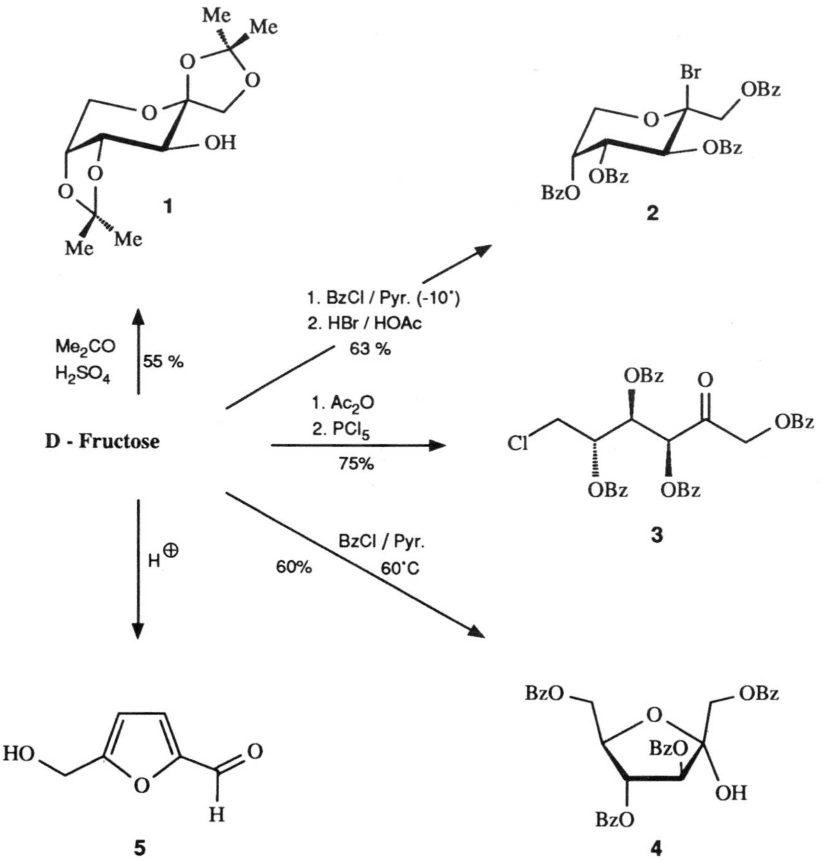

Another furanoid compound generated from fructose by acid-induced elimination of three moles of water,[29,30] is hydroxymethylfurfural (**5**, "HMF"). Despite of lacking chiral centers, HMF – as one of the few "petrochemicals" readily accessible from regrowing resources – is a highly versatile six-carbon

With ketoses the situation is much less auspicious. Fructose, for example, not only elaborates the two pyranoid tautomers in solution, but the furanose forms as well (cf. Table 3), which amount to more than 50 % in solvents like dimethyl sulfoxide or pyridine.[22] This makes simple "entry reactions" for tautomeric fixation considerably more complex, usually resulting in a mixture of derivatives of all five tautomers, the acyclic *keto* form included:

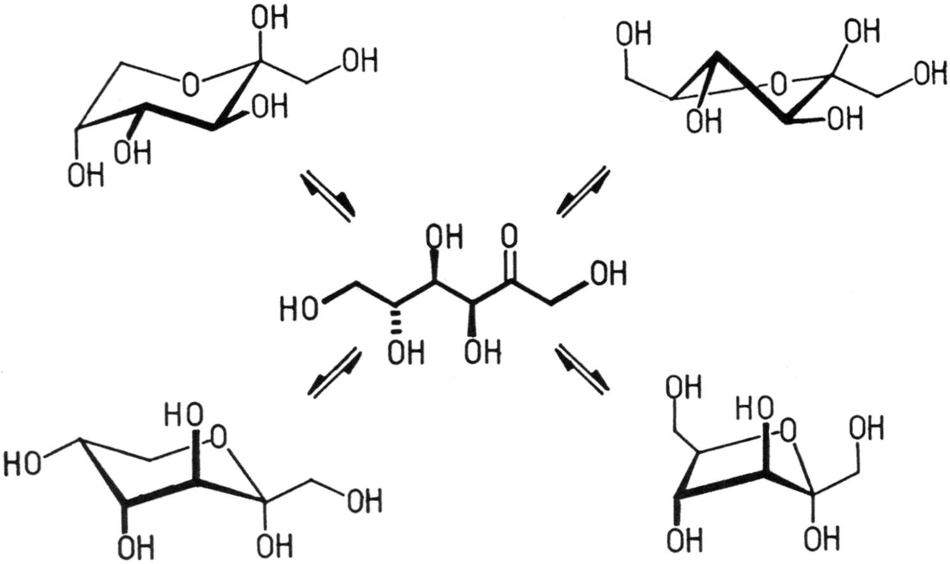

Table 3. Equilibrium composition in % of D-fructose tautomers in water, dimethylsulfoxide and pyridine[22]

Solvent	Temp.	β-*p*	α-*p*	β-*f*	α-*f*	keto
water	0 °C	80	2	15	3	
	25 °C	73	2	20	5	(0.5)
	50 °C	64	3	25	8	
	70 °C	56	4	30	10	
DMSO	20 °C	32	3	46	19	
	50 °C	21	4	51	24	
Pyridine	0 °C	60	4	27	9	
	20 °C	54	5	30	11	-
	60 °C	42	6	36	15	1

building block and, hence, of eminent industrial interest. As of now, however, its large-scale production is not effected although the pilot-plant size methodology appears to be available.[31,32]

Once, tautomeric fixation has been achieved, the mono- or disaccharide is to be converted into building blocks with useful functionality, such that the modern preparative armoury of organic chemistry can directly be applied. Thereby, it is an irrevocable necessity to strictly adhere to the criteria of practicality outlined above, if the products are to be of industrial significance.

This imposition of practical norms for the acquisition of furanoid, open-chain, and pyranoid building blocks from simple sugars reduces the number of possible reactions quite substantially, leaving comparatively few efficient, large-scale adaptable reaction sequences − *"reaction channels"*, so to say − which happen to be genuine for each specific mono- and disaccharide. The account given in the sequel outlines such "reaction channels" from simple sugars to recently developed building blocks of high versatility, and addresses their potential applications.

2. Prototype Reaction Channels to Building Blocks

One such highly efficient channel "sugar ⇒ versatile building block", feasible on a hectogram-scale, leads from D-glucose or D-maltose to pyranoid enediolones 7 and 9, i.e., dihydropyranones with two chiral centers at one side of the ring and useful functionality at the other.

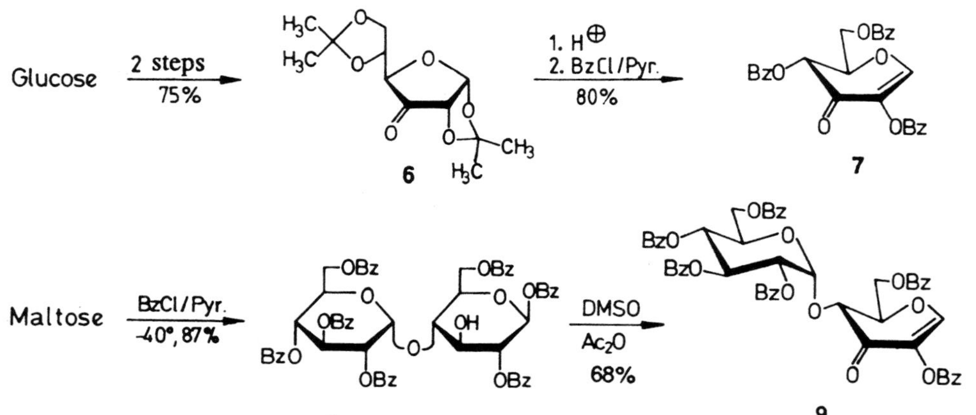

In the case of D-glucose it takes a 4-step procedure with a 60 % overall yield, involving acetonation, oxidation (→ **6**), acid removal of isopropylidene groups and benzoylation, the slightly basic conditions in the last step (pyridine) already sufficing to elicit β-elimination of benzoic acid.[33]

If one is prepared to accept a tetrabenzoylglucosyl residue as an acid-sensitive, alkali-stable blocking group – and a cheap one at that – the disaccharide-derived dihydropyranone **9** is even more directly accessible: low temperature benzoylation of maltose affords the heptabenzoate **8** in high yield, and subsequent DMSO / acetic anhydride oxidation effects both, oxidation at the 3-OH and β-elimination.[33]

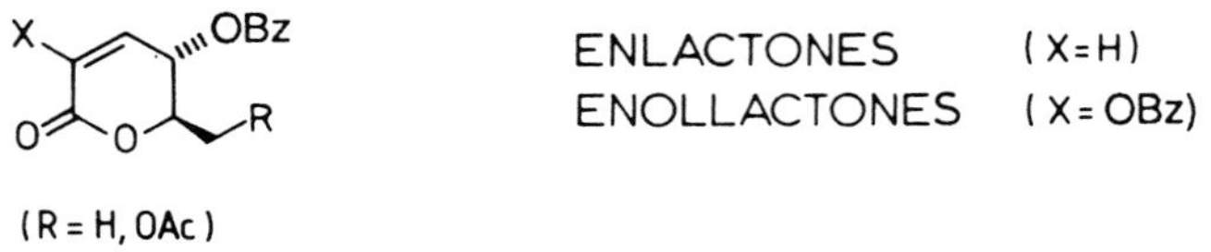

X
OBz
O
O
R
(R = H, OAc)
ENLACTONES (X=H)
ENOLLACTONES (X=OBz)

BF_3- catalyzed peroxidation of Glycal Esters

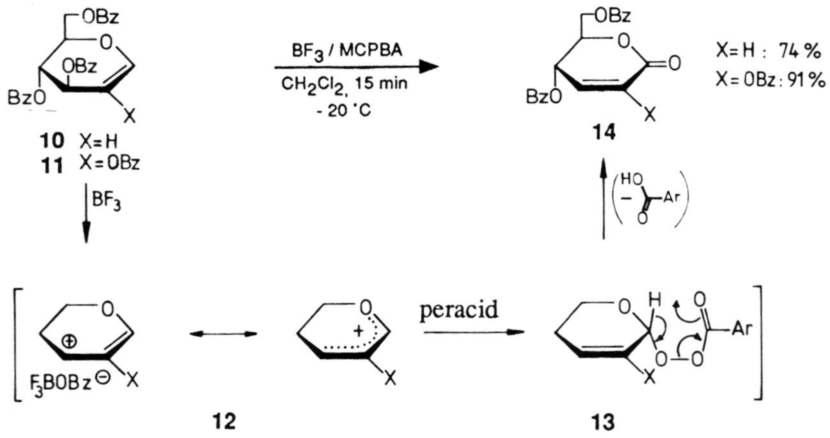

OBz
O
OBz
BzO
X
10 X=H
11 X=OBz
BF3
BF3 / MCPBA
CH2Cl2, 15 min
- 20 °C
OBz
O
O
BzO
X
14
X=H : 74 %
X=OBz : 91 %
HO
Ar
O
F3BOBz⊖
X
+
peracid
O
H
O
Ar
O
O
X
12
13

Another preparatively delightful reaction channel starting from glycal (**10**) or hydroxyglycal esters (**11**) provides enantiomerically pure dihydropyranones with the reverse functionality in the ring, i.e. enelactones or enolester lactones of type **14**.[34]

Initiated by a BF_3-induced removal of the allylic acyloxy function to form the allylcarboxonium ion **12**, the m-chloroperbenzoic acid (MCPBA) attacks solely at C-1 as expected from a hard nucleophile; the resulting perester intermediate **13** then undergoes fragmentation as indicated by the arrows to yield enelactones or enollactones of type **14**. This most readily effected conversion (15 min at -20 °C) can be applied to any mono- and disaccharide-derived glycal ester, and, thus, provides a whole series of dihydropyran-2-ones in excellent yields.[34]

Surprisingly, when changing the conditions slightly, i.e. by simply performing the peroxidation at room temperature rather than at -20 °C, the glycal esters, e.g. triacetyl-glucal **15** or its galactal analog, do not give the respective enelactones, but the acyclic pentenal **19** via oxidative cleavage of the olefinic double bond.[34]

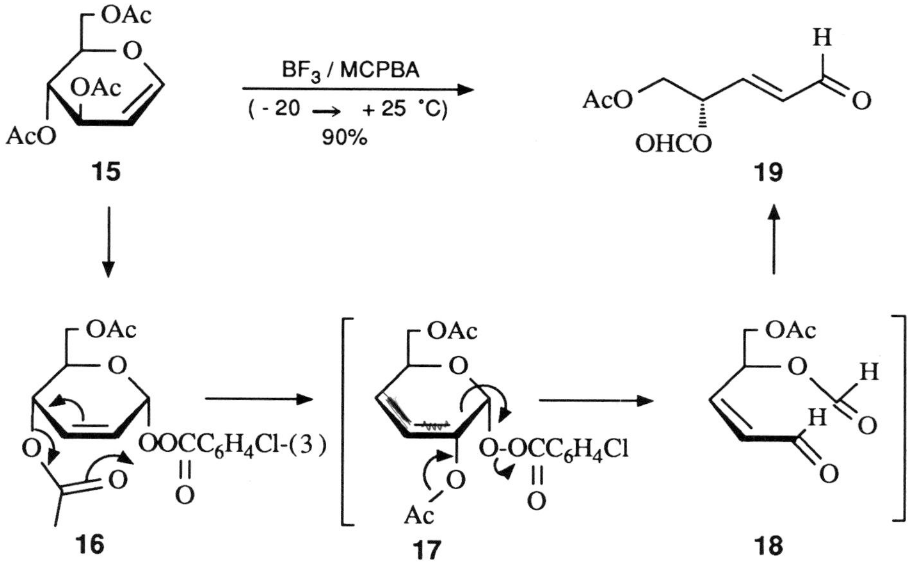

Mechanistically, this unusual conversion, elicited under comparatively mild conditions, comprises a glycol-type oxidative cleavage between C-1 and C-2, which is best rationalized in the following terms: the perester **16**, formed initially, does not undergo fragmentation to the enelactone under these conditions, but gives way to a $C_4 \rightarrow C_2$- rearrangement of the allylic acetoxy group instead, i.e. **16** → **17**. A Grob-type fragmentation in **17** (cf. arrows) then results in cleavage of the C_1-C_2-carbon bond to form **18** in which the *cis*-olefinic double bond is isomerized to the *trans*-isomer **19** under the influence of the Lewis acid present.[34]

This novel, remarkably smooth conversion of the six carbon glycal esters into five carbon pentenals of type **19**, which has only very few analogies in literature so far,[35,36] adds another useful "reaction channel" from hexoses to a versatile five-carbon building block via glycals as the key intermediate; others – aside from the enelactone conversion **15 → 20** – being their mercuric ion-catalyzed ring opening, e.g. **15 → 21**, to 3,4,5-trihydroxyhexenals.[37]

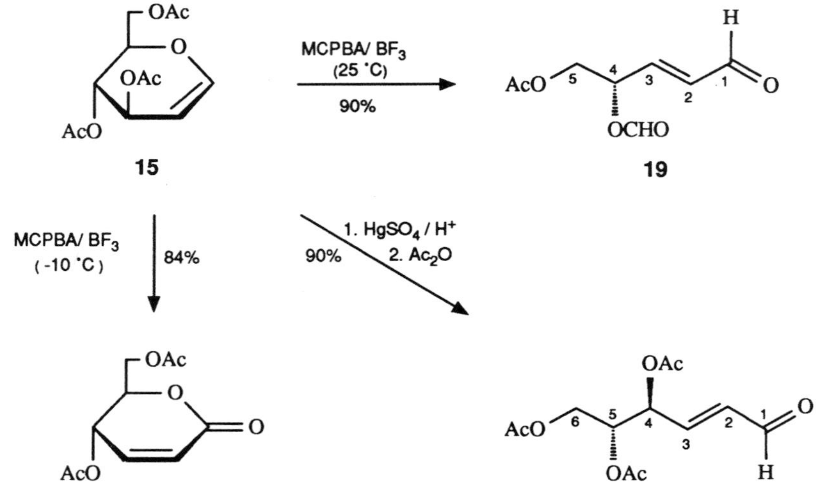

These examples amply demonstrate the ease and efficiency with which sugars may be converted into five- or six-carbon building blocks via practical, large-scale adaptable reaction channels. Rather than enumerating other such channels, in the sequel, I want to focus on potential target molecules, i.e. classes of biologically interesting compounds, that may advantageously be approached with sugar-derived building blocks.

3. Pyranoid Soft Coral Constituents

Intense interest in marine natural products have uncovered a large number of structurally novel compounds within the last decade, among them palythazine,[38] a heterocycle with an unusual dipyranopyrazine skeleton, and bissetone,[39] a branched pyranoid system :

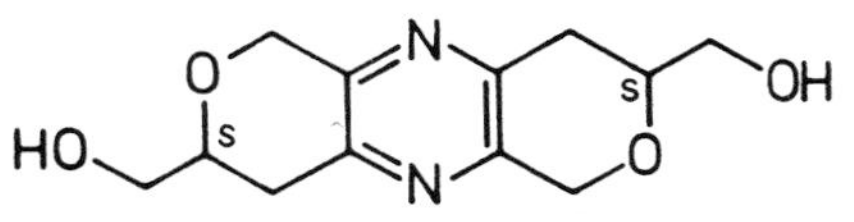

PALYTHAZINE   BISSETONE

from *Palythoa tuberculosa*   from *Briareum polyanthes*

(salt water invertebrate)   (Gorgonian soft coral)

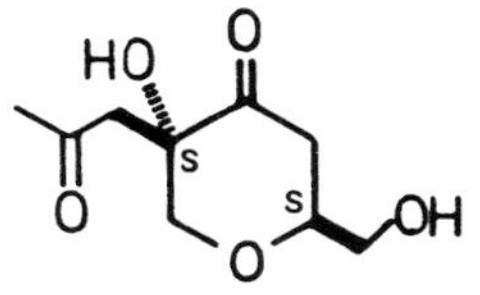

"optically active" $[\alpha]_D^{20}$ −43.6° (EtOH)

In either case, a straightforward synthesis could be developed from D-glucose,[40,41] the key compound being building block **25**, a (6*R*)-dihydropyranone, readily accessible in three high-yielding steps from the hydroxyglucal ester **11**. Liberation of the enol ester-blocked carbonyl function cannot be achieved directly, but via hydroxylaminolysis to the stable, highly crystalline oxime of 1,5-anhydro-D-fructose tribenzoate (**23**). Gratifyingly, this hydroxylaminolysis procedure does not affect primary or secondary ester groups elsewhere in the molecule, enables excellent yields, and appears to be generally applicable, a whole series of acylated 1,5-anhydro-ketoximes having been prepared via this most delightful new reaction channel.[42] Subsequent deoximation (**23** → **24**), and mild base-induced elimination of benzoic acid provides the dihydropyranone **25** in nicely crystalline, enantiomerically pure form.

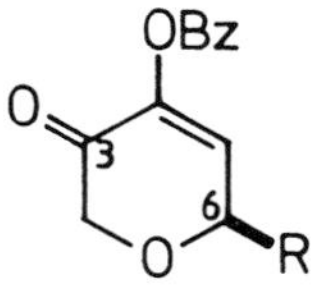

(6R) - DIHYDROPYRANON -3

($R = CH_2OBz, CH_3$)

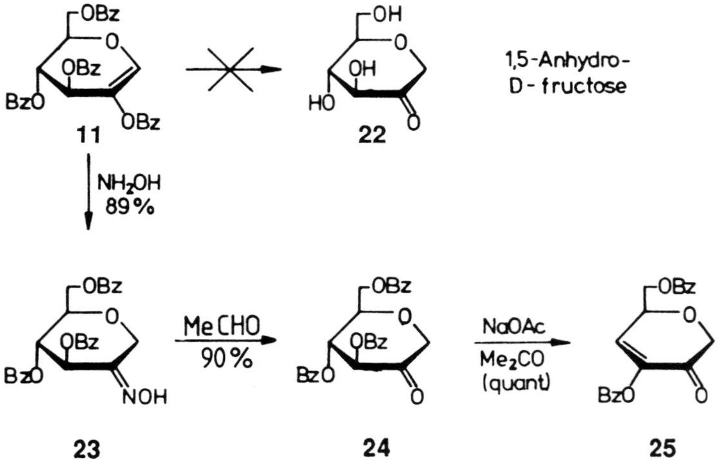

1,5-Anhydro-
D-fructose

11   **22**

NH$_2$OH
89%

23   **24**   **25**

Me CHO
90%

NaOAc
Me$_2$CO
(quant)

   The efficiency with which this building block can be elaborated from
D-glucose is noteworthy: the six-step reaction sequence involved can be reduced to
two hectogram-adaptable one-pot procedures comprising the conversion
D-glucose → hydroxyglucal ester (85 %, vide supra), and the sequence
hydroxylaminolysis → deoximination → elimination (84 %, when performed in one
continuous operation).[41]

   The elaboration of (S,S)-palythazine (**29**) from building block **25** was effected
in a simple, high-yielding reaction sequence: conversion into oxime **26**, liberation of
the second carbonyl function by debenzoylation (→ **27**), and controlled catalytic
hydrogenation to aminoketone **28**, which dimerizes at pH 9; the concluding step is
an air oxidation of the dihydropyrazine initially formed.[40]

SYNTHESIS OF S,S-PALYTHAZINE

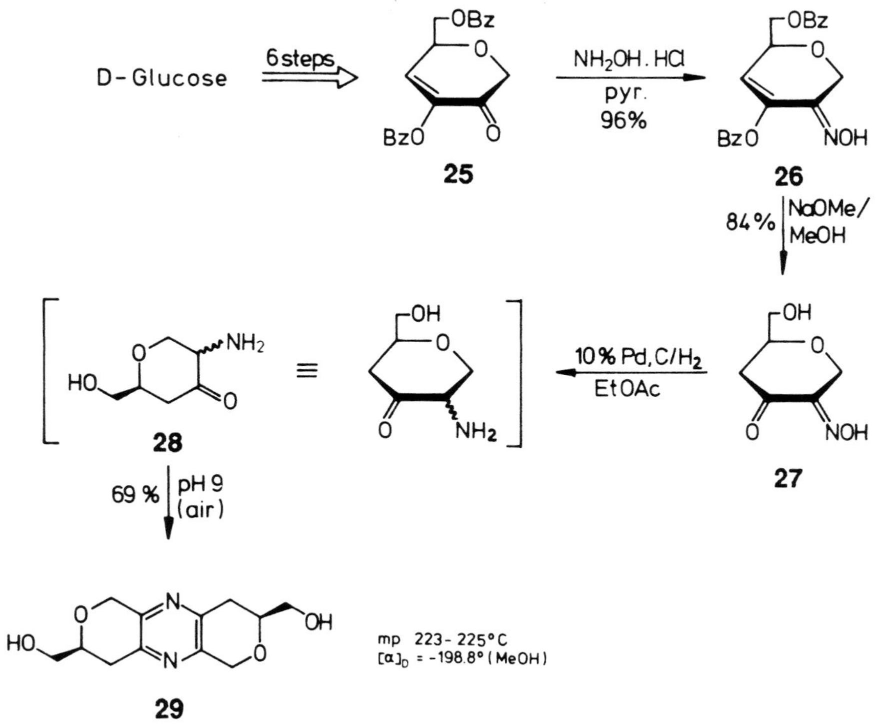

(S,S)-Bissetone (**30**), when traced back to building block **25**, only lacks the 3-carbon branch, i.e. acetone. Indeed, the lithium enolate of acetone proved to be a suitable three-carbon synthon attacking the carbonyl function with a 4:1 preference from the pro-axial side (**25 → 31**). The benzoyl group shift directly following the attack elaborates the desired 2-oxopropyl-branched tetrahydropyranone **32**, the dibenzoate of bissetone, in fact, from which the parent compound **30** is generated simply by de-*O*-benzoylation.[41]

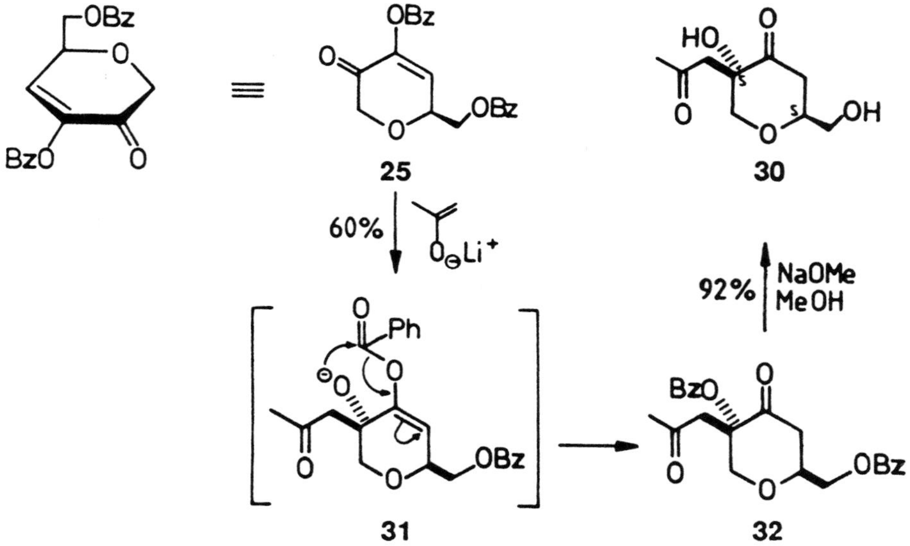

Thus, two soft-coral-derived constituents that have been obtained from their natural sources in so minute amounts (up to 50 mg each) as to preclude their biological evaluation, can now be acquired from D-glucose via highly efficient, large-scale adaptable syntheses, requiring 8 and 9 steps, respectively, with overall yields of 37 and 26 %.

4. Medium Ring Size Lactones

With *macrodialides* of the phorocantholide and diplodialide type – the former a defensive secretion from the metasternal gland of the Eucalypt longicorn beetle *Phorocantha synonyma*,[43] the latter a metabolite of a plant pathogenic fungus *Diplodia pinea* of high hydroxylase inhibitory activity[44] – it may be preposterous at first sight to think of sugars as suitable starting materials.

Diplodialides and Phoracantholides

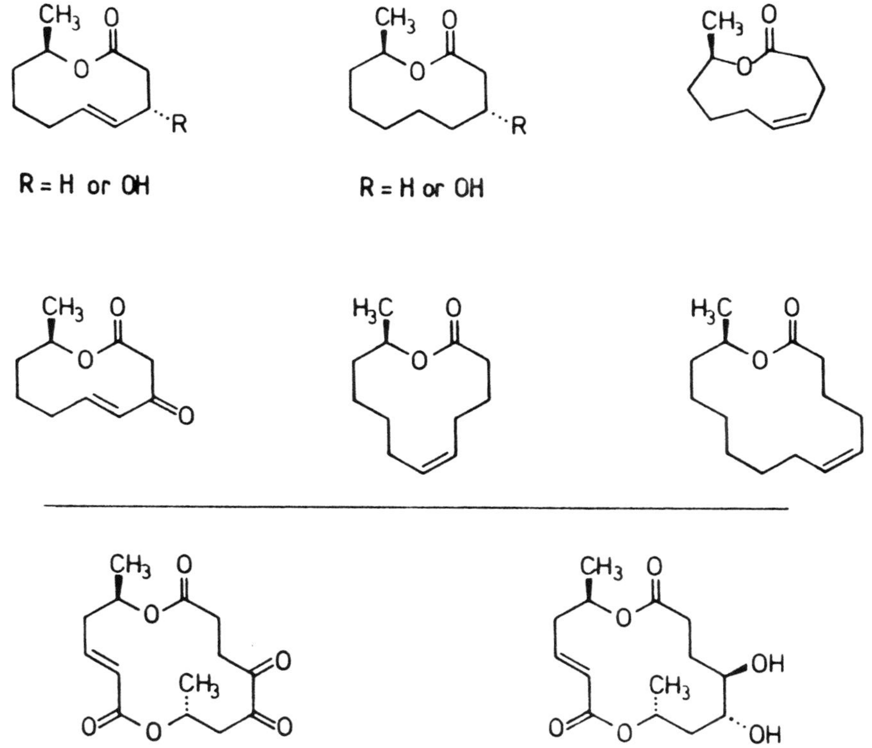

R = H or OH

R = H or OH

(−)−Grahamimycin A
(R,R)

(+)− Colletodiol
(all−R)

The mostly singular chiral center along the carbon chain comprises an (R)-hydroxy group tied up as a lactone. Elaboration of such structural elements from carbohydrate precursors will only be reasonable, if highly efficient reaction channels are available or may be developed for the conversion of a monosaccharide into a building block of the (5R)-hydroxyhexanal or (5R)-hydroxyhexenal type. Both compounds are known in racemic form, of the latter even, that on distillation isomerization takes place to the *trans*-olefin, which as such corresponds structurally to the entire left half of the fungal metabolites grahamimycin and colletodiol.

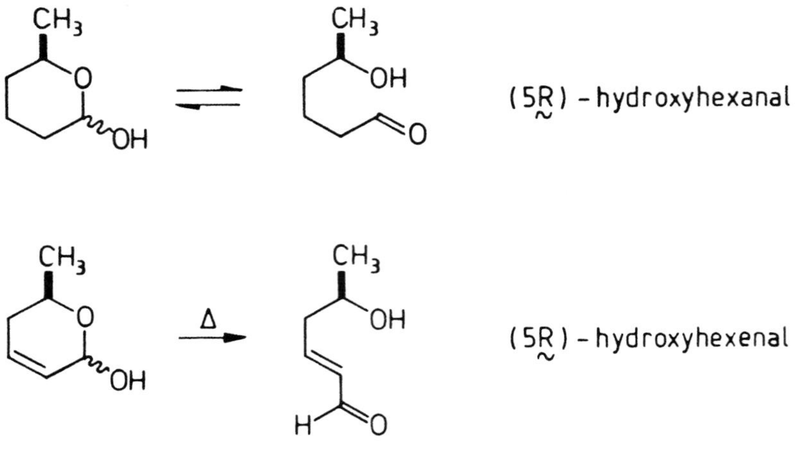

(5R) – hydroxyhexanal

(5R) – hydroxyhexenal

An efficient, practical access to enantiomerically pure (5R)-hydroxyhexenal – in the form of its pyranoid ethyl acetal **35** – could indeed be developed starting from D-glucose.[45,46] The initial steps comprise the well-elaborated conversion of D-glucose into triacetylglucal **15**, and, from thence, to the α,β-unsaturated hexenoside **33**.

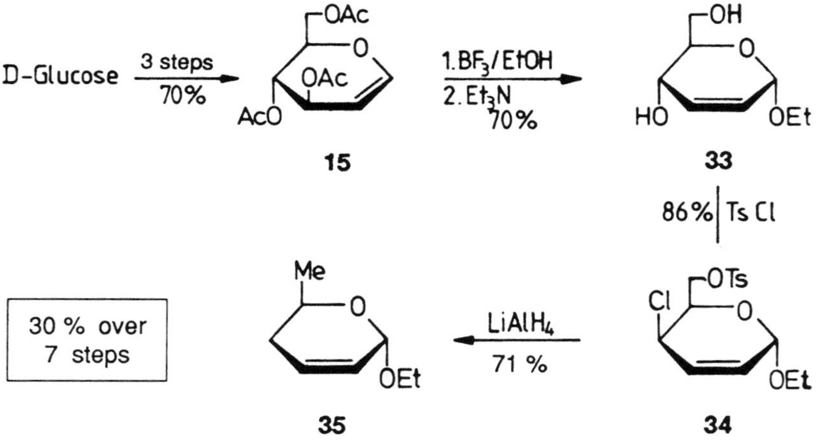

Under standard tosylation conditions, **33** is converted into the 4-chloro compound **34** rather than the di-*O*-tosyl derivative, since the chloride ions formed during the reaction displace the activated allylic 4-tosyloxy function in situ. Reductive removal of tosyloxy- and chloro groups, i.e. **34** → **35**, can be effected in one operation to afford the desired (5*R*)-hexenal synthon **35**. The overall yield of 30 % for the seven steps from D-glucose[45,46] appears quite satisfactory.

The high versatility of **35** as a key intermediate towards macrodialides is amply demonstrated by its smooth hydrolysis to the respective hexenal **37** or, when preceeded by hydrogenation, to the (5*R*)-hydroxyhexanal (**36**), whilst peroxidation yields (*R*)-parasorbic acid (**38**) – all enantiomerically pure six-carbon building blocks that represent major segments of macrolides.

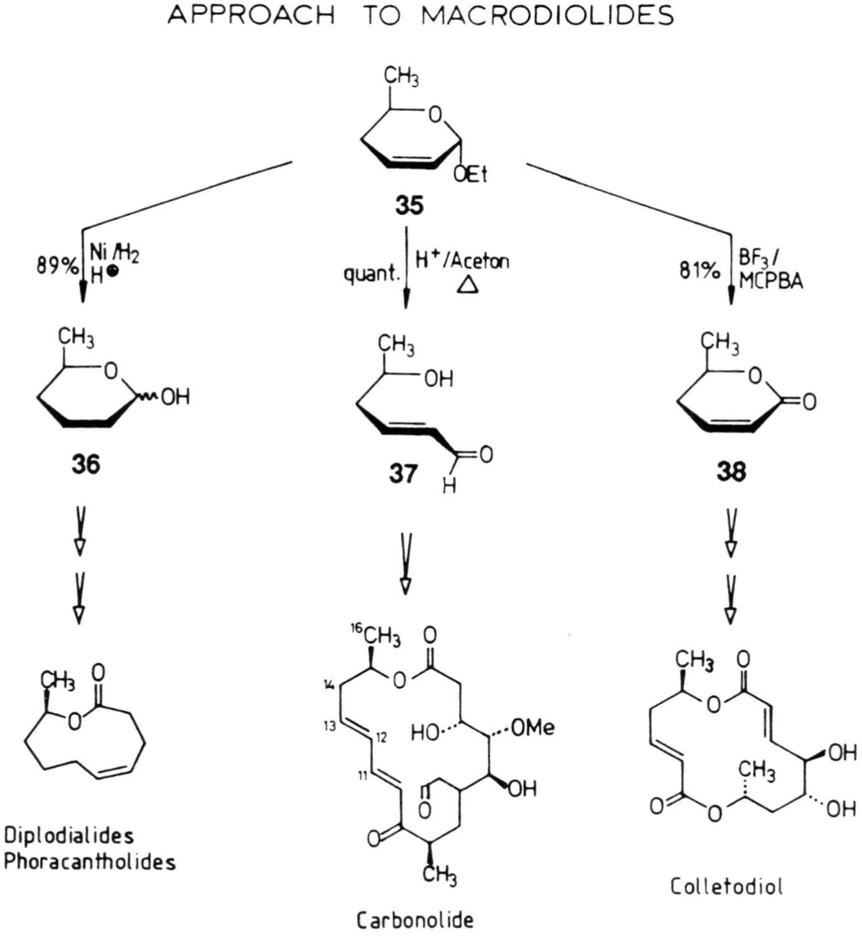

APPROACH TO MACRODIOLIDES

35

89% | Ni /H₂ H•   quant. | H⁺/Aceton △   81% | BF₃/ MCPBA

36   **37**   **38**

Diplodialides
Phoracantholides

Carbonolide

Colletodiol

Using this approach, the phoracantholides I and J, as well as the diplodialides A, B, and C have been synthesized in enantiomerically pure form.[45,46] The reaction sequences involved being practical and efficient, these biologically interesting compounds are thus available for the first time in amounts that allow a thorough biological evaluation.

Incidentically, the enantiomeric (5S)-hydroxy-hexenal analogs are also readily accessible from L-rhamnose in an even more expeditious reaction sequence.[47]

5. Dioxane-annellated Antibiotics and Cardiac Glycosides

Another category of biologically relevant natural products that may successfully be approached via sugar-derived pyranoid building blocks are those having the tetrahydropyrano-dioxan ring system annellated onto a cyclohexane or steroid. The broad spectrum antibiotic spectinomycin[48] belongs to this class, as does a group of cardiac glycosides,[49] isolable from the latex and leaves of *Calotropis procera*, indegenious in wide parts of Africa and India.

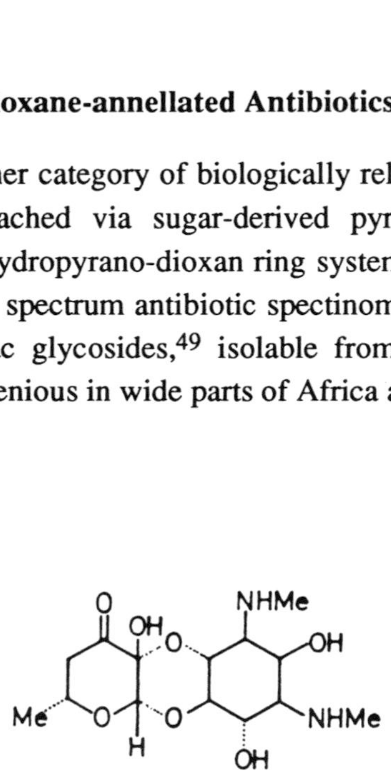

SPECTINOMYCIN 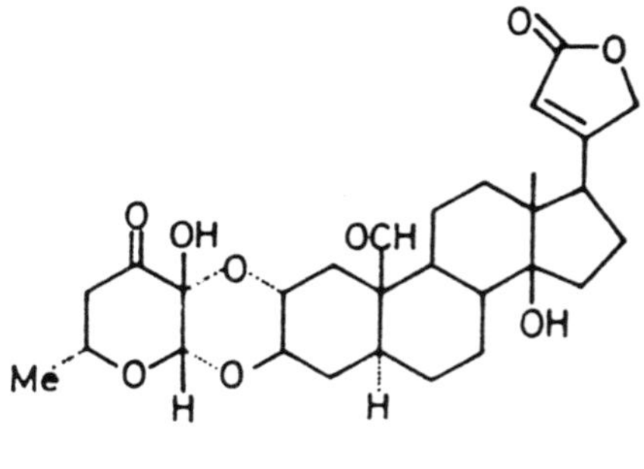   USCHARIDIN

SUGAR COMPONENT :

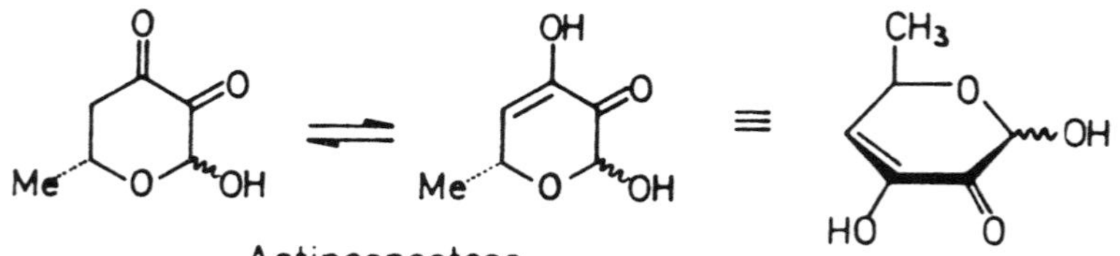

Actinospectose

The pyranoid tricarbonyl sugar, that is fused to the cyclohexandiol-type aglycon by both a β-glycosidic bond and a hemiketal linkage, is a 4,6-dideoxy-D-*glycero*-hexos-2,3-diulose, that has been designated actinospectose.[48] Due to its sensitivity towards acidic and basic conditions it could only be "characterized", though, as an intractable tar.

Stable acylated derivatives of actinospectose, i.e. 2,6-dihydro-3-pyranones with a 6-hydroxymethyl or 6-methyl substituent of matching configuration, can readily be obtained via another preparatively useful reaction channel starting from hydroxyglycal esters, namely by a chlorination → hydrolysis → elimination sequence:[50,51]

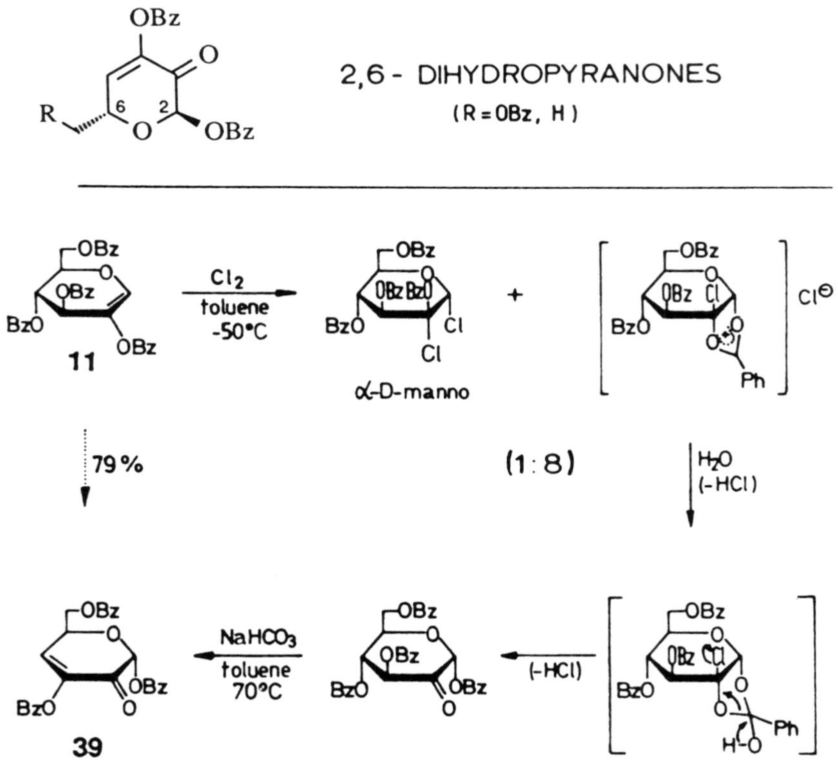

2,6-Dihydropyranones of type **39** provide a very prolific ensuing chemistry, particularly with respect to hydride addition, C-branching with Grignard or cuprate reagents,[52] and Diels-Alder type cycloaddition induced by either Lewis acids or high pressure.[53] O-Nucleophiles attack at the carbonyl function from the sterically and electronically more favored pro-axial face to provide upon the usually ensuing benzoyl group migration tetrahydropyrones of type **40**, which are bis-acetal derivatives of actinospectose in the form present in natural products.[51,54]

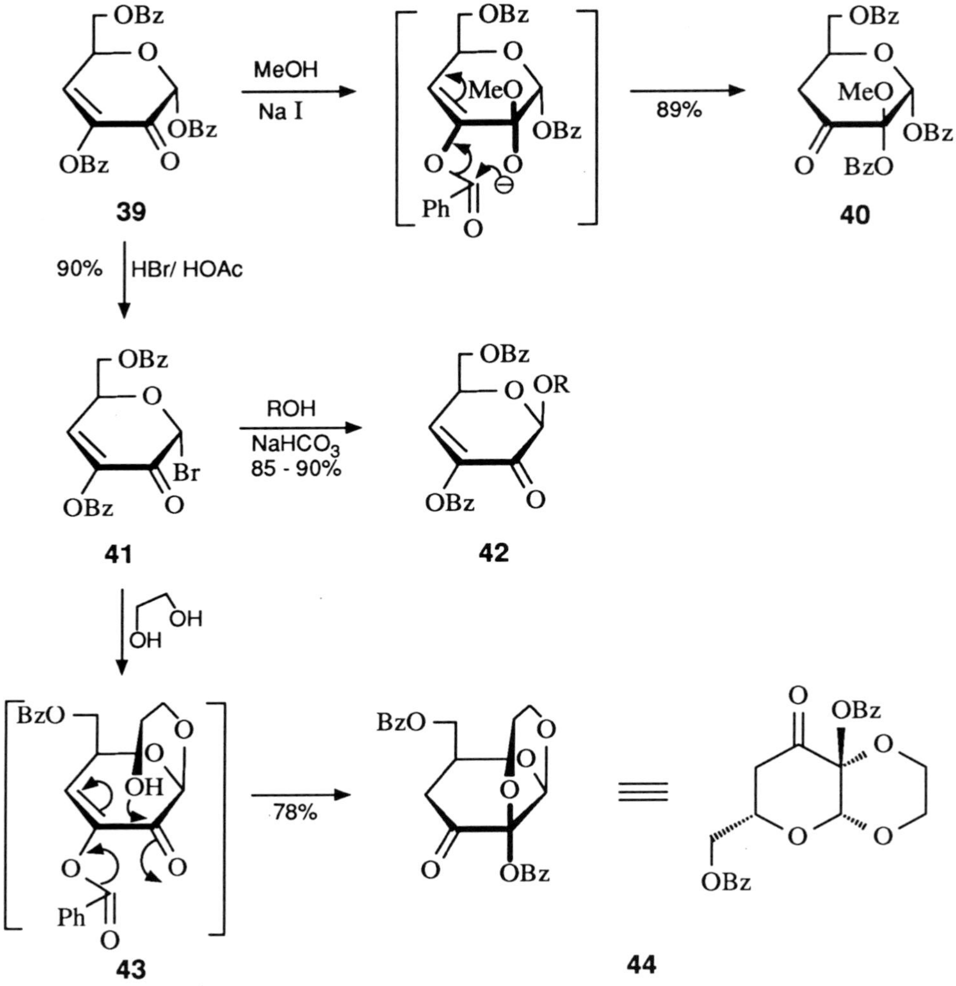

An even closer resemblance in this respect exhibits the pyrano[2,3-b]dioxane **44** preparable from **39** in two high-yielding steps. Being remarkably insensitive towards acidic conditions, **39** is easily converted into the respective 1-halides, as shown by its transformation into **41** on treatment with HBr / acetic acid.[53] Alcoholysis of such enolone α-D-bromides may be performed in the presence of simple acid scavengers, and give the β-glycosides in yields well over 80 %.[53] Similarly, when subjected to glycolysis with ethanediol, the bromide **41** smoothly elaborates the dioxane ring-annellated pyrone **44**, the β-selective glycosylation obviously being followed by cyclization to the hemiketal and subsequent benzoyl migration (arrows in **43**) with liberation of the C-3 carbonyl function.[54]

Thus, from the stage of a hydroxyglycal ester of type **11**, which is only a one-pot reaction away from the parent bulk sugar, a surprisingly simple, highly practical reaction sequence, i.e. **11** → **39** → **41** → **44**, is available with which dioxane-annellated glycosides can be prepared.[54-56]

Model experiments with the enantiomeric (*R,R*)- and (*S,S*)-cyclohexandiols substantiated the feasibility of this approach towards natural products of the spectinomycin- and uscharidine-type: brief reaction of the (*R,R*)-isomer with the pyranoid bromide **41** at ambient temperature gently elaborated the tricyclic product **45**, characterized as its highly crystalline monoetherate, of which an X-ray structural analysis proved its linkage configuration.[54]

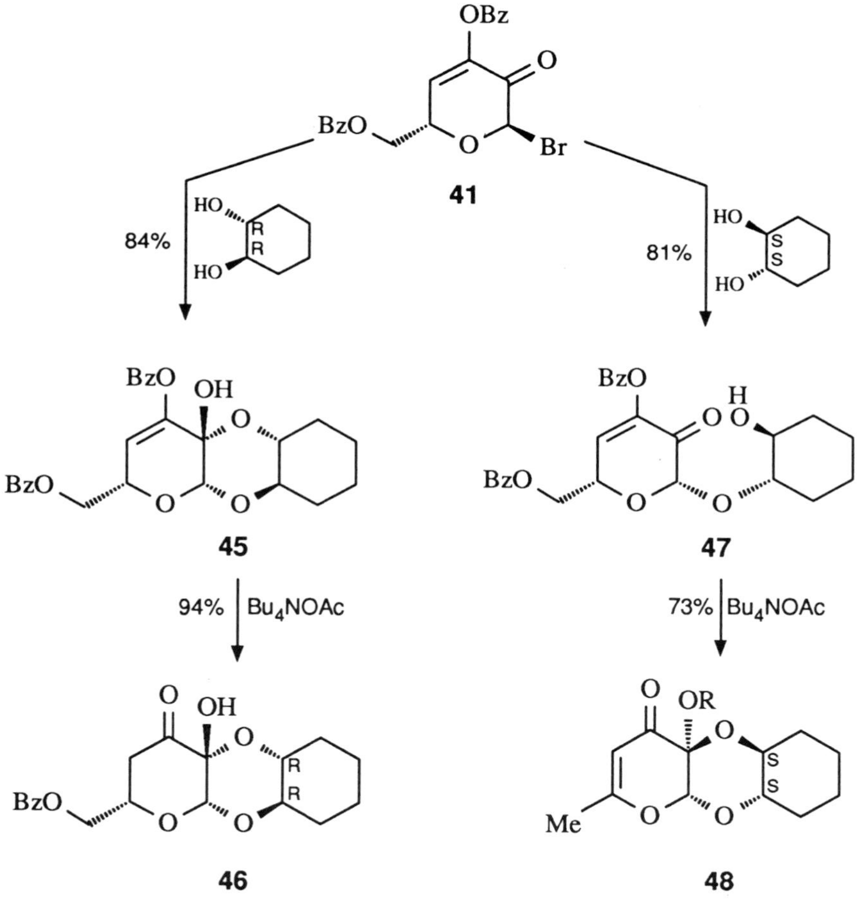

Liberation of the enolester-blocked carbonyl function in **45** expectedly was inducible by traces of alkali to provide **46**, a tricyclic structure, that has all essential functional elements as well as the correct stereochemistry of spectinomycin, except for the benzoyloxy group in the pyranoid half and some OH and methylamino substituents lagging in the cyclohexane portion.

By consequence, a practical synthetic strategy for spectinomycin called for correspondingly adapted cyclohexanoid and pyranoid halves, whereby the former is most readily provided by the bis-carbobenzoxy derivative of actinamine, a *myo*-inosadiamine, that has been synthesized several times, not the least by us some twenty years ago via useful reaction channel starting from *myo*-inositol:[57]

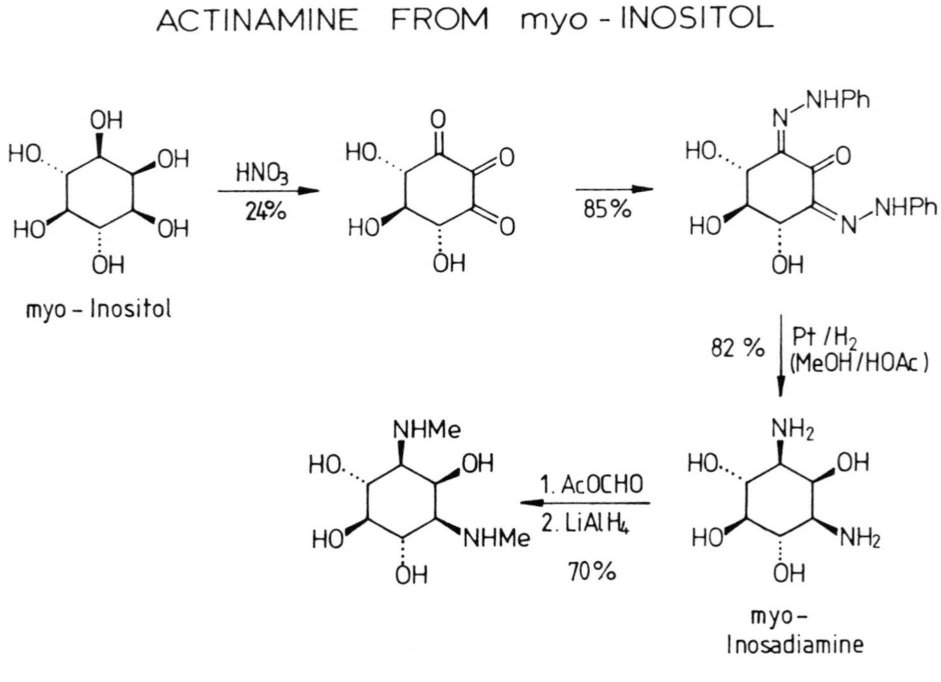

ACTINAMINE FROM myo - INOSITOL

The *R*-methyl group in the pyranoid portion of spectinomycin pointed towards a 6-deoxy-hexose as the chiral educt and – given the ready elaboration of dihydropyranones of type **41** from hydroxyglycal esters, outlined above – for an acylated 6-deoxy-D-hydroxyglucal as the actual starting material. The tribenzoate **49** was chosen for this purpose due to its expeditious preparation from methyl α-D-glucopyranoside in five large-scale adaptable, high yielding steps.[51] When subjected to low-temperature-chlorination, hydrolysis, elimination and treatment with HCl – in a manner analogous to the conversion **11 → 39 → 41** – the crystalline actinospectosyl chloride **50** is readily obtained in acceptable overall yield.[51]

Exposing actinospectosyl chloride (**50**) and bis-carbobenzoxy-actinamine (**51**) to suitable glycosylation conditions (silver triflate in THF), the desired β-selective glycosylation of the sterically less hindered 5-OH of actinamine is effected, whereafter ketalization and benzoyl group migration – in a manner analogous to **41 → 44** (cf. above) – yielded the bis-carbobenzoxy-spectinomycin benzoate **52** in respectable yield.[56] Since de-*O*-benzoylation is readily accomplished (K_2CO_3/methanol) and hydrogenolysis is proceeding smoothly, this sequence constitutes a most facile, efficacious total synthesis of spectinomycin, requiring 12 steps from D-glucose with an overall yield of 9.9 %, averaging 80 % per step.[56]

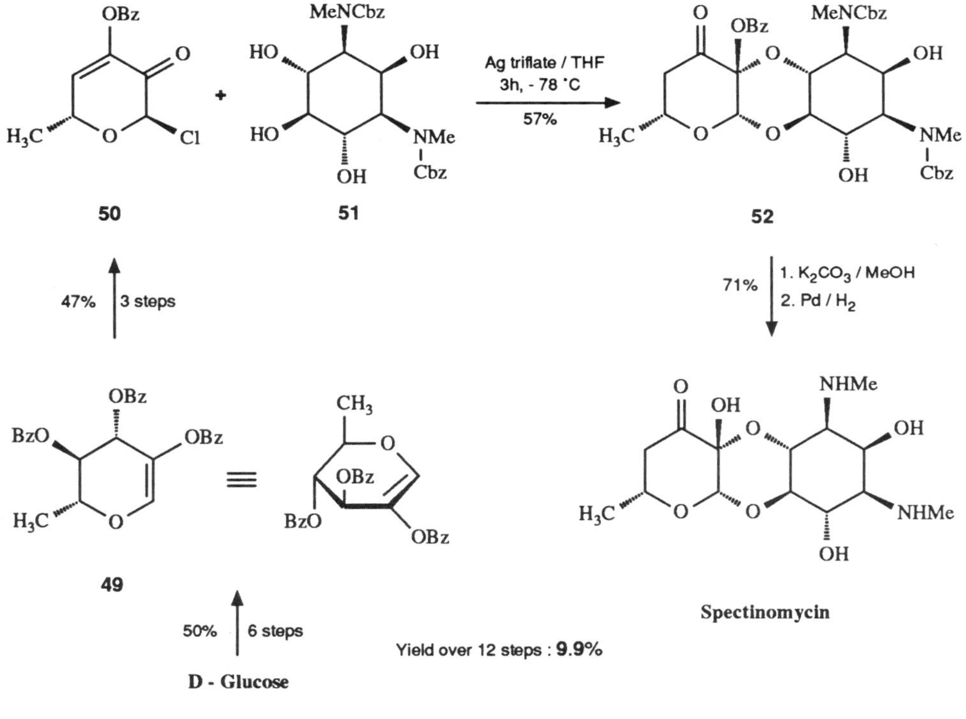

This expeditious synthetic route to spectinomycin not only compares most favorably with two previous syntheses of this antibiotic, one elaborating the pyranoid half from D-glucose in 23 steps,[58] the other requiring nine from the rather incommodiously accessible L-glucose.[59] This approach, in addition, is versatile enough to lend itself to the synthesis of uscharidine type cardiac glycosides from steroidal diols. Since calotropagenine, the aglycone of uscharidine, is accessible only with difficulty, model experiments have been carried out with the more readily available cholestan-2α,3β-diol **54**. On silver carbonate-mediated reaction with actinospectosyl bromide **53**, two isomers formed smoothly in an approximate 3:1 ratio, of which the major one – as evidenced by ^{1}H- and ^{13}C-NMR data, correborated by NOE experiments[60] – proved to be the "unnatural" annellation product **56**. The minor product **55**, however, could be readily debenzoylated by treatment with butylammonium acetate in aqueous acetonitril to afford the uscharidine analog **57** in crystalline form.[56,60]

SYNTHESIS OF USCHARIDIN -TYPE
CARDIAC GLYCOSIDES

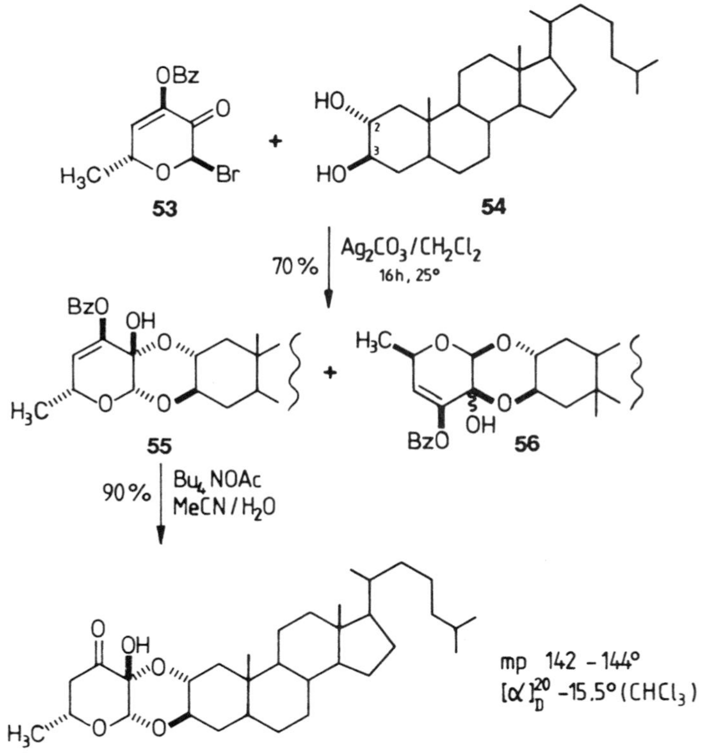

The 3 : 1 glycosylation selectivity observed clearly points towards a higher reactivity of the 2-OH group in the diol educt – a finding that similarly enables the effective introduction of an O^2-blocking group, whereafter reactions with actinospectosyl halides **50** or **53** are uniformly directed to the "natural"

glycosylation site. This obvious preparative solution has been carried out with 2-*O*-benzyl protected **54**, providing after deprotection the cardenolide analog **57** in highly satisfactory yield.[60]

6. Disaccharide Building Blocks for β-D-Mannose- and β-D-Mannosamine-containing Oligosaccharides

Oligosaccharides composed of more than one type of sugar unit such as the human blood group determinants and the bacterial antigens are major carriers of biological information which necessitates their regio- and stereocontrolled synthesis on a preparative scale.

The impressive advances made toward this end within the past decade usually comprise the stepwise construction of oligosaccharides from suitably blocked monosaccharide components, i.e. consecutive elongation of a saccharide chain by one sugar unit.[61]

The readily accessible bulk disaccharides such as lactose, maltose or cellobiose, in which one intersaccharidic linkage is already preformed, have not been systematically exploited for this purpose. A promising recent development along this vein is their transformation into 2-oxo- and 2-oximino glycobiosyl bromides, i.e. disaccharide building blocks suitably functionalized for direct glycobiosylation.[62,63] The methodology elaborated starts with the conversion of the basic disaccharides by benzoylation, HBr treatment, and dehydrobromination into their 2-hydroxyglycal esters as demonstrated below for lactose (→ **58**); it is followed by NBS-promoted methanolysis that effectively delivers the hexa-*O*-benzoyl-lactosulosyl bromide **58**. Alternatively, application of the high-yielding, three-step sequence hydroxylaminolysis → benzoylation → photobromination provides the 2-oximino-lactulosyl bromide **60**.[63]

The broad utility of these novel disaccharide building blocks with a non-participating group next to the anomeric center resides in the ease and uniformity with which stereocontrol over glycosidations can be effected: silver carbonate-induced alcoholysis with simple primary and secondary hydroxyl components exclusively provides the β-D-lactosidulose **61** or its oximino analog **62**.

Reduction of the 2-oxo (NaBH$_4$) and 2-benzoximino functions (diborane or LiBH$_4$/Me$_3$SiCl) proceed with stereoanomeric control in such that hydride addition takes place from the side opposite to the anomeric substituent; accordingly disaccharides with β-D-mannose (**63**) and β-D-mannosamine units (**64**) are generated in a most satisfactory way.[64-66]

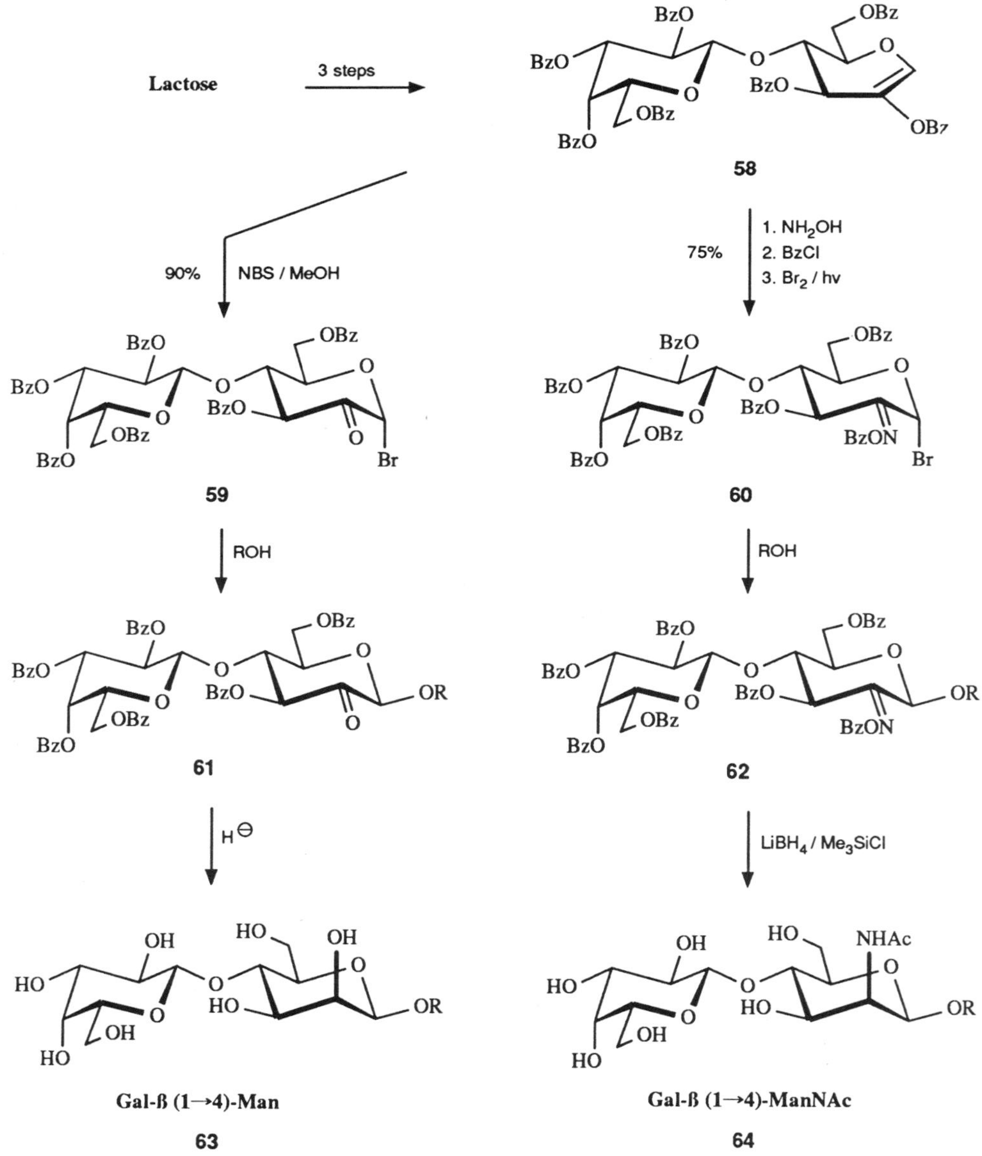

63
Gal-ß (1→4)-Man

64
Gal-ß (1→4)-ManNAc

Accordingly, these disaccharide building blocks are expected to lend themselves to the efficient attachment of α-linked lactosamine, maltosamine, and cellobiosamine units to any given sugar hydroxyl. The method's prime strength, though, would appear to lie in the annelation of glycosyl-(1→4)-β-D-mannose and glycosyl-(1→4)-β-D-mannosamine blocks onto a sugar chain inasmuch as difficulties still prevail with the direct incorporation of β-D-mannose or 2-acetamido-2-deoxy-β-D-mannose residues. The latter, for example, are important constituent sugars of pneumococcal vaccines, yet, as of now, no preparatively satisfactory methods are available.

Various further ramifications in the use of these disaccharide building blocks for the construction of biologically relevant hetero-oligosaccharides are currently under investigation, their application to the synthesis of trisaccharides with central β-D-mannose, α-D-glucosamine, and β-D-mannosamine units having already been worked out successfully.[64-66]

7. Glucosyl-α(1→5)-arabinonic Acid and Glucosyloxymethyl-furfural (GMF) from Isomaltulose

Isomaltulose (**65**), having become available on an industrial scale recently (cf. Table 2), makes it a particularly lucrative target for developing reaction channels to disaccharide-type building blocks with potential industrial application profiles.

Of the many chemical modifications that have been investigated, two high-yielding reaction channels free of protecting groups, have emerged, which meet to a high degree the criteria of practicality outlined in the introduction. The first is the air oxidation of isomaltulose in strongly alkaline (KOH) solution, which − via the enediolate intermediate **66** − generates the potassium salt of the next lower aldonic acid, i.e. glucosyl-α(1→5)-D-arabinonate (**67**) in a surprisingly clean reaction.[67-69] Unlike the air oxidation of D-fructose under analogous conditions, which is encumbered by many side products such as glycolic, glyceric, lactic and erythronic acids,[70] the oxidative degradation of isomaltulose is essentially free of such undesired products, enabling yields of around 80-90 %.[68,69]

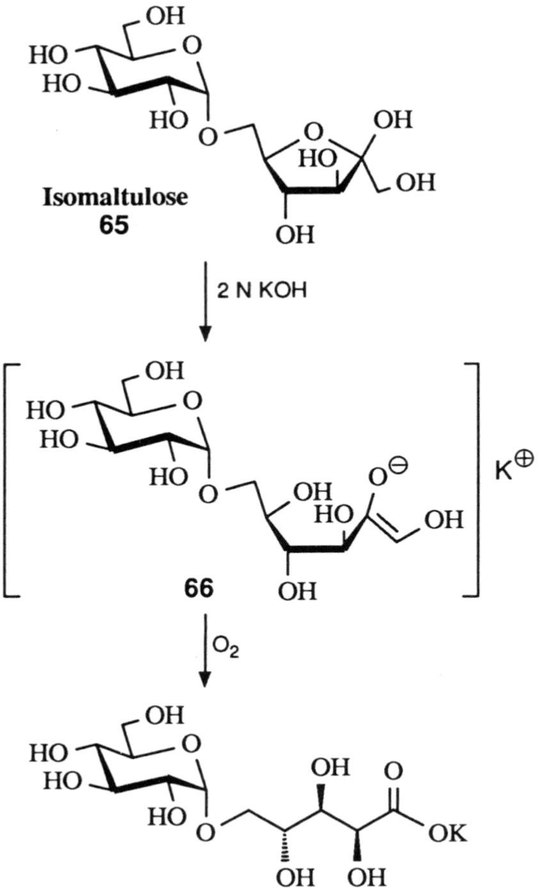

Isomaltulose
65

2 N KOH

66

O₂

Glucosyl-α(1→5)-arabinonate
67

The ensuing chemistry,[67,68] feasible with the potassium glucosyl-arabinonate **67**, comprises its essentially quantitative conversion into the lactone **68** simply by stirring its aqueous solution with a strongly acidic ion exchange resin. DIBAL reduction generates the glucosyl-α(1→5)-D-arabinose (**69**), characterized as such or its peracetate **70** – a disaccharide that previously was accessible in minute amounts via *Protaminobacter rubrum*-induced glucosyl transfer from sucrose to D-arabinose.[71] Alternately, on exposure of **67** to methanol/trimethoxymethane in the presence of an acidic ion exchange resin, the methyl ester **72** is obtained in highly crystalline form and satisfactory yield, which can either be converted into its amide by ammonolysis, i.e. **72 → 71**, or on reduction into glucosyl-α-(1→5)-D-arabinitol

73, a disaccharide-alcohol that has about the same sweetness as its six-carbon alditol-chain analog Palatinit®.

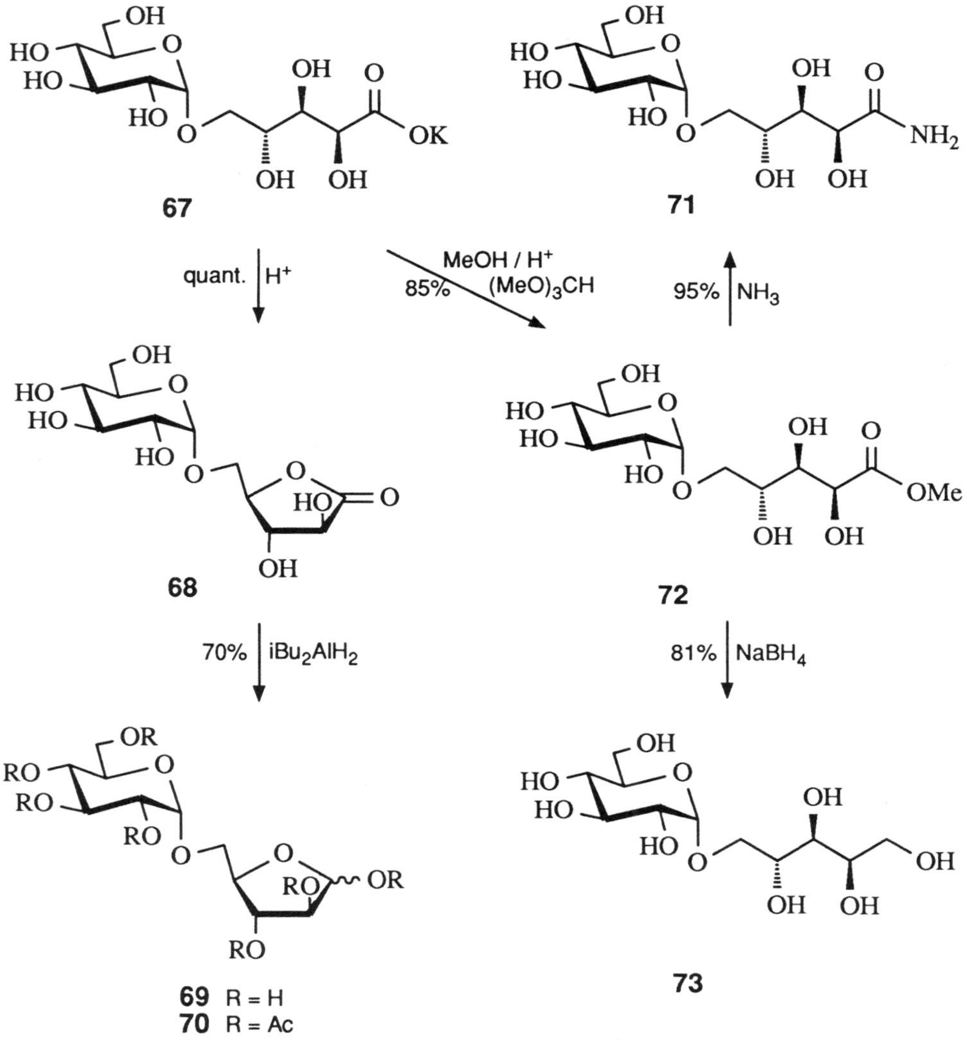

Of the five α-D-glucosyl-D-fructoses isomeric with sucrose, all of which, in solution, have different tautomeric distributions,[72] only the α(1→6)-linked isomaltulose (**65**) has the potential of elaborating a furfural structure from its fructose portion. If the intersaccharide linkage survives the acidic conditions required for the threefold dehydratization involved, the product would be α-D-**Glucopyranosyloxy-Methyl-Furfural** (**74**, "GMF"), a glucosylated, and, hence, substantially hydrophilic analog of HMF (**5**):

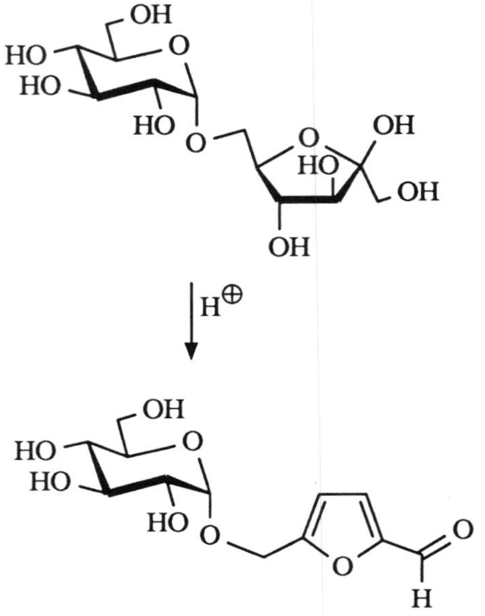

Isomaltulose
(**65**)

Glucosyloxymethyl -
furfural

("GMF", **74**)

The conversion isomaltulose → GMF could indeed be realized in a preparatively satisfactory way. In adaption of work on the conversion of fructose into HMF,[29,73] a series of conditions were evaluated, the best being heating for 3 h in dimethyl sulfoxide solution with a strongly acidic ion exchange resin.[74-76] From the resulting mixture of GMF (ca. 80 %), isomaltulose dimers (10 %), HMF, and glucose, the GMF can be isolated in up to 70 % yield in the form of well-shaped prisms, exhibiting a strongly positive rotation (+131° in methanol).[75] Aside from a batch conversion, a continuous process has also been elaborated using a flow reactor,[74] thus making GMF a well accessible product of bulk scale, if required.

The potential uses of GMF in ensuing reactions are multifold, yet, as of now, have not been fully exploited. Some modifications already performed[76] are :

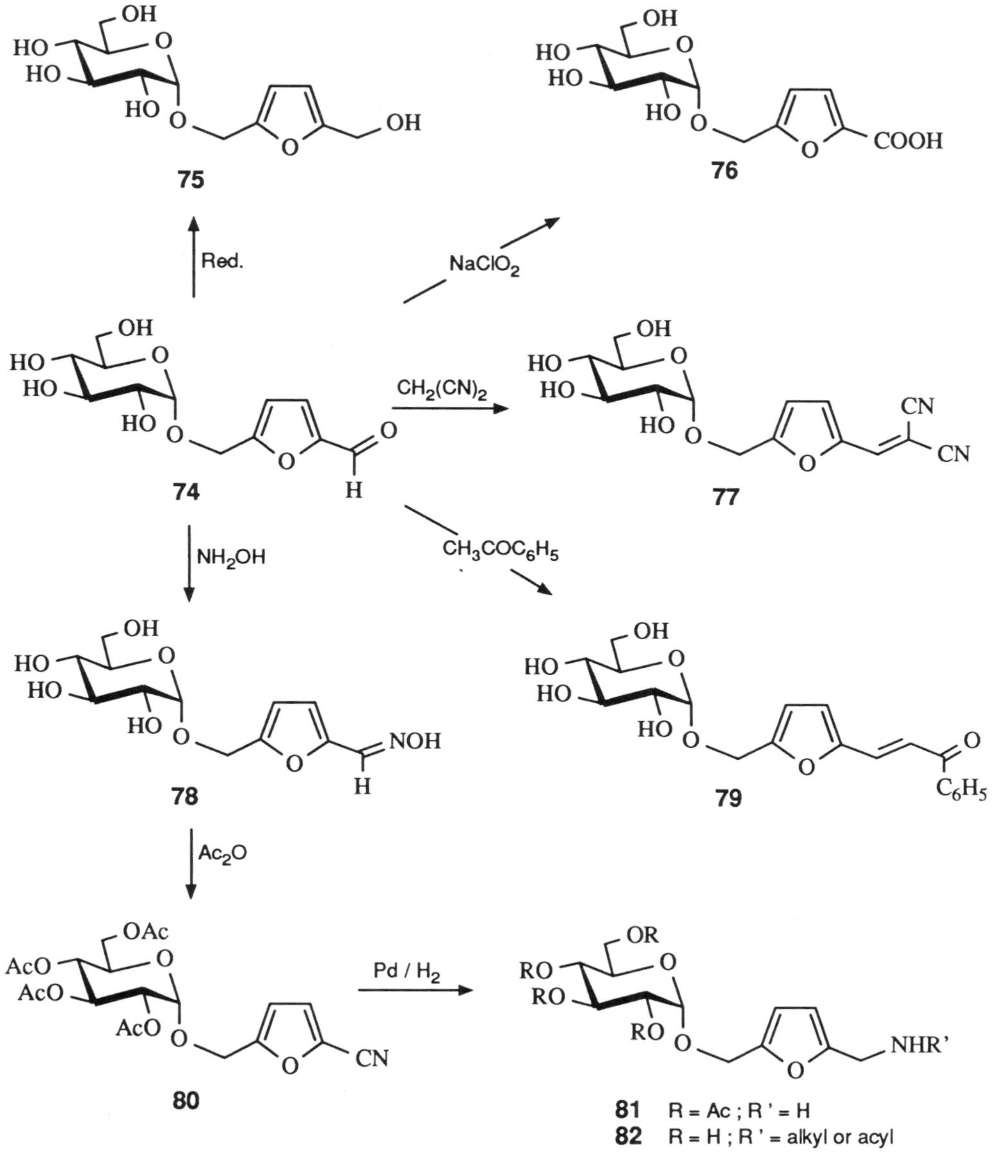

75

Red.

NaClO$_2$

76

COOH

74

H

CH$_2$(CN)$_2$

77

CN
CN

NH$_2$OH

CH$_3$COC$_6$H$_5$

78

NOH
H

79

O
C$_6$H$_5$

Ac$_2$O

80

CN

Pd / H$_2$

NHR'

81 R = Ac ; R' = H
82 R = H ; R' = alkyl or acyl

Of these products, the alcohol **75** and the acid **76** – on esterification with fatty acids and long-chain alcohols, respectively – should yield novel non-ionic surfactants in which the hydrophilic and hydrophobic portions are separated by a quasi-aromatic spacer of type **82**. The same holds for compounds resulting from GMF-amine **81**, which is expected to give cationic surfactants via alkylation or quaternization of the amino group, or non-ionics on amidation with fatty acids, all products with excellent environmental acceptability.

82

Similarly, the aldol-type condensation products of type **77** and **79**, that are smoothly obtained in high yield[76] are surmised to be useful photochromics in optical information processing.

8. Future Prospects

The examples described above provide a conceptual framework by which the chemistry of low-molecular weight carbohydrates, accessible in bulk quantities, can be moulded towards the practical elaboration of products with industrial application profiles. The synthetic potential, however, inherent in cheap, bulk-scale available, enantiomerically pure carbohydrate raw materials is huge and is far from being exhausted.

Thus, particularly in view of the comparatively few reaction sequences meeting process chemistry demands, there is an urgent necessity to further develop, practical, large-scale adaptable *reaction channels* from sugars to versatile building blocks – a task that can successfully be achieved only if the present chemical methodology is utilized to its fullest and the increasingly emerging biotechnological procedures as well. All of this, unambiguously, points towards broad-scale, practicality-oriented basic research to be performed not only in academic institutions, but also in industrial laboratories, most effectively, of course, if both cooperate closely. In short, the challenges, at least in outline form, are clear. The capacity to develop vibrant and inciteful collaborations between academic and industrial institutions on a European level is likely to emerge as one of the new frontiers of the utilization of the organic raw materials, which Nature offers us on an annual basis.

Acknowledgements. We would like to express our sincere appreciation to those colleagues that, over the years, have been members of our "Nachwachsende Rohstoffe"-Team, and whose contributions are documented in the references: Manfred BREHM, Susanne HAHN, Pan JARGLIS, Eisuke KAJI, Franz KLINGLER, Roger KLIMESCH, Peter KÖHLER, Uwe KREIS, Karlheinz NEFF, and Sabine SCHWIDETZKY, neé WEPREK. Our thanks are also due to the *Deutsche Forschungsgemeinschaft*, the *Fonds der Chemischen Industrie*, the *Bundesministerium für Forschung und Technik, Bonn,* and the *Südzucker AG, Mannheim/Ochsenfurt*, for substantial support of these investigations.

References

1. F. O. Licht: Estimation of World Sugar Production 1989/90. *Zuckerind.* **115** (1990) 408.

2. C. E. James, L. Hough, R. Khan: Sucrose and its Derivatives. *Progr. Chem. Org. Natl. Products* **55** (1989) 117-184.

3. W. A. Roelfsema, B. F. M. Kuster, H. Pluim, M. Verhage: Lactose and Derivatives. *Ullmann's Encyclop. Ind. Chem.*, 5th Edition, **A15** (1990) 107-114.

4. G. Tegge: Maltose und Maltotriose. *Ullmanns's Encyclop. Ind. Chem.*, 4th Edition, **24** (1983) 770-772.

5. L. A. W. Thelwall: Developments in the Chemistry and Chemical Modification of Lactose. *Developments in Dairy Chemistry, 3* (P. F. Fox, Ed.), Elsevier Appl. Science Publ., London, 1985, pp. 35-67.

6. R. Khan: The Chemistry of Maltose. *Adv. Carbohydr. Chem.* **39** (1981) 213-278.

7. R. Weidenhagen, S. Lorenz: Palatinose, ein neues bakterielles Umwandlungsprodukt der Saccharose. *Angew. Chem.* **69** (1967) 641; *Z. Zuckerind.* **7** (1957) 533-534; *Chem. Abstr.* **52** (1958) 6823h. – R. Weidenhagen, S. Lorenz (Süddeutsche Zucker AG): Verfahren zur Herstellung von Palatinose. *Ger. Offen.* 1,049,800 (1957); *Chem. Abstr.* **55** (1961) 2030b.

8. F. H. Stodola, H. I. Koepsell, E. S. Sharpe: A New Disaccharide Produced by *Leuconostoc mesenteroides.* *J. Am. Chem. Soc.* **78** (1956) 2514-2518.

9. W. Crueger, L. Draht, M. Munir (Bayer AG): Continuous Fermentation with Simultaneous Conversion of Sucrose to Isomaltulose. *Ger. Offen.* 2,741,197 (1977) and 2,806,216 (1978); *Chem. Abstr.* **90** (1979) 184932t and **91** (1979) 156068t.

10. C. Bucke, P. S. J. Cheetham (Tate & Lyle): Isomaltulose. *Brit. Pat. Appl.* 79/38,563 (1979); *Eur. Pat. Appl.* 28,900 (1981); *Chem. Abstr.* **95** (1981) 95468.

11. J. Shimizu, K. Susuki, Y. Nakajima (Mitsui Sugar Co.): Palatinose Production Using Immobilized α-Glucosyltransferase. *Jap. Pat. Appl.* 80/113 982 (1980); *Chem. Abstr.* **97** (1982) 53994m.

12. C. Kutzbach, G. Schmidt-Kastner, H. Schutt (Bayer AG): Immobilized Sucrose-mutase and its Use for the Production of Isomaltulose. *Ger. Offen.* DE 3,038,218 (1980); *Eur. Pat.* EP 49,801 (1982); *Chem. Abstr.* **97** (1982) 90444c.

13. M. Munir (Süddeutsche Zucker AG): Isomaltulose Production Using Immobilized Bacterial Cells. *Ger. Offen.* DE 3,038,219 (1980) and 3,213,107 (1982); *Chem. Abstr.* **97** (1982) 4680x and **100** (1984) 66611q.

14. D. Schwengers, H. Benecke (Pfeifer & Langen): Sweeteners and its Use. *Ger. Offen.* DE 3,446,380 (1984); *Chem. Abstr.* **105** (1986)77815p.

15. F. Reiff: Vitamin C. *Ullmann's Encyclop. Ind. Chem.*, 4th Edit., **23** (1983) 685-692.

16. O. Raunhardt: Xylit. *Ullmann's Encyclop. Ind. Chem.*, 4th Edit., **24** (1983) 777-778.

17. F. Reiff: Sorbitol and Mannitol. *Ullmann's Encyclop. Ind. Chem.*, 4th Edit., **24** (1983) 772-777.

18. G. Schulz: Gluconsäure. *Ullmann's Encyclop. Ind. Chem.*, 4th Edit., **24** (1983) 783-787.

19. J. Jurczak, S. Pikul, T. Bauer: (*R*)- und (*S*)-Isopropylidene-glyceraldehyde in Stereoselective Organic Synthesis. *Tetrahedron* **42** (1986) 447-488.

20. S. Hanessian: Total Synthesis of Natural Products. Pergamon Press, Oxford, 1983.

21. For Useful Preparative Procedures see: *Methods Carbohydr. Chem.* **2** (1963) 318-325, 326-328, 405-408, 411-414, 427-430.

22. B. Schneider, F. W. Lichtenthaler, G. Steinle, H. Schiweck: Distribution of Furanoid and Pyranoid Tautomers of D-Fructose in Water, Dimethyl Sulfoxide and Pyridine. *Liebigs Ann. Chem.* **1985**, 2443-2453.

23. R. F. Brady, Jr.: Synthesis of the Isomeric Di-*O*-isopropylidene-β-D-fructopyranoses. *Carbohydr. Res.* **15** (1971) 35-40.

24. S. Hahn: Ungesättigte Synthesebausteine aus D-Fructose. *Doctoral Dissertation*, Techn. Hochschule Darmstadt, 1989.

25. P. Brigl, R. Schinle: Über Benzoyl- und Benzal-Derivate der Fructose. *Ber. Dtsch. Chem. Ges.* **66** (1933) 325-330.

26. R. K. Ness, H. G. Fletcher, Jr.: Crystalline Tetrabenzoyl-β-D-fructopyranosyl Bromide and its Reduction. *J. Am. Chem. Soc.* **75** (1953) 2619-2623.

27. D. H. Brauns: Crystalline Chlorotetraacetyl-fructose and Related Derivatives. *J. Am. Chem. Soc.* **42** (1920) 1846-1854.

28. P. Brigl, R. Schinle: Tetrabenzoyl-fructofuranose. *Ber. Dtsch. Chem. Ges.* **67** (1934) 127-130; J. W. van Cleve, *Methods Carbohydr. Chem.* **2** (1963) 237.

29. H. E. van Dam, A. P. G. Kieboom, H. van Bekkum: The Conversion of Fructose and Glucose in Acidic Media. Formation of Hydroxymethylfurfural. *Starch* **38** (1986) 95-101.

30. M. J. Anatal, Jr., W. S. L. Mok, G. N. Richards: Mechanism of Formation of 5-Hydroxymethyl-2-furaldehyde from D-Fructose and Sucrose. *Carbohydr. Res.* **199** (1990) 91-109.

31. H. Schiweck, K. Rapp, M. Vogel: Utilization of Sucrose as an Industrial Bulk Chemical. *Chem. Ind. (London)* **1988**, 228-234.

32. B. J. Luberoff: Sucrochemistry, An Outsider's View. *Sucrochemistry* (J. L. Hickson, Ed.), *ACS Symp. Series No. 41*, American Chemical Society, Washington, 1977, p. 359 ff.

33. F. W. Lichtenthaler, S. Nishiyama, T. Weimer: 2,3-Dihydropyranones with Contiguous Chiral Centers. *Liebigs Ann. Chem.* **1989**, 1163-1170.

34. F. W. Lichtenthaler, S. Rönninger, P. Jarglis: An Expedient Approach to Pyranoid Ene and Enol Lactones by BF_3-Catalyzed Peroxidation of Glycal Esters. *Liebigs Ann. Chem.* **1989**, 1153-1161.

35. I. J. Borowitz, G. Gopnis: The Synthesis and Subsequent Oxidation of Tetrahydrochroman. A New Lactone Synthesis. *Tetrahedron Lett.* **1964**, 1151-1155.

36. D. N. Kirk, J. M. Wiles: Competing Reactions in the Peroxy-acid Oxidation of 3-Alkoxy-steroidal 3,5-Dienes. *J. Chem. Soc., Chem. Commun.* **1970**, 1015-1016.

37. F. Gonzales, S. Lesage, A. S. Perlin: Catalysis by Mercuric Ion of Reactions of Glycals with Water. *Carbohydr. Res.* **42** (1975) 267-274.

38. D. Uemura, Y. Toya, I. Watanabe, Y. Hirata: Palythazines from *Palythoa tuberculosa*. *Chem. Lett.* **1979**, 1481-1483.

39. J. H. Cardellina II, R. L. Hendrickson, K. P. Manfredi, S. A. Strobel, J. Clardy: Bissetone, a Unique Antimicrobial Pyranone from the Gorgonian *Briareum polyanthes*. *Tetrahedron Lett.* **28** (1987) 727-730.

40. P. Jarglis, F. W. Lichtenthaler: Stereospecific Synthesis of (*S,S*)-Palythazine from D-Glucose. *Angew. Chem.* **94** (1982) 140; *Angew. Chem., Int. Ed. Engl.* **21** (1982) 141.

41. M. Brehm, W. G. Dauben, P. Köhler, F. W. Lichtenthaler: Proof of the (*S,S*)-Configuration of (-)-Bissetone by Synthesis from D-Glucose. *Angew. Chem.* **99** (1987) 1318; *Angew. Chem., Int. Ed. Engl.* **26** (1987) 1271.

42. P. Jarglis, F. W. Lichtenthaler: Selective Deacylation of Enol Esters with Hydroxylamine. *Tetrahedron Lett.* **21** (1980) 1425-1428.

244   *F. W. Lichtenthaler, et al.*

43. B. P. Moore, W. V. Brown: Chemistry of the Metasternal Gland Secretion of the Eucalypt Longicorn. *Aust. J. Chem.* **29** (1976) 1365-1368.

44. T. Ishida, K. Wada: A Steroidal Hydroxylase Inhibitor from *Diplodia pinea*: Diplodialide A. *J. Chem. Soc., Chem. Commun.* **1975**, 209-210.

45. F. D. Klingler: Sechs Kohlenstoff-Syntheseblocks mit (5*R*)- und (5*S*)-Hydroxyl Gruppe – Darstellung aus Monosacchariden und Verwendung zur Synthese enantiomerenreiner Diplodialide. *Doctoral Dissertation*, Techn. Hochschule Darmstadt, 1985.

46. K. H. Neff: Darstellung enantiomerenreiner Sechs-Kohlenstoff-Bausteine des 5-Hydroxyhexanal-Typs und ihre Anwendung zur Synthese von Parasorbid und (-)-Phoracantolid I und J. *Doctoral Dissertation*, Techn. Hochschule Darmstadt, 1988.

47. F. W. Lichtenthaler, F. D. Klingler, P. Jarglis: Simple Synthesis of *S*-Parasorbic Acid and Other (5*S*)-Hydroxy Six-carbon Synthons from L-Rhamnose. *Carbohydr. Res.* **132** (1984) C1-C5.

48. W. Rosenbrook, Jr.: Chemistry of Spectinomycin. *J. Antibiotics* **32** (1979) S211-S227, and literature cited there.

49. F. Büschweiler, K. Stöckel, T. Reichstein: *Calotropis* Glycoside, vermutliche Teilstruktur. *Helv. Chim. Acta* **52** (1969) 2276-2303.

50. F. W. Lichtenthaler, U. Kraska: Preparation and Reactions Benzoylated 4-Deoxy-hex-3-enosuloses. *Carbohydr. Res.* **58** (1977) 363-377.

51. F. W. Lichtenthaler, A. Löhe, E. Cuny: An Approach to Actinospectose-type 2,3-Diketo Sugars. *Liebigs Ann. Chem.* **1983**, 1973-1985.

52. F. W. Lichtenthaler, S. Nishiyama, P. Köhler, H. J. Lindner: Anomeric Stereocontrol in Addition Reactions to Hexose-derived Dihydropyranones. *Carbohydr. Res.* **136** (1985) 13-26.

53. W. G. Dauben, B. A. Kowalczyk, F. W. Lichtenthaler: Diels-Alder Reactions of Pyranoid Enolone Esters with Cyclopentadiene. *J. Org. Chem.* **55** (1990) 2391-2398.

54. F. W. Lichtenthaler: Utilization of Enantiomerically Pure Building Blocks from Sugars in Natural Product Synthesis. *New Aspects in Organic Chemistry* (Z. Yoshida, Ed.), Verlag Chemie, Weinheim, 1989, p. 365 ff.

55. E. Cuny, U. Kreis, F. W. Lichtenthaler: Utilization of 2-Oxoglucosyl Bromides for the Efficient Generation of Pyranodioxan Type Natural Products. *5th Europ. Carbohydr. Symp.*, Prague, August 1989, Abstract A-1.

56. E. Cuny, U. Kreis, F. W. Lichtenthaler: Glycosidation and Cycloacetalization of
     2-Oxohexosyl Halides with 1,2-Diols; an Efficient Route to Spectinomycin and
     Calotropis Cardenolides. *XVth internat. Carbohydr. Symp.*, Yokohama, August
     1990, Abstract AO17.

57. F. W. Lichtenthaler, H. Leinert, T. Suami: Eine einfache Synthese von Streptamin
     und Actinamin. *Chem. Ber.* **100** (1967) 2383-2388.

58. S. Hanessian, R. Roy: Chemistry of Spectinomycin; Total Synthesis,
     Stereocontrolled Rearrangement, and Analogs. *Can. J. Chem.* **63** (1985) 163-172.

59. D. R. White, R. D. Birkenmeyer, R. C. Thomas, S. A. Mizsatz, V. H. Wiley: The
     Stereospecific Synthesis of Spectinomycin. *Tetrahedron Lett.* **1979**, 2737-2740.

60. E. Cuny, F. W. Lichtenthaler, unpublished results.

61. R. R. Schmidt: New Methods for the Synthesis of Glycosides and Oligosaccharides.
     Angew. Chem. **98** (1986) 213-226; *Angew. Chem., Int. Ed. Engl.* **25** (1986) 212-225.

62. F. W. Lichtenthaler, E. Cuny, S. Weprek: A Facile, Efficient Synthesis of Acylated
     Glycosyl Bromides. *Angew. Chem.* **95** (1983) 906. *Angew. Chem., Int. Ed. Engl.* **61**
     (1983) 891.

63. F. W. Lichtenthaler, E. Kaji, S. Weprek: Disaccharide-derived 2-Oxo- and
     2-Oximino-glycosyl Bromides. *J. Org. Chem.* **50** (1985) 3505-3515.

64. F. W. Lichtenthaler, E. Kaji: A Facile Access to Hetero-trisaccharides with Central
     β-D-Mannose and β-D-Mannosamine Units. *Liebigs Ann. Chem.* **1985**, 1659-1668.

65. E. Kaji, F. W. Lichtenthaler, T. Nishino, A. Yamane, S. Zen: Practical Synthesis of
     Immunologically Relevant β-Glycosides of 2-Acetamido-2-deoxy-D-mannose. *Bull.
     Chem. Soc. Jpn.* **61** (1988) 1291-1297.

66. E. Kaji, F. W. Lichtenthaler, Y. Osa, S. Zen: Synthesis of N-Acetyl-lactosaminyl
     Donors from Lactose-derived Disaccharide Building Blocks. Paper presented at
     XVth Internat. Carbohydr. Symp., Yokohama, August 1990, Abstract AO85.

67. R. G. Klimesch: Darstellung α(1→6)- und α(1→5)-verknüpfter Glucosylalditole
     aus Isomaltulose. *Doctoral Dissertation*, Techn. Hochschule Darmstadt, 1983.

68. F. W. Lichtenthaler, R. G. Klimesch (Süddeutsche Zucker AG): Derivate und
     Reduktionsprodukte der D-Glucopyranosyl-α(1→5)-D-arabinonsäure. *Ger. Offen.*
     DE 3,248,404 (1982); *U.S. Pat.* 4,618,675 (1983); *Chem. Abstr.* **102** (1985) 7034x.

69. H. Röger, H. Puke, M. Kunz: Untersuchungen zur oxidativen Spaltung von
     reduzierenden Disacchariden. *Zuckerind.* **115** (1990) 174-181.

70. J. Dubourg, P. Naffa: Oxidation of Reducing Hexoses by Oxygen in Alkaline
     Medium. *Bull. Soc. Chim. Fr.* **1959**, 1353-1362.

71. W. Mauch, F. El Aama: 5-*O*-α-Glucopyranosyl-(1→5)-D-arabinofuranose, a Bacterioenzymatically Synthesized Disaccharide. *Z. Zuckerind.* **26** (1976) 21-25 and 711-713; *Chem. Abstr.* **84** (1976) 166575h and **86** (1977) 102209t.

72. F. W. Lichtenthaler, S. Rönninger: α-D-Glucosyl-D-fructoses: Distribution of Furanoid and Pyranoid Tautomers in Water, Dimethyl Sulphoxide, and Pyridine. *J. Chem. Soc., Perkin Trans. 2*, **1990**, 1489-1497.

73. A. Gaset, L. Rigal, G. Paillasa, J.-P. Salome, G. Fleche (Roquette Fr.): Procédé de Fabrication du 5-Hydroxymethylfurfural. *Fr. Demande* FR 2.551.754 (1983); *Chem. Abstr.* **103** (1985) 21571t.

74. F. W. Lichtenthaler, D. Martin, T. A. Weber, H. Schiweck (Südzucker AG): 5-(α-D-Glucopyranosyloxy)-furan-2-carboxaldehyd, Darstellung und Folgereaktionen. *Ger. Pat. Appl.* 1989.

75. T. A. Weber: Untersuchungen zum chemischen Verhalten der Isomaltulose. *Doctoral Dissertation*, Techn. Hochschule Darmstadt, 1990.

76. D. Martin: Darstellung von 5-(α-D-Glucosyloxymethyl)-furfural aus Isomaltulose. *Diploma Thesis*, Techn. Hochschule Darmstadt, 1990.

Hydrogen Fluoride,
Solvent and Reagent for Carbohydrate Conversion Technology

Jacques Defaye[†] and Christian Pedersen[‡]

[†] Département de Recherche Fondamentale, Centre d'Etudes Nucléaires de
Grenoble, and Centre National de la Recherche Scientifique,
F-38041 Grenoble, France,
[‡] Department of Organic Chemistry, Technical University of Denmark,
DK-2800 Lyngby, Denmark.

Summary. Hydrogen fluoride is an excellent solvent for carbohydrates combining good hydrogen acceptor abilities, low degradative properties with an exceedingly high stabilization capacity for ions. Especially this last feature may be exploited for glycosidations proceeding via reactive glycosyl fluoride intermediates or the corresponding carboxonium ion pairs. Using pyridinium poly(hydrogen fluoride) as solvent and reagent, even fatty alkyl and thioalkyl glycosides can be prepared, exhibiting detergent properties.

Another reaction pathway, utilizing the activating property of HF, is the formation of sugar anhydrides. Thus, 2-hexuloses and 2-hexulosans are quantitatively converted into mixtures of hexulose dianhydrides, their structure being determined mainly by anomeric and steric effects. Alditols may be transformed into their corresponding dianhydrides, affording isomannide and isosorbide, respectively. Treatment of D-glucono- and D-mannono-lactone with hydrogen fluoride catalyzed by formic or acetic acid yields the respective 3,6-anhydro-lactone.

Introduction

Current economic considerations suggest that decisive developments in the use of carbohydrates as organic raw materials will be, at least in the intermediate term, dependant on the availability of appropriate technologies, based on reasonably low-cost reagents and a limited number of steps, for conversion of agricultural products into speciality chemicals having reasonable added value.

Hydrogen fluoride (HF) is an inexpensive reagent that has found extensive industrial use. It is unique among mineral acids in the possibilities that it offers to the chemist. It melts at -83°C and boils at 19.5°C, enabling reactions to be carried out in HF-solution over a wide range of temperatures and to be terminated by evaporation, a point of economic interest as it permits facile recovery of the reagent. It is a strong protonating agent, with a Hammett acidity function of -10.2, which explains its solvent properties for oxygen-containing compounds. Many such compounds dissolve and are not appreciably degraded, even though HF is also a strong dehydrating agent, probably because extensive protonation prevents dehydration in HF-solution. This point is of importance for carbohydrates, which are readily dehydrated to furan compounds by mineral acids. Furthermore, the extensive hydrogen-bonding network of HF makes it a highly polar solvent, having a dielectric constant $\varepsilon = 84$, similar to that of water, suggesting that tautomeric equilibria based on this parameter might not differ greatly between water to HF, a further point of importance for carbohydrates.

More than a century ago, Gore[1] reported that "sugar of milk", cane sugar, and gum arabic dissolved freely and quickly in HF, and Ville and Mestrezat,[2] as early as in 1910, indicated the potential of HF for the hydrolysis of cellulose with the further objective of its transformation into ethanol. Fredenhagen and Cadenbach[3] found later that polysaccharides could be extracted from wood by this reagent, and patents were issued in the meantime by the I.G. Farbenindustrie[4] for a process of continuous extraction of wood using HF, an objective recently completed by the Hoechst company[5] in Germany. HF has furthermore attracted attention in recent years for its versatility in the cleavage of glycosidic linkages and an account of its application to the structural analysis of polysaccharides and glycoconjugates has appeared recently.[6] The increasing interest in the specific conversion of carbohydrates into substances having added value, using HF and HF-based reagents, is therefore not unexpected. The present contribution, which covers reactions at both anomeric and non-anomeric positions, sets out to summarize some results in the field.

Hydrogen Fluoride-induced Glycosidations

As HF may be expected to enhance the concentration of oxocarbenium ions at the anomeric position, glycoside formation is a common aspect of the reactivity of carbohydrates in HF, and alcohols of low molecular weight react readily, even at temperatures as low as -20°C, to form alkyl glycopyranosides.

Studies of the reaction kinetics by ^{13}C-NMR spectroscopy suggests a mechanism involving the formation of a glycosyl fluoride, presumably in equilibrium with the corresponding cyclic oxocarbenium ion or its ion pair (scheme 1, Pathway A). The reaction thus appears different from the classical Fischer acid-catalyzed glycosidation process, whereby the kinetically favoured glycofuranosides are formed initially. The latter reaction scheme is also in effect when an alcohol containing hydrogen fluoride is used as the glycosylating agent (scheme 1, Pathway B). Advantage has been taken of this reactivity for the preparation of methyl D-mannopyranosides, which are obtained in almost quantitative yield in an α,β-ratio of about 6 : 4 via Pathway A, whereas conditions of Fischer glycosidation yield, under mild conditions, complex mixture of furanosides and pyranosides.[7] A process has been claimed for the conversion of carbohydrates and polysaccharides, and specifically D-glucose, cellulose, and starch, into alkyl glycosides (anomeric ratio α,β ~ 7 : 3) in HF under economical conditions that involve ambient temperature and almost complete recovery of the reagents.[8]

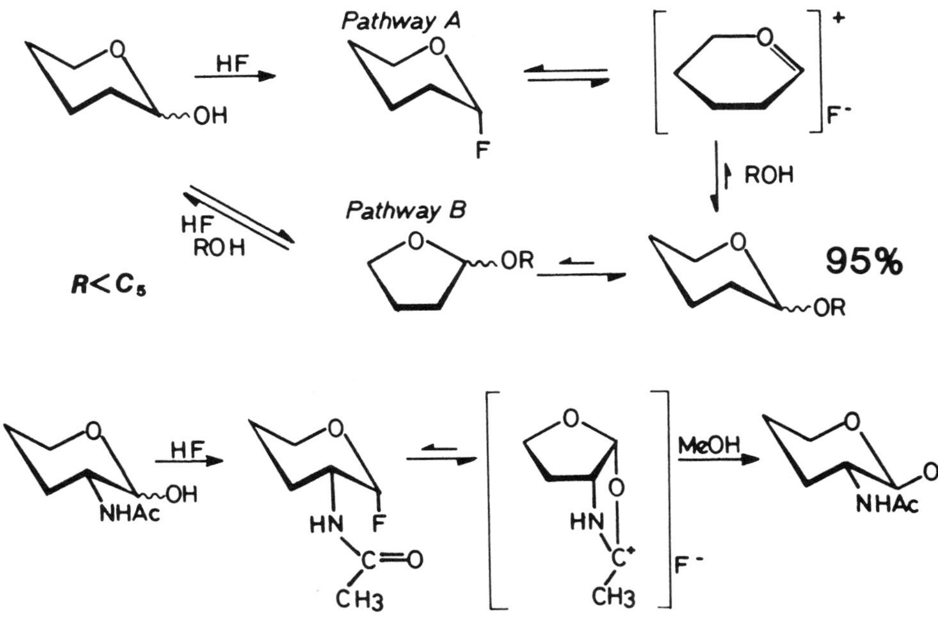

Scheme 1. Reaction of carbohydrates with alcohols in HF

Participation reactions may be of importance, as encountered with 2-acetamido sugars, as the β-anomers, resulting from a transient glycofuranosyl-oxazolinium ion detectable by ^{13}C-NMR, are formed[9] exclusively (scheme 1, bottom). Limitations in glycosidation reactions in HF are the concentration of solute, being critical if the formation of oligosaccharide reversion products is to be avoided, and the aglycon chain length, which may not exceed five carbons for aliphatic alcohols. This latter factor is probably related to the low solubility of the alcohol in HF and the formation of positively charged micelles, which decreases the nucleophilicity of the protonated species.

Glycosides and 1-thioglycosides having long-chain fatty alkyl aglycons, of interest as amphiphiles and nonionic detergents, have nevertheless[10-12] been obtained by use of an HF-based reagent, pyridinium poly(hydrogen fluoride). This stable 7 : 3 HF – pyridine complex was originally introduced by Olah[13] as a fluorinating agent and applied by Noyori[14] and Szarek[15] for the preparation of glycosyl fluorides from partially acylated or benzylated sugars. It has now been found to be an efficient glycosylation agent, effecting the smooth conversion of unprotected sugars into glycosides and 1-thioglycosides in the presence of a stoichiometric proportion of an alcohol or a thiol.

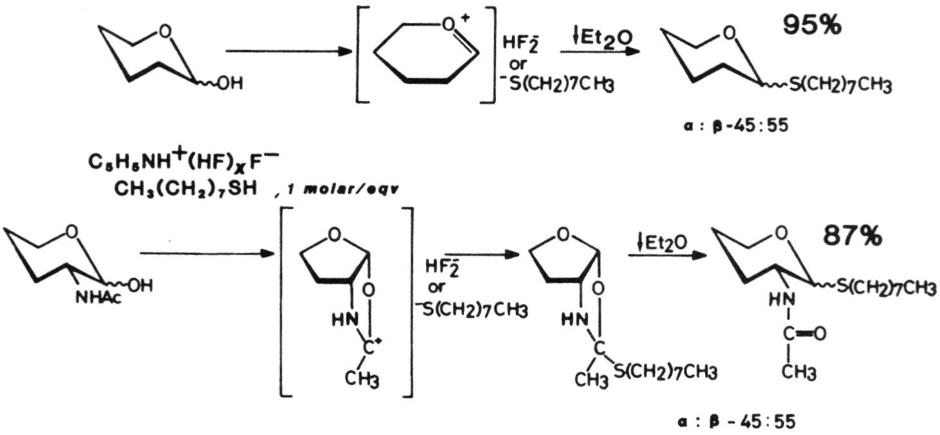

Scheme 2. Reaction of carbohydrates with long chain alkanethiols in pyridinium poly(hydrogen fluoride)

Although of general application, the method is particularly suited for the preparation of fatty alkyl 1-thioglycosides, which are obtained in almost quantitative yield (scheme 2, top) with an α,β-ratio of about 45 to 55 for heptyl or octyl 1-thio-D-glucopyranosides. Surprisingly, a similar anomeric mixture was obtained with 2-acylamido sugars, and this result has been explained on the basis of an intermediate thio-oxazoline, which opens during the work-up procedure in water (scheme 2, bottom).

The possibility of preparing long-chain alkyl 1-thio-oligosaccharides using this procedure is particularly attractive. Thus, reaction of 1-octanethiol with either cellobiose, lactose, or maltose resulted in the formation of the corresponding octyl 1-thiodisaccharides with an α,β anomeric ratio of about 1 : 1 (scheme 3). Concomitant hydrolysis of the interglycosidic linkage, which would have been expected either in HF or under Fischer glycosidation conditions, does not occur.

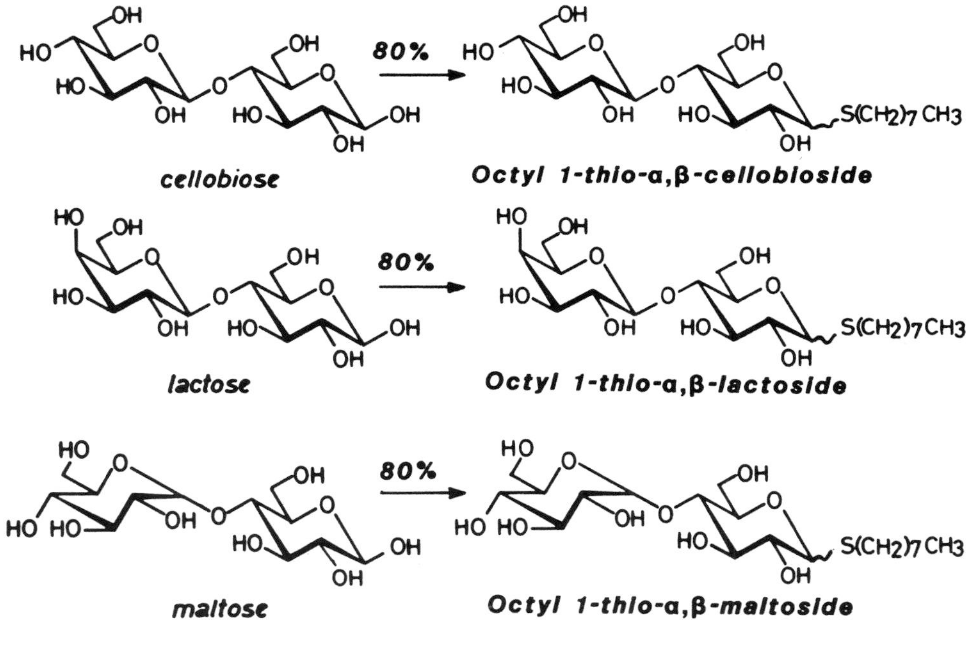

Scheme 3. Pyridinium poly(hydrogen fluoride)-mediated synthesis of fatty alkyl 1-thiodisaccharides.

The amount of the thiol reagent used in the HF – pyridine thioglycosidation is critical for the outcome of the reaction as the formation of a dithioacetal is a competing reaction when a stoichiometric excess is used.

The *critical micellization concentration* (CMC) is defined as the limit of detergent concentration above which micelles are formed and constitutes a good indication of detergency properties. The values observed for these glycosides are not very different from those found for pure β-anomers (Table 1). They appear quite uniform for fatty alkyl 1-thiodisaccharides and appreciably higher for alkyl glucosides, reflecting variations in the balance between the hydrophobic tail and the hydrophilic carbohydrate moiety of the molecule.

Table 1 : Comparative CMC values for alkyl 1-thioglycosides prepared by the HF – pyridine glycosylation process and data from the literature.

1-Thioglycoside	C.M.C. (mM)
Heptyl 1-thio-α,β-D-glucopyranoside	4
Heptyl 1-thio-β-D-glucopyranoside[16]	30
Octyl 1-thio-α,β-D-glucopyranoside	3.1
Octyl 1-thio-β-D-glucopyranoside[16]	9
Octyl 1-thio-α,β-cellobioside	0.2
Octyl 1-thio-α,β-lactoside	0.2
Octyl 1-thio-α,β-maltoside	0.2

The potential of this class of detergents, now readily available in one step from the free sugar, has been evaluated for the extraction of cytochrome P-450$_{17\alpha}$, an enzymic protein located in bovine adrenal microsomes, which is of biotechnological interest because of its of involvement in the hydroxylation of steroids.

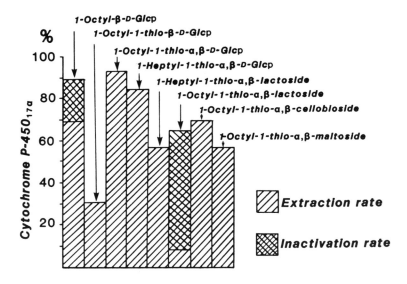

Fig. 1: Comparative extraction rates of cytochrome P-450$_{17\alpha}$ from bovine adrenal microsomes with various 1-alkylthio nonionic detergents.

Comparisons were made with commercial samples of octyl β-D-glucopyranoside and its 1-thio analog that had been prepared by the classical approach from the acetohalogeno sugar[16,17] precursor. Compounds prepared in one step by the HF – pyridine procedure show an excellent efficiency and an absence of inactivation of the enzyme, except for the octyl 1-thiolactoside (cf. Fig. 1).

A further point of interest of this simple technique for the preparation of fatty alkyl glycosides and 1-thioglycosides is that it makes a wide variety of nonionic detergents readily available. As Fig. 1 shows, there may be appreciable differences in extraction rates from one detergent to another, a point well known to membrane biochemists. The interest in having such a convenient technique of access to a large variety of detergents is thus evident.

Autocondensation ("reversion") is another aspect of interest in relation to the anomeric reactivity of carbohydrates. The propensity for reversion in a monosaccharide is closely related to its concentration and its ability to develop a positive charge at the anomeric position. As may be expected from the ionizing properties of HF, reversion is an important aspect of the reactivity of carbohydrates in this solvent. A highly branched oligo- and poly-saccharide mixture of average DP about 10 - 15 is thus obtained by slow evaporation of solutions of cellulose or starch in HF.[18] From ^{13}C-NMR and glc./ms methylation-analysis studies, the branching is determined as being α-(1→6) > α-(1→2) > α-(1→3) predominantly.

Interest in such products lies in their possible use as food additives and bulking agents, as shown for close analogs prepared by heat treatment of a starch-citric acid mixture and marketed under the name of "Polydextrose" by Pfizer Corp. in the U.S.A..[19] The enzymic efficiency of intestinal "brush-border" sucrase-isomaltase with this branched glucan, as expressed by the ratio V_m/K_m (~ 24), is about one-third of the value found for sucrose, which is indicative of a probable decreased rate for intestinal absorption for this compound.

Dianhydrides of Hexuloses

As may be anticipated from their tertiary structure at the anomeric position, 2-hexuloses and polysaccharides thereof are quite reactive in HF, and are immediately and quantitatively converted into mixtures of hexulose dianhydrides. Six main products have been characterized from D-fructose,[20] D- or L-sorbose,[21] and inulin.[20] In every instance the anomeric[22] and exoanomeric[23] effects, as well as conflicting influences that result from the known ring and conformational stabilities of 2-hexuloses in solution, appear to be the main factor governing the thermodynamic stability of these tricyclic spiro-dioxane systems.[24] The reaction of racemic sorbose in HF – sulfur dioxide, which yields almost quantitatively the symmetrical α-D/α-L-sorbopyranose 1,2':2,1'-dianhydride (scheme 4), a mixed dianhydride in which the 1,4-dioxane ring adopts a rigid chair conformation and both pyranose rings are configured in such a way as to conform with the most favorable anomeric and exoanomeric effect. This is a striking example of the importance of electronic factors in the formation of hexulose dianhydrides:[24]

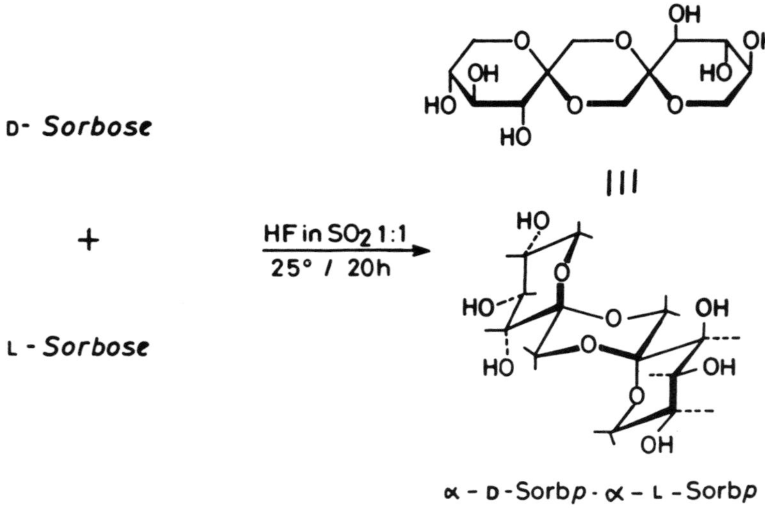

D- *Sorbose*

+

L - *Sorbose*

HF in SO$_2$ 1:1
25° / 20h

III

α – D -Sorb*p* · α – L -Sorb*p*

Scheme 4.   Reaction of racemic sorbose in HF.

From kinetic studies with D-fructose in HF – SO$_2$ solution at low temperatures, it appears that di-β-D-fructofuranose 1,2':2,1'-dianhydride is formed initially and rearranges quickly into the conformationally more stable α,β- isomer and then is converted mainly into the thermodynamically favored α-D-fructofuranose β-D-fructopyranose 1,2':2,1'-dianhydride, where such relative energy factors as the anomeric and exoanomeric effects, ring strain, and steric effects are most effectively balanced (scheme 5).

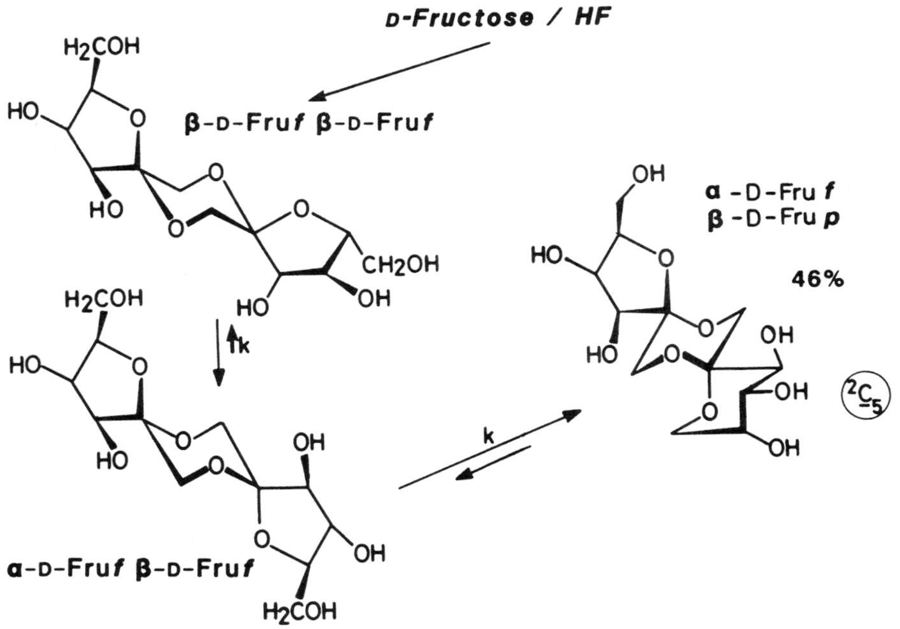

Scheme 5. Steric and electronic effects in the formation of D-fructose dianhydrides.

Apart from their theoretical interest as further experimental evidence for the exoanomeric effect, these results indicate that selectivity may be exercised with HF or HF-containing reagents for the preparation of hexulose dianhydrides. Several difructose dianhydrides have been claimed in recent patents[25] as valuable low-calorie additives in food and for preventing dental caries. It may furthermore be expected that such oxygenated, rigid, and relatively thermostable structures may be of interest in polymer chemistry.

With this objective in mind, α-D-fructofuranose β-D-fructofuranose 1,2':2,1'-dianhydride, which is about half as sweet as sucrose, was prepared[26,27] in excellent yield from acetylated inulin by taking advantage of the stability of acyloxy substituents under the reaction conditions (HF diluted in liquid SO_2) (cf. scheme 6). As well, the isomeric α-L-sorbofuranose β-L-sorbofuranose 1,2':2,1'-dianhydride, along with the 2,1':3,2' di-α-L-dianhydride, have been obtained by use of temporary acetal protecting groups (scheme 7):

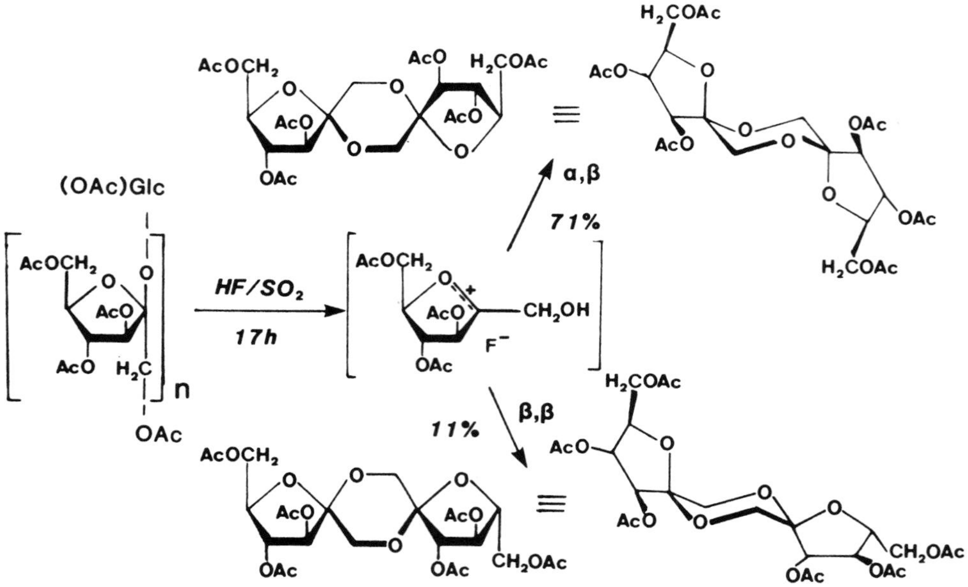

Scheme 6. Ester protecting groups in the HF-catalyzed synthesis of α-D-fructofuranose β-D-fructofuranose 1,2':2,1'-dianhydride from inulin peracetate.

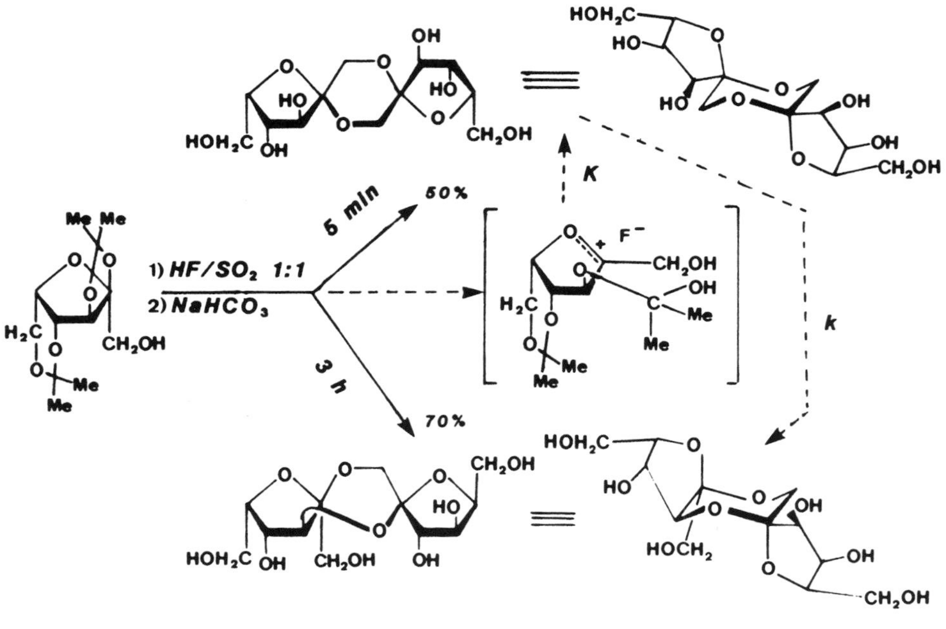

Scheme 7. Acetal protecting groups in the HF-catalyzed synthesis of di-L-sorbofuranose dianhydrides.

Difructofuranose dianhydrides have furthermore been prepared from sucrose, which reacts quantitatively in HF, leading to glucosylated difructose dianhydrides and thus opening up new opportunities for the transformation of sucrose into low-calorie food additives or polymeric materials:[28]

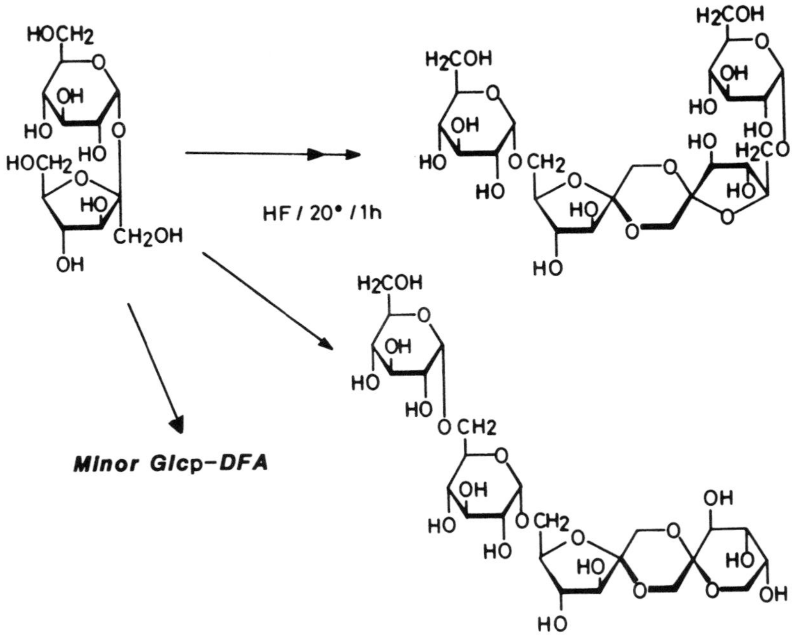

Scheme 8. Reaction of sucrose in HF.

Internal Anhydride Formation of Polyols

Glycosyl fluorides, useful as carbohydrate synthons[29-31] and as glycosyl donors in enzymic synthesis involving glycosyl transferases,[32] are readily formed in HF[9,20,21,33,34] (cf. scheme 1). However their isolation from HF-solutions presents technical difficulties, unless electron-withdrawing groups are present vicinal to the anomeric fluoride.[29-35] Thus, although preparation of α-D-glucopyranosyl fluoride by fluorolysis of starch has been claimed,[36] its isolation is hampered by the simultaneous formation of reversion oligosaccharides. However, 2-amino-2-deoxy-α-D-glucopyranosyl fluoride hydrofluoride and the corresponding glucosaminyl oligosaccharides are readily obtained by fluorolysis of chitosan.[37]

The reactions of carbohydrates at nonanomeric positions are another promising aspect of the reactivity of carbohydrates in HF. It is already known that carboxylic acids in HF are protonated, leading to the formation of acylium ions, which readily react with alcohols to form esters. Such reactivity is also found with carboxylic acid halides and anhydrides, and a common intermediate in such reactions being an acyloxonium ion readily detectable by ^{13}C-NMR[38](see scheme 9). Isomerizations at the central chiral carbon of *cis-trans*-1,2,3-triol triesters has been shown to occur in HF and advantage of this reactivity has been taken for the preparation of rare cyclitols.[39] Extension of the acylation concept has been recently applied to the preparation of carbohydrate fatty esters of interest as surfactants, fat substitutes, and thickeners.[40]

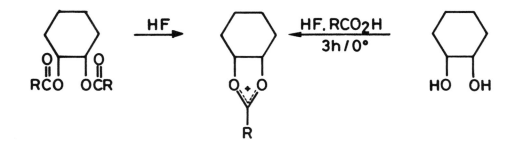

Scheme 9. Reactions of carboxylic acids with *cis*-diols and of *cis*-diol esters in HF.

Interestingly, in our hands, when D-glucitol and D-mannitol were treated with 1 molar equivalent of acetic or formic acid in HF, internal anhydride formation was found to occur exclusively. By using appropriate conditions, 1,4:3,6-dianhydro-D-glucitol ("isosorbide") and 1,4:3,6-dianhydro-D-mannitol ("isomannide") were obtained in excellent to almost quantitative yields, at room temperature[41] (cf. scheme 10).

The reactions, which were monitored by use of ^{13}C-NMR spectroscopy,[42] have been shown to involve dioxolenium ion intermediates. The subsequent formation of anhydrides by intramolecular opening of these electrophilic species takes place rapidly with the formoxonium ions issued from the formic acid catalysis, resulting in the formation of internal dianhydrides.[42] When acetic acid is used as catalyst, the reaction is slower and 3,6-anhydro-D-glucitol as well as 1,4-anhydro-D-mannitol may be prepared from D-glucitol and D-mannitol respectively in excellent yields, especially when two molar equivalents of acetic acid are used. It is noteworthy that, although the dianhydrides, "isosorbide" and "isomannide" are prepared industrially by dehydration of D-glucitol and D-mannitol using aqueous sulfuric acid or sulfonic ion-exchange resins at high temperature, the foregoing monoanhydrides cannot be obtained from such reactions.

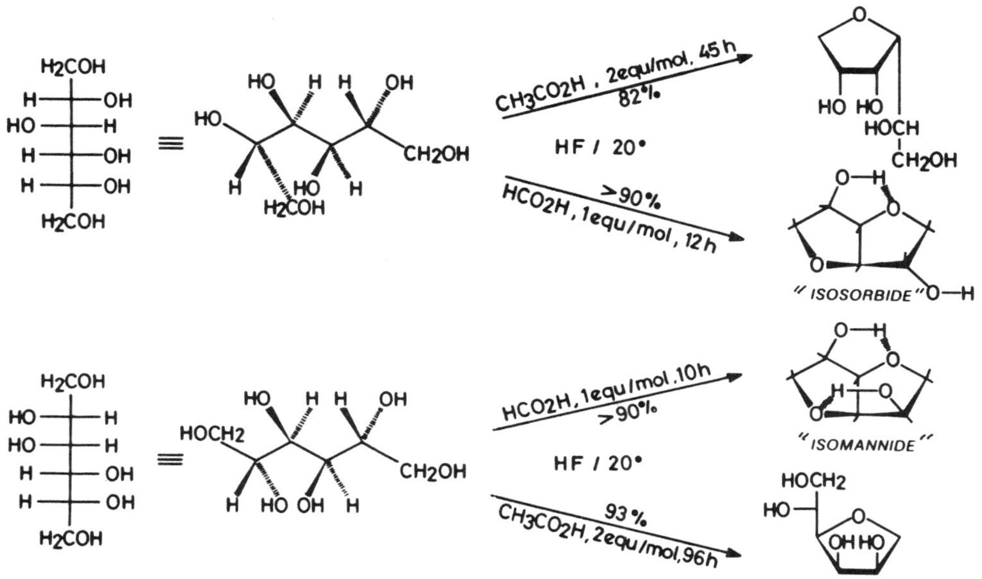

Scheme 10. HF – carboxylic acid-catalyzed synthesis of anhydropolyols from polyols.

"Isosorbide", "isomannide", and the corresponding monoanhydrides are valuable building blocks for polymer chemistry if they can be prepared economically, and they are widely used in form of their nitrate esters in the pharmaceutical industry and (as fatty acid esters) in the area of surfactants today.[43]

This reaction concept has been extended to the preparation of alkyl 3,6-anhydro-D-glucofuranosides from D-glucose and such glucans as cellulose and starch (cf. scheme 11), and to the preparation of 2-acylamido-3,6-anhydro-2-deoxy-D-glucose and also for the corresponding alkyl β-D-glycofuranosides from 2-acetamido-2-deoxy-D-glucose and chitin.[42]

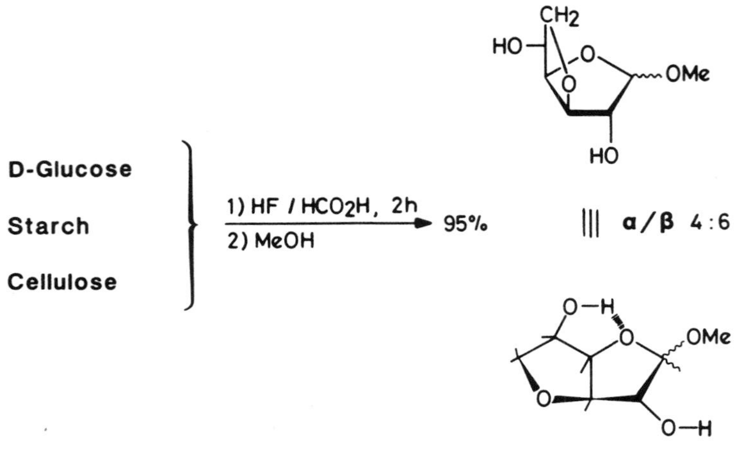

Scheme 11. HF – carboxylic acid-catalyzed synthesis of methyl 3,6-anhydro-
D-glucofuranoside from D-glucose and glucans.

3,6-Anhydro-D-mannono-1,4-lactone and 3,6-anhydro-D-glucono-1,4-lactone,
which may find applications as metal chelating agents, are similarly prepared by
treatment of the corresponding lactones in HF in the presence of stoichiometric
amounts of formic acid:[41-43]

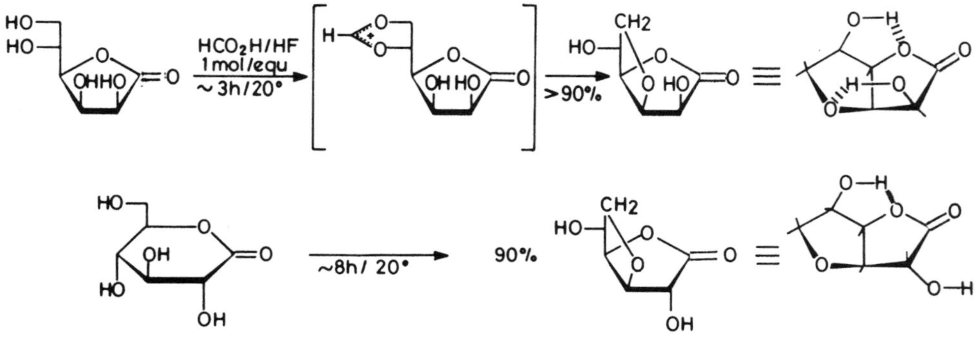

Scheme 12. HF – carboxylic acid-catalyzed synthesis of 3,6-anhydrohexonolactones.

Conclusions

From the preceding examples, which are not intended to cover all current applications, it is clear that hydrogen fluoride has an enormous potential as a solvent and reagent for the transformation of carbohydrate raw materials in industry.

The main advantage over classical reactions of carbohydrates, which involve charge displacements at the anomeric and nonanomeric positions and frequently use multi-step sequences, elevated temperatures, and costly reagents, is that they can be carried out in HF under energetically favoured conditions and, most often, with recovery of the solvent-reagent.

Interestingly, a measure of selectivity may be exercised in reactions with HF, and such reaction conditions as temperature, or the use of convenient removable protecting groups, may enhance this selectivity without need for more-sophisticated chemistry.

References

1. G. O. Gore: On Hydrofluoric Acid. *J. Chem. Soc.* **22** (1869) 368-406.

2. J. Ville, W. Mestrezat: Sur l'Hydrolyse Fluorhydrique de la Cellulose.
    C. R. Acad. Sci. **150** (1910) 783-784.

3. K. Fredenhagen, G. Cadenbach: Der Abbau der Cellulose durch Fluorwasserstoff und ein neues Verfahren der Holzverzuckerung durch hochkonzentrierten Fluorwasserstoff. *Angew. Chem.* **46** (1933) 113-124.

4. G. Pfleiderer, E. Koch (I.G. Farbenindustrie A.G.): Verfahren zur Behandlung fester oder flüssiger Stoffe mit Gasen oder Dämpfen. *Ger. Pat.* 585 318 (1933);
    Chem. Abstr. **28** (1934) 1426^4.

5. a) R. Erckel, R. Franz, M. Schlingmann (Hoechst A.G.): Recovery of Water-soluble Saccharide from Cellulose-containing Material. *Ger. Offen.* DE 3,040,850 (1982); *Chem. Abstr.* **96** (1982) 219565d.

    b) R. Erckel, R. Franz, R. Woernle, T. Riehm (Hoechst A.G.): Digestion of Cellulose-containing Material with Gaseous Hydrogen Fluoride. *Ger Offen.* DE 3,142,214, 3,142,215, and 3,142,216 (1981); *Chem. Abstr.* **98** (1983) 217530f, 217529n, and 217531g.

6. A. Knirel, E. V. Vinogradov: Application of Anhydrous Hydrogen Fluoride for the Structural Analysis of Polysaccharides. *Adv. Carbohydr. Chem. Biochem.* **47** (1989) 167-202.

7. D. F. Mowery, Jr.: Methyl D-Mannosides. *Methods Carbohydr. Chem.* **2** (1963) 328-331.

8. J. Defaye, E. Wong, C. Pedersen (Beghin-Say S.A.): Glycosides. *Fr. Demande* FR 2,567,891 (1986); *Wo. Pat.* 86-00,906 (1986); *Chem. Abstr.* **105** (1986) 227221g.

9. J. Defaye, A. Gadelle, C. Pedersen: Hydrogen Fluoride Catalyzed Formation of Glycosides. Preparation of Methyl 2-Acetamido-2-deoxy-β-D-gluco- and β-D-galacto-pyranosides, and of β-(1→6)-Linked 2-Acetamido-2-deoxy-D-gluco- and D-galacto-pyranosyl Oligosaccharides. *Carbohydr. Res.* **186** (1989) 177-188.

10. J. Defaye, A. Gadelle, C. Pedersen: Nouveau Procédé de Préparation d'Alkyl-1-thioglycosides et d'Alkyl-glycosides; Nouveaux Mélanges d'anomères Obtenus par ce Procédé et leur Application comme Détergents non Ioniques. *Fr. Demande* FR 10,301 (1989).

11. A. Gadelle, G. Defaye, J. Defaye, C. Pedersen: Hydrogen Fluoride Mediated Synthesis of Fatty Thioalkyl Glycosides and Oligoglycosides as Detergents for the Solubilisation of Membrane Proteins. *5th Eur. Symp. Carbohydr.* Prague 1989, Abstr. A-26.

12. J. Defaye, A. Gadelle, C. Pedersen: Pyridinium Poly(hydrogen fluoride), a Smooth and Efficient Glycosylation Agent. *15th Int. Carbohydr. Symp.*, Yokohama 1990, Abstr. A49, p. 85.

13. G. A. Olah, J. T. Welch, Y. D. Vankar, M. Nojima, I. Kerekes, J. A. Olah: Synthetic Methods and Reactions. 63. Pyridinium Poly(hydrogen fluoride) (30 % Pyridine / 70 % Hydrogen Fluoride): a Convenient Reagent for Organic Fluorination Reactions. *J. Org. Chem.* **44** (1979) 3872-3881.

14. M. Hayashi, S.-I. Hashimoto, R. Noyori: Simple Synthesis of Glycosyl Fluorides. *Chem. Lett.* **1984**, 1747-1750.

15. W. A. Szarek, G. Grynkiewicz, B. Doboszewski, G. W. Hay: The Synthesis of Glycosyl Fluorides using Pyridinium Poly(hydrogen Fluoride). *Chem. Lett.* **1984**, 1751-1754.

16. S. Saito, T. Tsuchiya: Synthesis of Alkyl-β-D-thioglucopyranosides, a Series of New Nonionic Detergents. *Chem. Pharm. Bull.* **33** (1985) 503-508.

17. S. Saito, T. Tsuchiya (Dojin Kagaku Kenkyusho K.K.): Preparation of Alkyl Thioglycosides. *Jpn. Kokai Tokkyo Koho* JP 61-07,288 (1986); *Chem. Abstr.* **105** (1986) 134292s.

18. A. Bouchu, J. Chedin, J. Defaye, E. Wong (Beghin-Say S.A.): Preparation of Branched Oligo- and Poly-saccharides, especially with Starch as Starting Material. *Fr. Demande* FR 2,597,872 (1987); *Wo. Pat.* 87-06,592 (1987); *Chem. Abstr.* **108** (1988) 149149m.

19. R. E. Smile: Sweeteners 4. Applications of Polydextrose. *Food Technol.* **1986**, 129-130.

20. J. Defaye, A. Gadelle, C. Pedersen: The Behaviour of D-Fructose and Inulin towards Anhydrous Hydrogen Fluoride. *Carbohydr. Res.* **136** (1985) 53-65.

21. J. Defaye, A. Gadelle, C. Pedersen: The Behaviour of L-Sorbose towards Anhydrous Hydrogen Fluoride. *Carbohydr. Res.* **152** (1986) 89-98.

22. R. U. Lemieux, S. Koto: The Conformational Properties of Glycosidic Linkages. *Tetrahedron* **30** (1974) 1933-1944.

23. a) R. U. Lemieux, A. A. Pavia, J. C. Martin, K. A. Watanabe: Solvation Effects on Conformational Equilibria. Studies Related to the Conformational Properties of 2-Methoxytetrahydropyran and Related Methyl Glycopyranosides. *Can. J. Chem.* **47** (1969) 4427-4439.

    b) R. U. Lemieux, S. Koto, D. Voisin: The exo-anomeric Effect. *Am. Chem. Soc. Symp. Ser.* **87** (1979) 17-29.

24. K. Bock, C. Pedersen, J. Defaye, A. Gadelle: Steric and Electronic Effects in the Formation of Dihexulose Dianhydrides. Reaction of Racemic Sorbose in Anhydrous Hydrogen Fluoride and a Facile Synthesis of D-Sorbose. *Carbohydr. Res.* (1991) in press.

25. a) J. Defaye, A. Gadelle, C. Pedersen (Beghin-Say S.A.): Cyclodehydration of Ketoses. *Fr. Demande* FR 2,550,535; *Wo. Pat* 85-00,814 (1985); *Chem. Abstr.* **104** (1986) 110115q.

    b) K. Haraguchi, M. Kishimoto, S. Kobayashi, K. Seki, K. Nagata, K. Pponbo, M. Kadoma, K. Kainuma (National Food Research Institute; Nippon Denpun Kogyo Co. Ltd): Preparation of Difructose Anhydride III by Fermentation of Inulin with *Arthrobacter globiformis*. *Jpn. Kokai Tokkyo Koho* JP 62-275,694 (1988); *Chem. Abstr.* **109** (1988) 21920j.

    c) K. Tamura, T. Kuramoto, S. Kitahata (Maruzen Chemical Co., Ltd): Manufacture of Di-D-fructofuranose 2,6':6,2'-Dianhydride from Levan or Phlein with *Pseudomonas*. *Jpn. Kokai Tokkyo Koho* JP 01-91,793 (1989); *Chem. Abstr.* **111** (1989) 76554q.

    d) S. Kobayashi, K. Seki, K. Haraguchi, M. Kishimoto, K. Nagata, K. Honbo, K. Kainuma, M. Kadoma (Norin Suisansho Shokuhin Sogo Kenkyusho Nippon Denpun Co., Ltd): Food Ingredients containing Difructose Dianhydrides. *Jpn. Kokai Tokkyo Koho* JP 63-269,962 (1988); *Chem. Abstr.* **111** (1989) 56162g.

    e) K. Tamura, T. Kuramoto, S. Kitahata (Maruzen Chemical Co., Ltd): Sweeteners containing Di-D-fructosylfuranose 2,6':6,2'-Dianhydride for Cavity Control. *Jpn. Kokai Tokkyo Koho* JP 63-214,160 (1988); *Chem. Abstr.* **110** (1989) 153057p.

26. J. Defaye, A. Gadelle, C. Pedersen, D. Lafont (Beghin-Say S.A.): Preparation of Dianhydrides of Fructofuranose and their Use as Sweetening Agents. *Eur. Pat. Appl.* EP 252,837 (1988); *Chem. Abstr.* **110** (1989) 75973t.

27. J. Defaye, A. Gadelle, C. Pedersen: Acetal and Ester Protecting-groups in the Hydrogen fluoride-catalyzed Synthesis of D-Fructose and L-Sorbose Difuranose Dianhydrides. *Carbohydr. Res.* **174** (1988) 323-329.

28. A. Bouchu, J. Chedin, J. Defaye, D. Lafont, E. Wong (Beghin-Say S.A.): Preparation of Branched Oligo- and Polyglycosides, especially from Sucrose. *Fr. Demande* FR 2,599,040; *Wo. Pat* 87-07,275 (1987); *Chem. Abstr.* **109** (1988) 95053a.

29. P. J. Card: Synthesis of Fluorinated Carbohydrates. *J. Carbohydr. Chem.* **4** (1985) 451-487.

30. a) K. C. Nicolaou, R. E. Dolle, D. P. Papahatjis, J. L. Randall: Practical Synthesis of Oligosaccharides, Partial Synthesis of Avermectin B_{1a}. *J. Am. Chem. Soc.* **106** (1984) 4189-4192.

    b) K. C. Nicolaou: New Strategies and Methods for the Synthesis of Complex Oligosaccharides. *15th Int. Carbohydr. Symp.* Yokohama (1990), Abstract P.L. 10, p. 11.

31. J. Thiem, W. Fritsche-Lang, M. Schlingmann, H. M. Mathias, M. Kreuzer (Hoechst A.G.): Preparation of Glycosides from Glycosyl Fluorides. *Ger. Offen.* DE 3,426,074 (1986); *Chem. Abstr.* **105** (1986) 191564f.

32. a) J. E. Hehre, T. Sawrai, C. F. Brewer, M. Nakano, T. Kanda: Trehalose: Stereocomplementary Hydrolytic and Glucosyl Transfer Reactions with α- and β-D-Glucosyl Fluoride. *Biochemistry* **21** (1982) 3090-3097.

    b) E. J. Hehre, K. Mizokami, S. Kitahata: Cyclodextrin Glycosyltransferase. Catalysis of Irreversible C-F Glycosylic Bond Cleavage and *de novo* Maltosidic Linkage Synthesis. *Denpun Kagaku* **30** (1983) 76-82.

33. J. Defaye, A. Gadelle, C. Pedersen: The Behaviour of Cellulose, Amylose, and β-D-Xylan towards Anhydrous Hydrogen Fluoride. *Carbohydr. Res.* **110** (1982) 217-227.

34. C. Bosso, J. Defaye, A. Domard, A. Gadelle, C. Pedersen: The Behaviour of Chitin towards Anhydrous Hydrogen Fluoride. Preparation of β-(1→4)-Linked 2-Acetamido-2-deoxy-D-glucopyranosyl Oligosaccharides. *Carbohydr. Res.* **156** (1986) 57-68.

35. A. A. E. Penglis: Fluorinated Carbohydrates. *Adv. Carbohydr. Chem. Biochem.* **38** (1981) 195-285.

36. R. Franz, H. M. Deger, M. Schlingmann (Hoechst A.G.): Glycosyl Fluorides. *Ger. Offen.* DE 3,432,565 (1986); *Chem. Abstr.* **106** (1987) 5368t.

37. J. Defaye, A. Gadelle, C. Pedersen (CEA-CNRS): Nouveaux Fluorures d'Aldosaminyle, leur Préparation et leur Utilisation pour la Fabrication d'Oligosaccharides Aminés liés (1→4). *Fr. Demande* FR 88,16647 (1988).

38. I. Lundt, C. Pedersen: Reaction of some Derivatives of *cis*-1,2-Cyclohexanediol and *cis*-1,2-Cyclopentanediol with Anhydrous Hydrogen Fluoride. *Acta Chem. Scand.* **26** (1972) 1938-1946.

39. H. Paulsen: Cyclic Acyloxonium Ions in Carbohydrate Chemistry. *Adv. Carbohydr. Chem. Biochem.* **26** (1971) 127-195.

40. H. M. Deger, W. Fritsche-Lange, A. Reng, M. Schlingmann, C. J. Lawson (Hoechst A.G.): Carbohydrate Fatty Acid Esters, a Method for their Preparation, and Emulsions for Cosmetic Applications containing them. *Ger. Offen.* DE 3,639,878 (1988); *Chem. Abstr.* **111** (1989) 134679x.

41. J. Defaye, C. Pedersen (Beghin-Say S.A.): Process for Preparing Anhydrides of Hexitols, Hexonolactones, Hexoses, and Hexosides. *PCT Int. Appl. Wo.* 89-00,162 (1989); *Chem. Abstr.* **111** (1989) 39828m.

42. J. Defaye, A. Gadelle, C. Pedersen: Acyloxonium Ions in the High-yielding Synthesis of Oxolanes from Alditols, Hexoses, and Hexonolactones catalyzed by Carboxylic Acids in Anhydrous Hydrogen Fluoride. *Carbohydr. Res.* **205** (1990) 191-202.

43. G. Flèche, M. Huchette: Isosorbide. Preparation, Properties, and Chemistry. *Starch/Staerke* **38** (1986) 26-30.

13

Selective Oxidation of D-Glucose : Chiral Intermediates for Industrial Utilization

Harald Röper

Cerestar Research and Development, Gruppo Ferruzzi,
B-1800 Vilvoorde, Belgium

Summary. D-Glucose (dextrose) is a building unit of starch, cellulose, sucrose, and lactose. For technical and economical reasons D-glucose is produced on large industrial scale by enzymatic hydrolysis of starch. Main use is for nutrition, as feedstock for sorbitol and HFCS (high fructose corn syrup) and as carbon source for industrial fermentations.

For broader industrial use of D-glucose, problems of selectivity, pH- and temperature instability in organic solvents have to be overcome. Selective chemical and biotechnical transformations, preferably in water, are necessary and favourable intrinsic properties like biodegradability, biocompatibility, and chirality have to be exploited.

This paper reviews existing knowledge and experience on the possibilities to oxidize selectively the different hydroxyl functions at positions C-1 to C-6 of unprotected D-glucopyranose in water, using biotechnical catalysts (microbial cells, enzymes) or heterogenous chemical catalysts.

Transformations to the following oxidized D-glucose derivatives will be covered: Glucosone (2-keto-D-glucose), 6-aldehydo-D-glucose, D-gluconic acid, D-glucuronic acid, 2-keto-L-gulonic acid, D-glucaric acid, 2-keto-D-gluconic acid, 5-keto-D-gluconic acid, 2,5-diketo-D-gluconic acid, L-ascorbic acid (vitamin C), and kojic acid.

Some of these products are already produced on industrial scale, others show interesting potential to be used as intermediates for different industries and applications areas.

1. Introduction

More than 5 million tons of starch are annually produced in the EEC from raw materials like maize, wheat, barley, and potatoes.

Starch, modified starches, starch hydrolysates, crystalline dextrose, and derivatives are used in the food and non-food sector, and as fermentation feedstock.[1]

Advanced products obtained from starch using biotechnical and chemical processes have been reviewed. They comprise glucosides and glucans, biopolymers, oxidized carbohydrates, hydrogenated carbohydrates (polyols, aminopolyols), organic acids, and chiral building blocks (L-erythrulose).[2,3]

In this paper the focus will be on D-glucose, the building unit of starch, to ask and answer the following questions:

1. What is known about selective oxidation (dehydrogenation) of hydroxyl groups in unprotected D-glucose at positions C1 - C6 (cf. Fig. 1), using heterogenic catalytic oxidaton or biotechnical oxidation with whole cells or enzymes, and water as solvent ?

2. What is the potential of these products to be used as intermediates or finished products in different industrial sectors ?

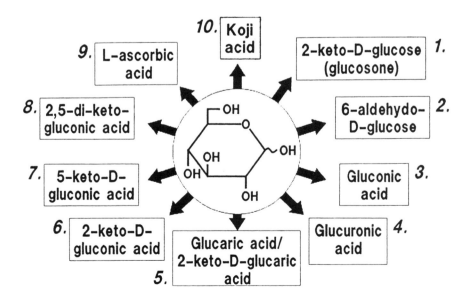

Fig. 1 Oxidation products of D-glucose via heterogeneous catalysis, enzymes, or fermentation

2. D-Glucosone (2-Keto-D-glucose)

2-Keto-D-glucose (1), which has also been designated D-glucosone, can be used as an intermediate in the production of D-fructose, D-mannitol, D-sorbitol, and 2-keto-D-gluconic acid. Chemical routes resulted in glucosone yields of < 50 % with many byproducts.

Microbial transformation of D-glucose utilizing the pyranose-2-oxidase in whole cells of *Aspergillus flavus, Saxidomus giganteous, Tridophycus flaccidum,* and *Oudemansiella mucida* resulted in low glucosone yields. Enzymatic conversion of D-glucose to D-glucosone has been performed using crude enzyme preparations from the red algae *Tridophycus flaccidum.*

The responsible enzyme, pyranose-2-oxidase (EC 1.1.3.10) has been purified and characterised.[4] Pure pyranose-2-oxidase enzyme can be obtained from the mycelium of *Coriolus versicolor* or *Lenzites betulinus*, free of glucosone-utilizing enzyme contaminants, e.g. pyranosone dehydratase. *Coriolus versicolor* NRRL 15152 pyranose-2-oxidase has a MW of 130 000 Daltons (2 subunits of 65 000 D) and covalently bound flavin. Optimum pH is 5. Pure pyranose-2-oxidase can also be obtained from *Polyporus obtusus,*[5,6] which, in the presence of oxygen and of catalase in buffered solution at 20 °C, converts D-glucose primarily into glucosone with very low byproduct formation.

A patent from Nabisco Brands Inc.[7] describes a screening procedure for glucose-2-oxidase producing microorganisms, based on the concominantly formed hydrogen peroxide, which is indicated by a colour-forming reagent. Using whole microbial cells for transformation, a membrane reactor is used to separate the glucosone formed from hydrogen peroxide. Glucosone is reduced to D-fructose, hydrogen peroxide is degraded by anions, cations, or enzymes (catalase, peroxidase).[8]

Under optimal conditions glucose-2-oxidase converts > 99 % of glucose into glucosone, which subsequently can be transformed into D-fructose using metal catalysts, e.g. Pd, Raney-Ni and hydrogen at elevated temperature and pressure, or homogenous $NaBH_4$ reduction:[9]

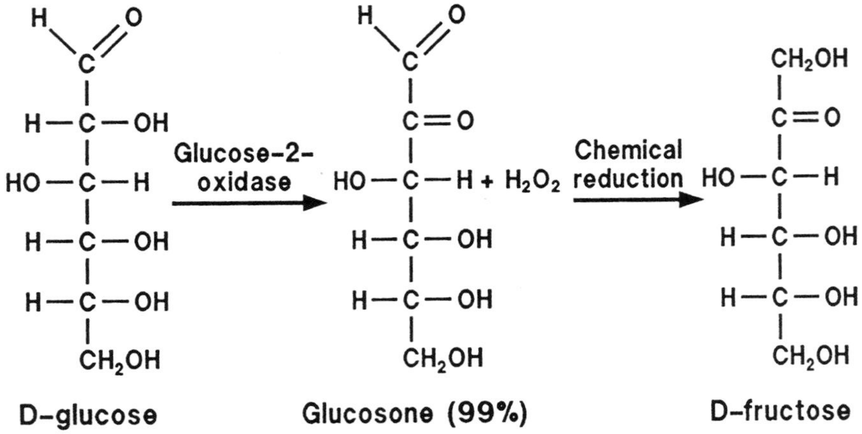

D-glucose   **Glucosone (99%)**   **D-fructose**

Glucose-2-oxidase from *Polyporus obtusus* or *Aspergillus oryzae* is used in batch or immobilized form in the presence of oxygen (pH 3 - 8, T: 15 - 66 °C) and a peroxide remover, e.g. catalase. H_2O_2 can also be reacted with propylene and halogenides in presence of enzymes to yield the halohydrin and epoxide:[9]

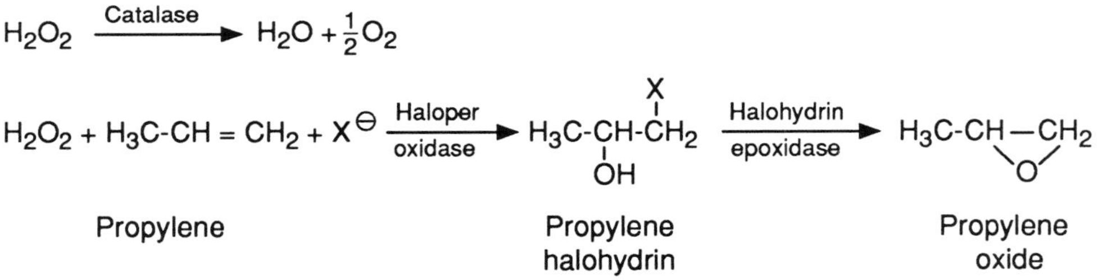

Propylene   **Propylene halohydrin**   **Propylene oxide**

(X$^{\ominus}$= Halogenid)

Glucosone is not produced on an industrial scale, but could be an interesting intermediate for catalytic reduction to D-fructose, for reductive amination to 1,2-diaminosorbitol / mannitol, and for the preparation of 4-deoxy-2,3-diketo-D-glucose using a pyranosone-dehydratase:

2-keto-D-glucose: potential applications

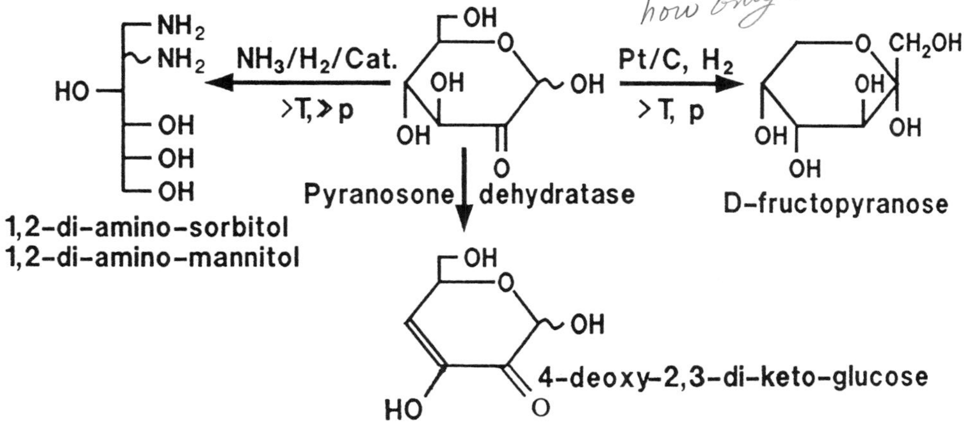

how only aldehyde ?

1,2-di-amino-sorbitol
1,2-di-amino-mannitol

NH$_3$/H$_2$/Cat.
>T,>p

Pyranosone dehydratase

Pt/C, H$_2$
>T, p

D-fructopyranose

4-deoxy-2,3-di-keto-glucose

3. 6-Aldehydo-D-glucose

The enzyme galactose-oxidase (EC 1.1.3.9.), in the presence of O_2, catalyzes the oxidation of D-galactose selectively in position 6 to 6-aldehydo-D-galactose and H_2O_2. The enzyme is neither electron-acceptor specific (O_2) nor electron-donor specific (D-galactose). Thus, it has the capability of oxidizing polyols and non-sugar alcohols to aldehydes:[10]

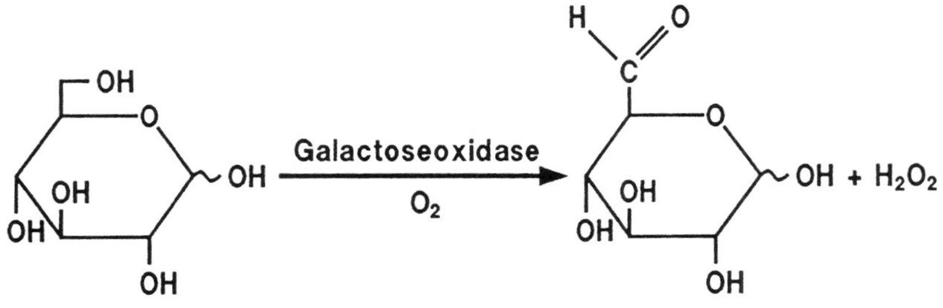

Galactoseoxidase
O_2

$+ H_2O_2$

Route only known for D-galactose

Galactose-oxidase oxidizes glycerol with absolute prochiral specificity to (*S*)-glyceraldehyde and oxidizes (*R*,*S*)-3-chloro-propane-1,2-diol enantio-specific to the corresponding (*S*)-aldehyde leaving the (*S*)-diol unaffected.[11] H_2O_2 has to be removed, e.g. by catalase reaction.

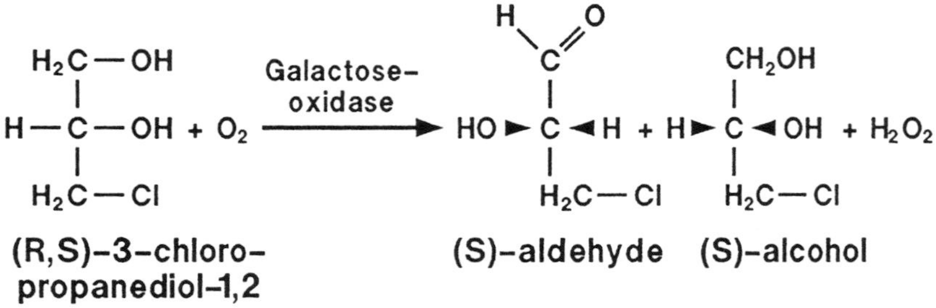

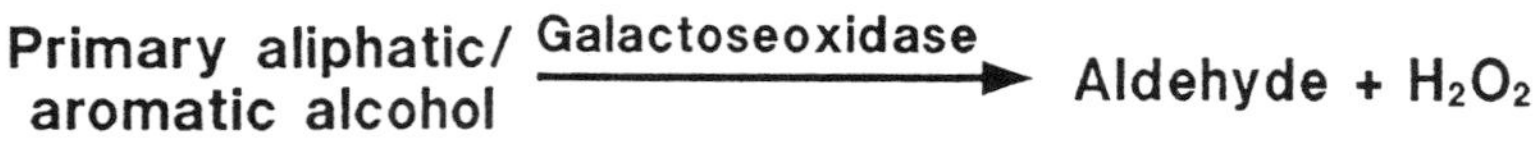

$$H_2C-OH$$
$$|$$
$$H-C-OH \quad + \; O_2 \quad \xrightarrow{\text{Galactoseoxidase}}$$
$$|$$
$$H_2C-OH$$

Glycerol (meso)

$$\underset{\text{H}}{\overset{\text{O}}{C}}$$
$$|$$
$$HO \blacktriangleright C \blacktriangleleft H \quad + \; H_2O_2$$
$$|$$
$$CH_2OH$$

(S)-glyceraldehyde

$$H_2C-OH$$
$$|$$
$$H-C-OH \; + \; O_2 \quad \xrightarrow{\begin{array}{c}\text{Galactose-}\\\text{oxidase}\end{array}}$$
$$|$$
$$H_2C-Cl$$

(R,S)-3-chloro-propanediol-1,2

$$\underset{\text{H}}{\overset{\text{O}}{C}} \qquad\qquad CH_2OH$$
$$| \qquad\qquad\qquad |$$
$$HO \blacktriangleright C \blacktriangleleft H \; + \; H \blacktriangleright C \blacktriangleleft OH \; + \; H_2O_2$$
$$| \qquad\qquad\qquad |$$
$$H_2C-Cl \qquad H_2C-Cl$$

(S)-aldehyde   (S)-alcohol

$$\text{Primary aliphatic/} \atop \text{aromatic alcohol} \quad \xrightarrow{\text{Galactoseoxidase}} \quad \text{Aldehyde} + H_2O_2$$

Advantages of this biocatalytic reaction in comparison to conventional chemical catalysis are as follows:

♦ alcohol oxidation stops at the oxidation stage of the aldehyde and does not proceed to the acid as in normal chemical oxidations

♦ the reaction is stereospecific / enantioselective.

Using galactose-oxidase, D-glucose most probably can be oxidized to 6-aldehydo-D-glucose, which shows potential as an intermediate for 1,6-diamino-sorbitol production via reductive amination:

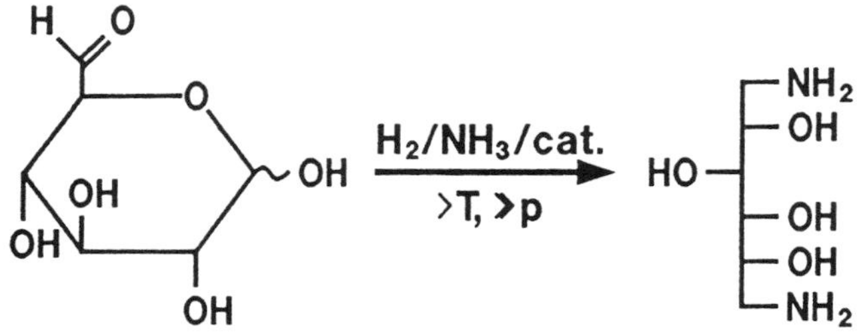

$$\xrightarrow[\text{>T, >p}]{\text{H}_2/\text{NH}_3/\text{cat.}}$$

$$\begin{array}{c} -NH_2 \\ -OH \\ HO- \\ -OH \\ -OH \\ -NH_2 \end{array}$$

4. D-Gluconic Acid

About 45.000 tons / year of D-gluconic acid are being produced worldwide by fermentative oxidation of D-glucose:

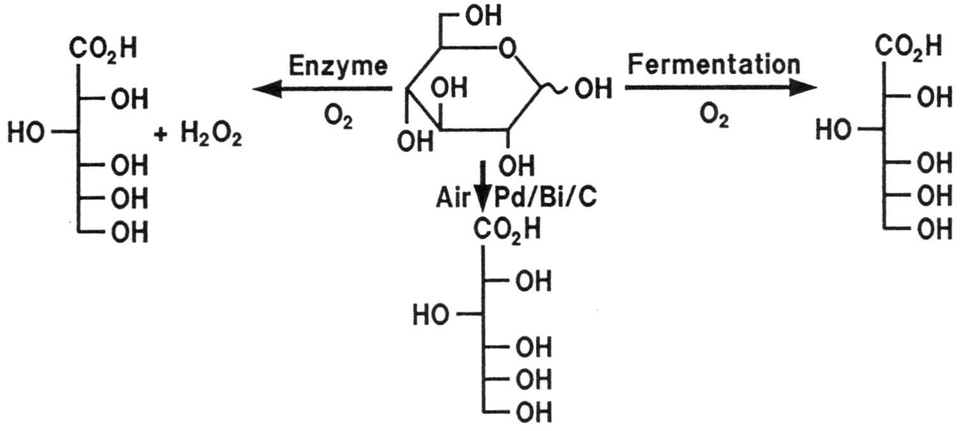

This oxidation of D-glucose in position 1 is achieved by microbial fermentation using *Acetobacter*, *Pseudomonas*, or *Penicillium* species. Conversion yields of > 97 % are obtained. Among the different gluconates, the sodium salt is the most important for application as metal chelating agent in conditions of extreme alkalinity, e.g. for bottle washing, cleaning, ferrous metal derusting etc. Also pharmaceutical applications and the use of glucono-δ-lactone as accidulant in food processing are important.

By using Pd / Bi / C as heterogenous catalyst and air (O_2) as hydrogen-acceptor, gluconate can be obtained from D-glucose in 99 % yield in very high purity[12] under alkaline conditions. The economics of this process are more favourable than with fermentation.

A patent by Roquette Frères[13] describes a similar process using Pd on charcoal / alumina / silica-alumina, $BaSO_4$, or TiO_2 as support with metals of Group IV, V, and VI (Sn, Pb, Sb, Bi, or Se) as promotors. Using a 5 % Pd / 3.5 % Bi / C catalyst, 98 % yield of D-gluconic acid was obtained after the 35th recycle.[13]

Composite catalysts are prepared by loading on inorganic supports as first catalysts Pt, Ru, Rh, or Pd; as second catalysts Sn, Bi or Sb and as third catalysts Se, Sn or Te at 0.1 - 20 % of the composite weight. These catalysts can be used to oxidize selectively the reducing end groups of starch hydrolysates and maltodextrins to oligobionic acids which show potential to be used as chelating agents, glass or metal cleaners, detergent builders, concrete admixtures, or as additives in food and pharmaceuticals. Maltose was catalytically oxidized to maltobionic acid in 99.5 % yield.[14]

Similar catalysts Pd, Pt, Rh, Os on inert supports like carbon, Al_2O_3, SiO_2, SiO_2-Al_2O_3, $BaSO_4$, TiO_2 containing promotors like Bi or Pb are used to oxidize starch hydrolysates (glucose and maltose syrups) and maltodextrins using O_2 (air) under alkaline conditions. High conversion yields, selectivity, and retention of interglucosidic bonds are reported.[15]

Co-immobilized coupled glucose oxidase / catalase on nylon mesh has been described for the enzymatic oxidation of glucose to gluconic acid: 3.2 g D-glucose (1.8×10^{-2} mmol) was dissolved in 400 ml water; glutardialdehyde treated coatings were added and the solution was stirred at room temperature for 24 h, keeping the pH between 6 - 8. Conversion was > 90 % based on consumed glucose, 4 g of gluconate was isolated.[16] Isolation, purification, and characterization of glucose oxidase from *Aspergillus niger* is described in detail.[17]

Performance of a rotating-disk reactor containing alginate beads with entrapped glucose oxidase / catalase was studied with regard to glucose / gluconic acid conversion at pH 5.6, 30 °C with 100 mmol feed of D-glucose, saturated with air in dependence of rotation speed, flow rate, submergence of the disk, and feed rate.[18]

A Boehringer patent describes a process for enzymatic oxidation of glucose with glucose-oxidase / catalase immobilized on hardened protein particles at temperatures < 10 °C and the freezing point of the solution. With 100 ml of commercial enzyme a total of 1 000 kg glucose was converted in 100 runs.[19] Another Boehringer patent describes a "Permeabilisiertes Schimmelpilzmycel mit intaktem immobilisiertem Glucoseoxidase-Katalase-System sowie dessen Herstellung und Anwendung".[20]

Linko et al. described the continuous production of free gluconic acid in a bioreactor using living cells of *Gluconobacter oxydans* immobilized on nylon at pH 2.5. Highest volumetric productivity with respect to glucose concentration was obtained at 175 g/l glucose, with about 120 g/l gluconic acid level. At maximum gluconic level, productivities of 12 - 15 g/l h at relatively high substrate feed rate of 0.166 l/l h and relatively low aeration rate of 0.5 l/l min were obtained which correspond to 80 % yield. Thereby, the formation of 2- and 5-keto-gluconic acid can be kept low by fermentation control.[21]

5. D-Glucuronic acid

D-Glucuronic acid (4), presently, is not produced on industrial scale. D-Glucose can be oxidized in position 6 by *Ustulina deusta*[22] or *Bacterium industrium var. hoshigaki*[23] under aerobic conditions:

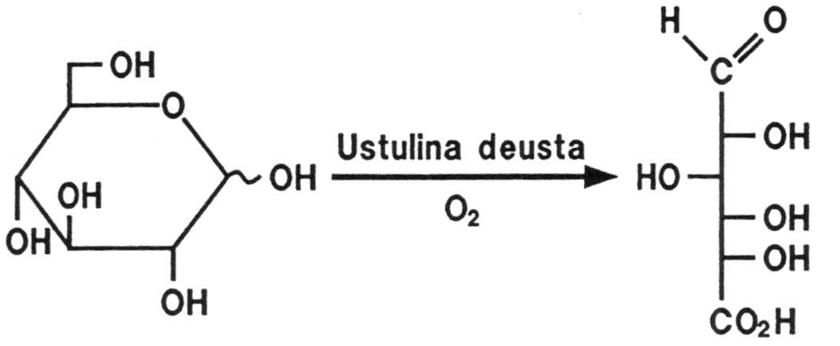

As of now, noble catalysts like Pt or Pd / C show very low selectivity for D-glucuronic acid starting with D-glucose.

Glucuronic acid is a building unit of hyaluronic acid. It can be obtained in low yields by hydrolysis of N_2O_4-oxidized starch and cellulose (polyglucuronic acid).[24,25] A microbial oxidation of cellulose to polyglucuronic acid using *Sporocytophaga myxococcoides* is also described in the literature.[26]

D-Glucuronic acid (4) could be used as starting material for L-gulonic acid, e.g. by selective chemical reduction of the aldehydo function. Reductive amination using ammonia, hydrogen, and a catalyst yields 6-amino-L-gulonic acid, an interesting building unit for fine chemicals, polymers and anionic detergents:

L-gulonic acid   **6-amino-L-gulonic acid**

L-Gulonic acid can be further transformed by gulonic acid dehydrogenase / NADP⁺ to 2-keto-L-gulonic acid, the precursor of vitamin C (L-ascorbic acid).

Starting with D-glucuronic acid, Kulbe et al.[27] describe an enzymatic membrane process with intrasequential cofactor regeneration via L-gulonic acid to 2-keto-L-gulonic acid. Enzymes are a hexonate dehydrogenase and gulonic acid dehydrogenase; cofactor is NADP⁺ / NADPH:[27]

D-glucuronic acid / vitamin C

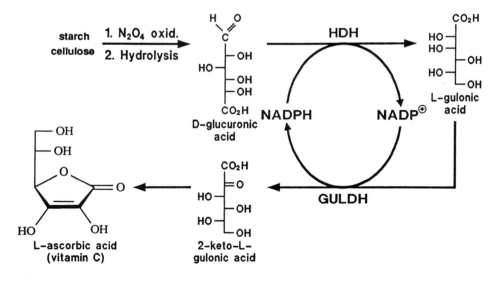

HDH = hexonate dehydrogenase   GULDH = gulonic acid dehydrogenase

This would be one possibility to shortcut the 8-step Reichstein synthesis for vitamin C:

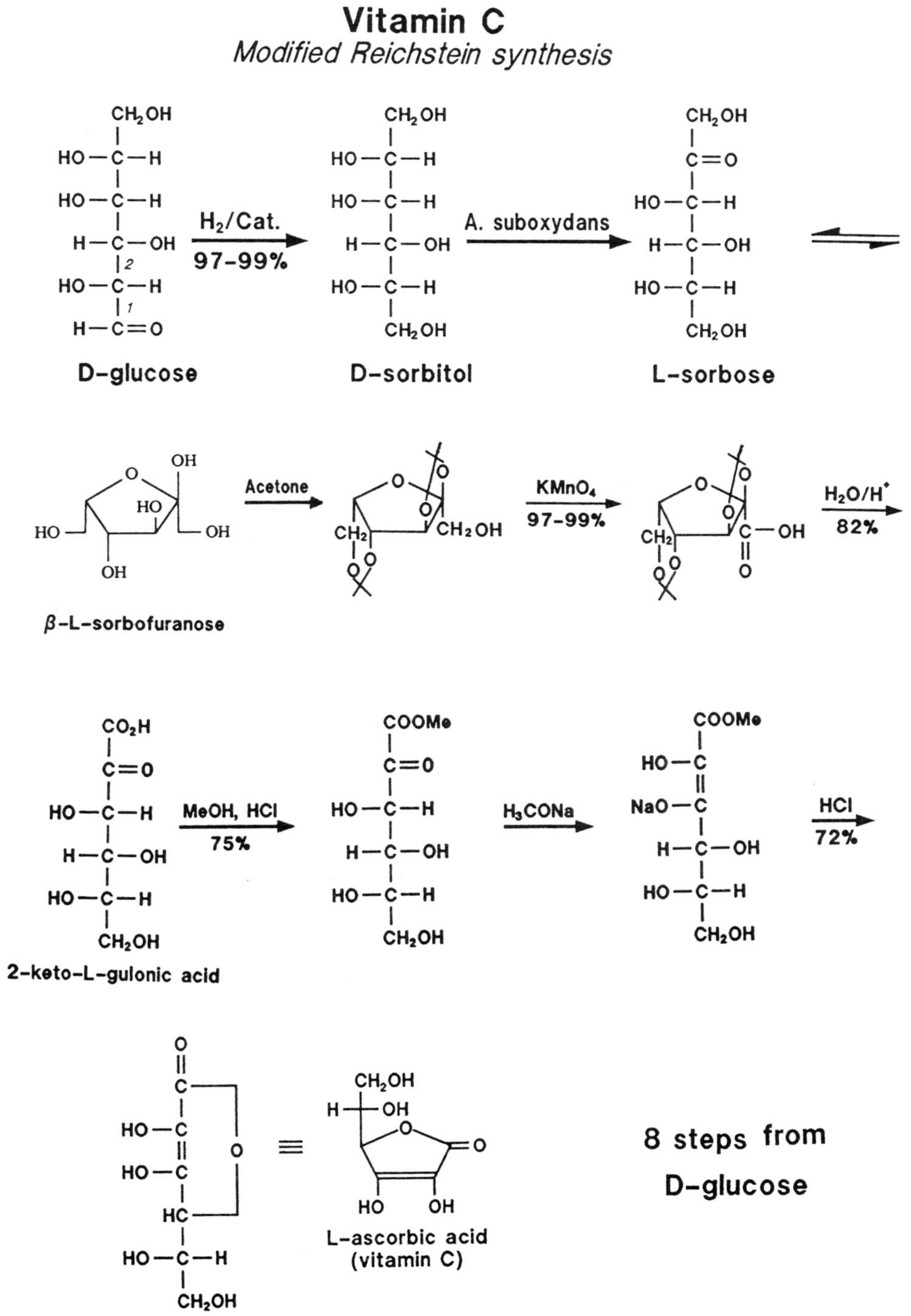

Vitamin C
Modified Reichstein synthesis

D-glucose →(H₂/Cat. 97–99%) **D-sorbitol** →(A. suboxydans) **L-sorbose** ⇌

β-L-sorbofuranose →(Acetone)→(KMnO₄ 97–99%)→(H₂O/H⁺ 82%)

2-keto-L-gulonic acid →(MeOH, HCl 75%)→(H₃CONa)→(NaO-C ... HCl 72%)

L-ascorbic acid (vitamin C)

8 steps from D-glucose

Using Pt / C, air, and NaHCO$_3$ as buffer, L-sorbose could be converted at 25 °C to 2-keto-L-gulonic acid with 60 % yield in 48 - 60 hours. The oxalic acid formed was precipitated as Ca-oxalate and the sodium-2-keto-L-gulonate was crystallized from the filtrate in 52 % yield.[28]

6. D-Glucaric acid

D-Glucaric acid can be obtained by oxidation of position 1 and 6 of D-glucose. Using HNO$_3$ as oxidizing agent, yields of about 30 % of crystalline potassium glucarate could be obtained:

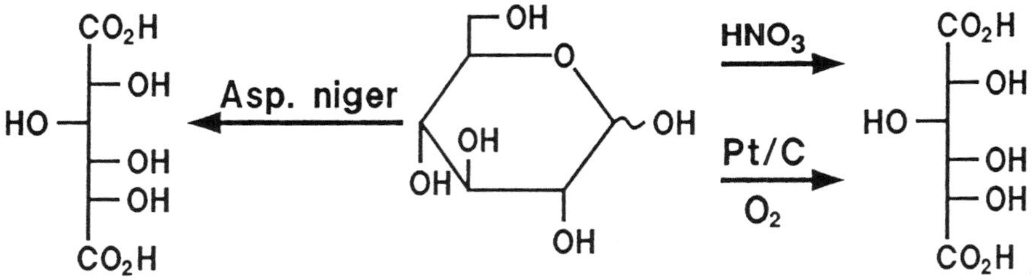

Using D-glucose or D-gluconic acid as the starting material, catalytic oxidation over Pt-on-carbon as catalyst yielded under optimized conditions about 40 % K-glucarate after crystallization. Byproducts are L-guluronic acid, 2-keto-D-gluconic acid, 5-keto-D-gluconic acid, glycolic acid, oxalic acid, and formic acid. Due to limited catalyst selectivity, an economic process for D-glucaric acid is not feasible yet.

An *Aspergillus niger* strain has been described converting D-glucose into D-glucarate (saccharic acid).[29]

Potassium glucarate with equimolar amounts of boric acid shows good Ca and Mg sequestering properties (151 mg Ca/g K-glucarate at 20 °C and 130 mg Ca/g K-glucarate at 60 °C). However complex stability at elevated washing temperatures is inferior compared to STPP. Thus the potential of glucarate to be used as builder in detergents is limited. However, it shows application potential as special sequestrant in cooling fluids and as an intermediate for special emulsifiers and polyesters. Catalytic reductive amination with ammonia under high pressure and temperature would yield 1,6-diamino-sorbitol, an interesting intermediate for polymers (PU; polyamides) and for biodegradable detergents:

Glucaric acid

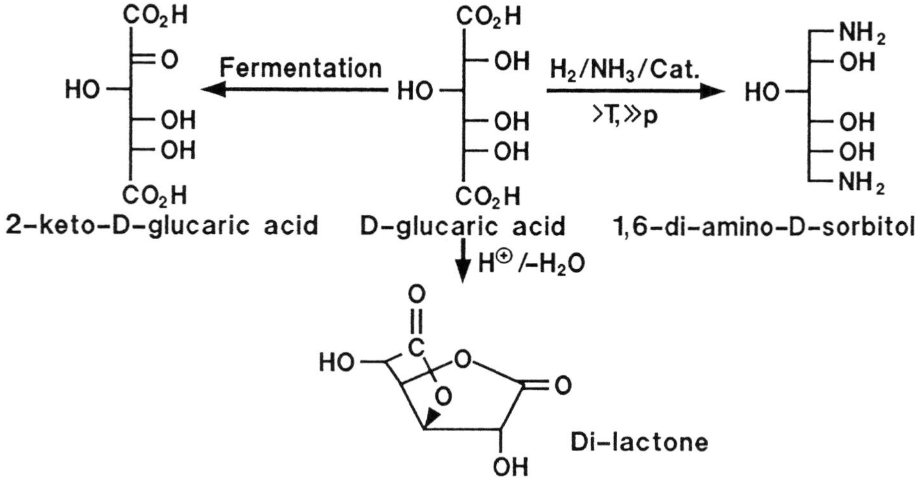

2-keto-D-glucaric acid   Fermentation   D-glucaric acid $H_2/NH_3/Cat.$ >T,≫p   1,6-di-amino-D-sorbitol

$H^{\oplus}/-H_2O$

Di-lactone

2-Keto-glucarate is being produced by contacting up to 20 % d.s. D-glucose or D-gluconic acid with a bacterium of the genus *Pseudogluconobacter*, e.g. *Pseudogluconobacter saccharoketogenes*.[30] Also *Pseudomonas suboxidans* transforms D-glucose, D-mannose, or D-fructose into 2-keto-D-glucarate (vide supra). This can be used as a calcium-enriched feedstuff additive, a detergent builder, cement plasticizer or as a reagent for metabolic studies.[31]

7. 2-Keto-D-gluconic acid

Using a Pt / Pb / C catalyst, D-glucose or D-gluconic acid are converted with 90 % selectivity to 2-keto-D-gluconic acid.[32] *Acetobacter, Pseudomonas, Xanthomonas, Serratia sp.*, and others convert D-gluconic acid (D-glucose) with up to 100 % yield into 2-keto-D-gluconic acid:[33]

2-keto-D-gluconic acid

Production methods: – microbial oxidation
– catalytic oxidation

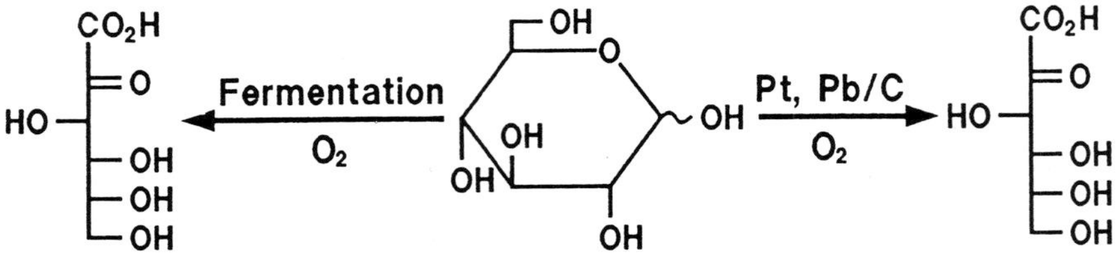

2-Keto-D-gluconic acid is precursor for the production of D-*arabino*-ascorbic acid (iso-vitamin C, erythrobic acid). It has a very low anti-scorbutic activity, but can be used as an antioxidant, e.g. in food.

Reaction of 2-keto-D-gluconic acid with HCN yields the cyanohydrin which upon saponification gives a mixture of 2-carboxy-D-gluconic acid and 2-carboxy-D-mannoic acid. These polyhydroxy-dicarboxylic acids could have interesting builder properties.

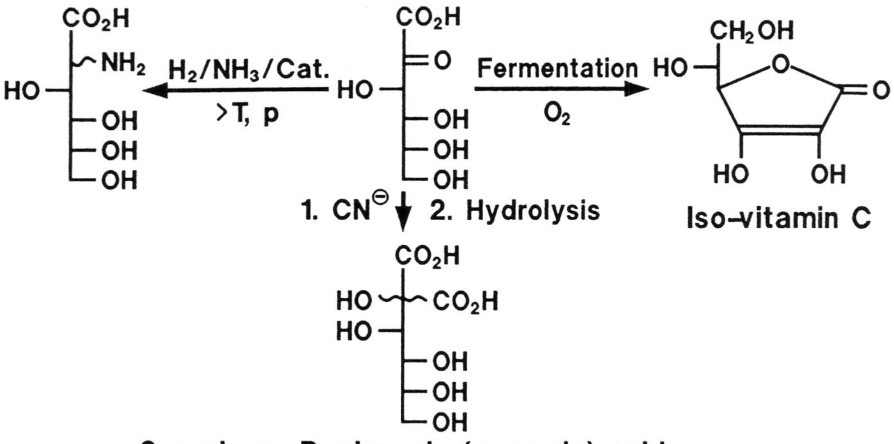

2–carboxy–D–gluconic (mannoic) acid

Reductive amination of 2-keto-D-gluconic acid yields a mixture of 2-amino-2-deoxy-D-gluconic acid and 2-amino-*manno* analog, interesting building blocks for surface active compounds and special polymers.

8. 5-Keto-D-gluconic acid

Using whole cells of *Acetobacter* species, D-glucose is oxidized to 5-keto-D-gluconate in very high yields (95 %):[33]

5-keto-D-gluconic acid

Preparation method: fermentation

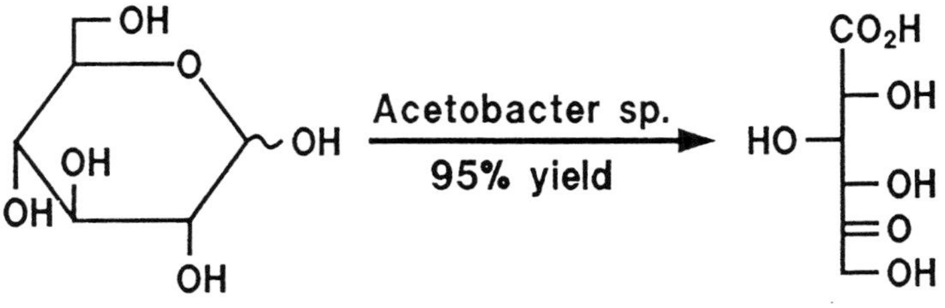

Selectivity of the presently available Pt / C catalysts for 5-keto-D-gluconic acid formation from D-glucose is very low.

The product is not produced on industrial scale. It could be used to produce 5-carboxy-D-gluconate (complexing agent; polyester component) or 5-amino derivatives of D-gluconic acid:

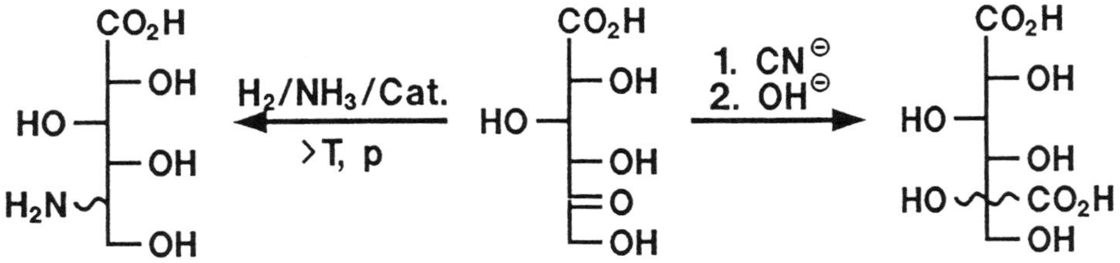

9. 2,5-Diketo-D-gluconic acid

Eriwinia sp. have been identified transforming D-glucose at 30 % d.s. with > 70 % yield to 2,5-diketo-gluconate.[34] Pfizer discloses a process where D-glucose at 20 - 30 % d.s. is oxidized by *Acetobacter cerineus* in the presence of 0.1 - 0.04 % cholin to yield 90 - 95 % of 2,5-diketo-gluconate in 40 - 50 h.[35]

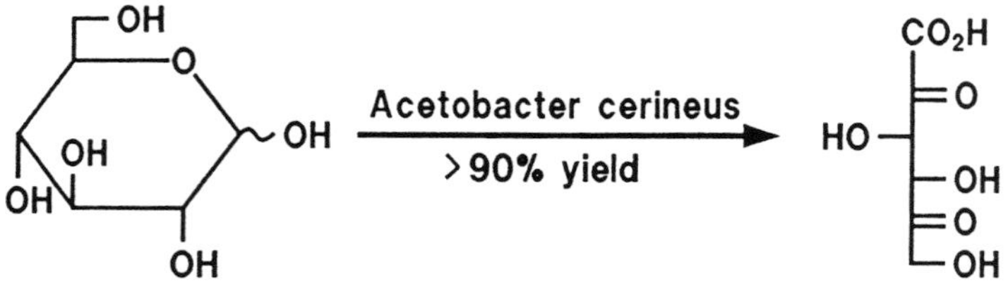

D - glucose   2,5 - diketo - D -
                                                  gluconic acid
                                                      (6)

Other strains like *Acetobacter melanogenum, Acetobacter aurantium, Gluconobacter rubiginosus, Gluconobacter liquefaciens,* and *Pseudomonas sesami* showed low productivity, low yields, and coloured pigments as byproducts.

2,5-Diketo-gluconate, by fermentative reduction, e.g. with *Corynebacterium mutants*, can selectively be converted into 2-keto-L-gulonate, a compound, that, chemically, can be transformed into vitamin C. This represents another short cut of the Reichstein vitamin C synthesis.

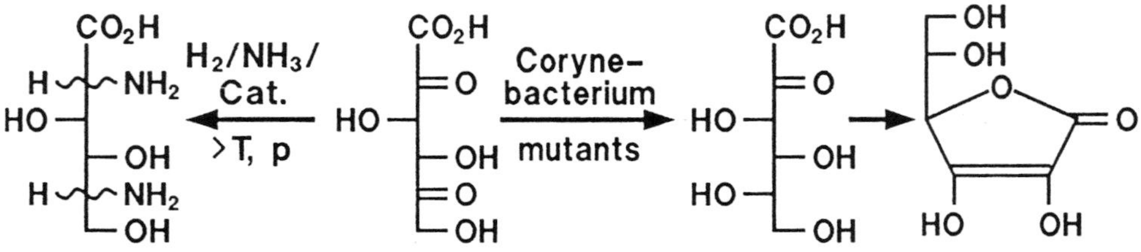

Another potential follow-up product of 2,5-diketo-gluconic acid is the 2,5-diamino derivative obtained by reductive amination.

10. L-Ascorbic Acid (Vitamin C)

Different approaches are being followed to short cut the traditional 8-step Reichstein synthesis:

(i)   Selective catalytic oxidation of L-sorbose to 2-keto-gluconic acid using Pt / C catalyst.[28]

(ii) Oxidation of L-sorbose, e.g. with *Pseudomonas sorbosoxidans* and *Gluconobacter oxidans* or L-sorbose dehydrogenase.[36]

(iii) Direct fermentation of D-glucose to 2,5-diketo-D-gluconic acid and 2-keto-L-gulonic acid involving mutants and genetically engineered microorganisms.

L-asorbic acid (vitamin C)

Production method: multi–step "Reichstein", synthesis starting from D–glucose, combining fermentation and chemical steps

Alternative potential production method: direct fermentation of D–glucose using genetically engineered microorganisms

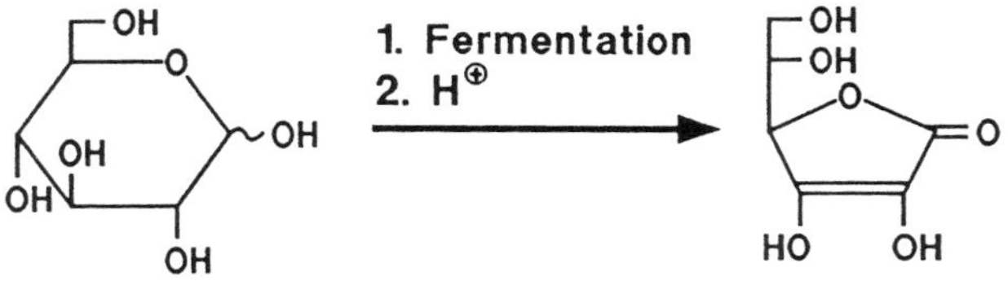

Genentech/Lubrizol/Pfizer

2,5–di–keto–D–gluconate reductase gene from Corynebacterium (2,5–di–keto–gluconic acid →2–keto–L–gulonic acid) cloned into E. herbicola (D–glucose → 2,5–di–keto–D–gluconic acid)

This approach can be realized in a two-step fermentation, starting with the *Erwinia*-induced conversion of D-glucose to 2,5-diketo-D-gluconic acid,[34,35] which is subsequently reduced by *Corynebacterium* to 2,5-L-gluconic acid, the direct precursor of vitamin C.

In the one-step fermentation with genetically engineered microorganisms (Genentech) the 2,5-diketo-D-gluconic acid reductase gene from *Corynebacterium* is cloned into plasmid and transformed by *Eriwinia* and excreted into the fermentation medium. The final cencentration of acid obtained in the broth, however, is low (20 g/l).[37-39]

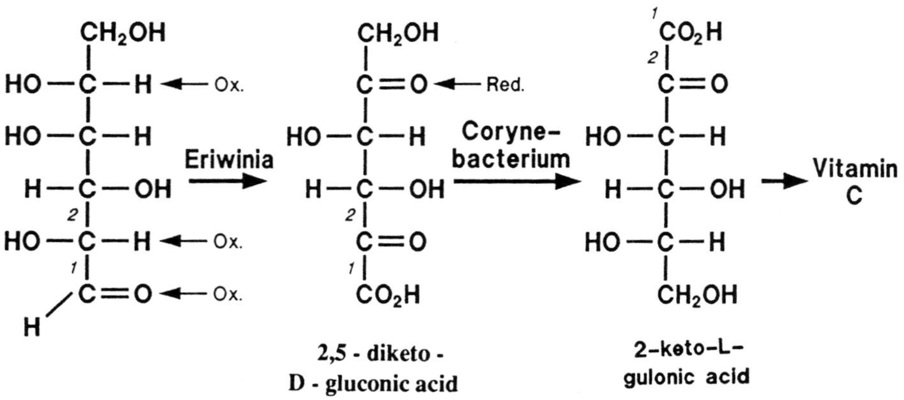

The Fischer projection structures at the top of the page show a 5 or 3 step synthesis of vitamin C:

First structure (D-glucose derivative):

CH₂OH
|
HO—C—H   ← Ox.
|
HO—C—H   Eriwinia →
|
H—C—OH
2 |
HO—C—H   ← Ox.
1 |
/ C=O   ← Ox.
H

Second structure (2,5-diketo-D-gluconic acid):

CH₂OH
|
C=O   ← Red.
|
HO—C—H   Coryne-bacterium →
|
H—C—OH
2 |
C=O
1 |
CO₂H

2,5 - diketo - D - gluconic acid

Third structure (2-keto-L-gulonic acid):

1 CO₂H
2 |
C=O
|
HO—C—H   → Vitamin C
|
H—C—OH
|
HO—C—H
|
CH₂OH

2–keto–L– gulonic acid

5 or 3 steps to vitamin C, starting from D - glucose

11. Kojic acid

By fermentation of D-glucose with *Aspergilli* very small quantities of kojic acid are being produced. It is used as an antibiotic, an insecticide, and as complexing agent. The γ-pyrone structure is derived from glucopyranose by eliminating two moles of water and oxidizing the 3-position to a keto function:

Kojic acid

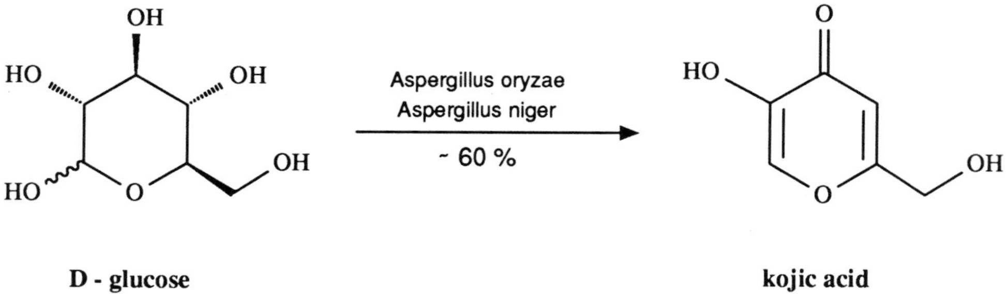

D - glucose   →   Aspergillus oryzae / Aspergillus niger   ~ 60 %   →   kojic acid

12. Conclusion

For broader industrial use of D-glucose, problems of selectivity, pH- and temperature instability, and insolubility in organic solvents have to be overcome.

It could be demonstrated, that chemical and biological catalysts are able to activate selective oxidation (dehydrogenation) reactions at different positions of unprotected D-glucose under mild reaction conditions, using water as solvent. A range of products is obtained. Some of them are already produced on industrial scale, others show potential to be used as intermediates for different industries and application areas.

Intensified R & D efforts are necessary to develop economical processes and to explore their application potential, especially to make use of their intrinsic properties like heterofunctionality, chirality (bioactivity), biocompatibility, and biodegradability.

References

1. H. Koch, H. Röper: New Industrial Products from Starch. *Starch/Stärke* **40** (1988) 121-131.

2. H. Röper, H. Koch: New Carbohydrate Derivatives from Biotechnical and Chemical Processes. *Starch/Stärke* **40** (1988) 453-464.

3. H. Röper: Neue Stärkeprodukte mit biotechnischen und chemischen Verfahren. D.I.L. Workshop: Verfahrenstechnik für nachwachsende Rohstoffe unter besonderer Berücksichtigung eines modernen Öl- und Kohlenhydratkonzeptes. Quakenbrück 28/29 Sept. 1988 (unpublished).

4. S. L. Neidleman, J. Geigert: The Preparation of D-Glucosone: a Case History Favoring Enzymatic over Chemical Synthesis using Pyranose-2-oxidase. Online International Inc., 989 Avenue of the Americas, New York, USA, *World Biotech. Rep.* **1986**, 2, 3, 49-54.

5. K. E. Koths, R. S. Halenbeck, D. B. Ring (Cetus Corp.): Fungal Pyranose-2-oxidase Preparations. *US Pat.* 4,568,645 (1984); *Chem. Abstr.* **104** (1986) 166931v.

6. K. E. Koths, R. S. Halenbeck, P. M. Fernandes (Cetus Corp.): Methods and Reagents for Pyranosone Production. *US Pat* 4,569,910 (1984); *Chem. Abstr.* **104** (1986) 184888r.

7. R. O. Horwath (Nabisco Brands Inc.): Screening Microorganisms for the Production of Glucose-2-oxidase. *US Pat.* 495,193 (1983); *Chem. Abstr.* **102** (1985) 44371a.

8. J. A. Maselli, R. O. Horwath (Standard Brands Inc.): Carbohydrate Process for Enzymic Glucosone Production. *Eur. Pat.* 56 038 (1984); *Chem. Abstr.* **96** (1982) 141215t.

9. S. L. Neidleman, W. F. Amon, J. Geigert (Cetus Corp.): Process for the Production of D-Glucosone and D-Fructose. *US Pat.* 4,246,347 (1981); *Chem. Abstr.* **94** (1981) 207379s.

10. G. A. Hamilton, J. De Jersey, P. K. Adolf: Oxidases and Related Redox Systems (T. E. King, H. S. Mason, M. Morrison, Eds.). Vol. 1, University Park Press, Baltimore 1973, 103 ff.

11. A. M. Klibanov, B. N. Alberti, M. A. Marletta: Stereospecific Oxidation of Aliphatic Alcohols Catalyzed by Galactose Oxidase. *Biochem. Biophys. Res. Commun.* **108** (1982) 804-808.

12. H. Saito, S. Ohnaka, S. Fukuda (Kawaken Fine Chemicals): Gluconic Acid. *Eur. Pat.* 142 725 (1985); *Chem. Abstr.* **103** (1985) 196366m.

13. P. Fuertes, G. Fleche (Roquette Frères S. A.): A Process for Oxidaton of Aldoses to Aldonic Acids in the Presence of Palladium Catalysts. *Eur. Pat.* 233 816 (1987); *Chem. Abstr.* **108** (1988) 187 206k.

14. H. Kimura, A. Kimura, Y. Mitsuta (Kao Corp.; Kawaken Fine Chemicals): Catalyst Compositions for Oxidation of Saccharides. *Jap. Pat.* 62/269 748 (1987); *Chem. Abstr.* **108** (1988) 206594d.

15. P. Fuertes, G. Fleche (Roquette Frères S. A.): Oxidation of Di-, Tri-, Oligo- and Polysaccharides to Polyhydroxycarboxylic Acids in the Presence of Carbon Supported Palladium - Bismuth and Palladium - Lead Catalysts. *Eur. Pat.* 232 202 (1986); *Chem. Abstr.* **108** (1988) 187205j.

16. B. A. Burdick, J. R.Schaeffer: Co-Immobilized Coupled Enzyme Systems on Nylon Mesh Capable of Gluconic and Pyruvic Acid Production. *Biotechnology Lett.* **9** (1987) 252-258.

17. J. H. Pazur: Glucose Oxidase from *Aspergillus niger*. Methods in Enzymology (W. A. Wood, Ed.), Vol. IX, Academic Press, 1986, pp. 82-87.

18. Ho Nam Chang, In Seong Joo, Young Sung Ghim: Performance of Rotating Packed Disk Reactor with Immobilized Glucose Oxidase. *Biotechnology Letters* **6** (1984) 487-492.

19. W. Hartmeier (Boehringer Ingelheim): Immobilized Glucosidase Preparation and its Use in Oxidizing Glucose. *Ger. Pat.* 2,911,192 (1980); *US. Pat.* 4,460,686 (1984); *Chem. Abstr.* **93** (1980) 217248s.

20. W. Hartmeier, T. Doeppner (Boehringer Ingelheim): Permeabilized Fungus with Intact Immobilized Glucose Oxidase-Catalase System and its Use. *Ger. Pat.* 3,301,992 (1984); *Chem. Abstr.* **101** (1984) 149809g.

21. P. Seiskari, Y.-Y. Linko, P. Linko: Continuous Production of Gluconic Acid by Immobilized *Gluconobacter oxydans* Cell Bioreactor. *Appl. Microbiol. Biotechnol.* **21** (1985) 356-360.

22. H. Wünschendorff, C. Killian: Über den Stoffwechsel von *Ulstulina vulgaris*. *C. R. Acad. Sci.* **187** (1928) 572-574.

23. T. Takahashi, T. Asai, *Zentralbl. Bakteriol. Parasitenk. Infektionskr. Abt. II* **84** (1931) 193.

24. G. Gräfe: D-Glucuronic Acid and its Preparation from Starch. *Stärke* **5** (1953) 205-209.

25. K. Heyns, G. Gräfe: Oxidative Umwandlungen an Kohlenhydraten VII. Synthese von D-Glucuronsäure über Carboxyl-Stärke. *Chem. Ber.* **86** (1953) 646-650.

26. M. S. Loicjanskaya: The Primary Stages in the Decomposition of Cellulose by *Spirocheta cytophaga*. *Compt. Rend. Acad. Sci. USSR* **14** (1937) 381; *Chem. Abstr.* **31** (1937) 6279(8).

27. K. D. Kulbe: Enzymtechnologie und nachwachsende Rohstoffe. *Nachr. Chem. Techn. Lab.* **36** (1988) 612-624.

28. K. Heyns: Oxydative Umwandlungen an Kohlenhydraten. Katalytische Oxydation von L-Sorbose zu 2-Keto-L-gulonsäure. *Liebigs Ann. Chem.* **558** (1947) 177-187.

29. F. Challenger, V. Subramaniam, T. K. Walker: Bildung von organischen Säuren aus Zuckern duch *Aspergillus niger*. *Nature* **119** (1927) 674.

30. H. Shirafuji, T. Yamaguchi, I. Nogami (Takeda Chemical Ind.): Fermentation Method for Producing 2-Keto-D-glucaric Acid. *Jap. Pat.* 85/294 577 (1985); *Eur. Pat.* 228 274 (1987); *Chem. Abstr.* **108** (1988) 36347f.

31. Takeda Chemical Ind.: Preparation of 2-Keto-D-glucaric Acid by Contacting *Pseudommonas sorbosoxidans* with D-Glucose, D-Fructose, or D-Mannose. *Jap. Pat.* 63/091 088 (1988).

32. P. C. C. Smith: The Selective Catalytic Oxidation of D-Gluconic Acid to 2-Keto-D-gluconic Acid or D-Glucaric Acid. *Doctoral Dissertation*, Technical University Eindhoven, NL, 1984.

33. K. Kieslich: Microbial Transformations of Non-steroid Cyclic Compounds. Georg Thieme Publishers, Stuttgart, 1976, pp. 273-274.

34. T. Sonoyama, S. Yagi, B. Kageyama: Facultatively Anaerobic Bacteria showing high Productivities of 2,5-Diketo-D-gluconate from D-Glucose. *Agric. Biol. Chem.* **52** (1988) 667-674.

35. D. A. Kita, D. M. Fenton (Pfizer Inc.): 2,5-Diketogluconic Acid. *Ger. Pat.* 30 36 413 (1981); *Chem. Abstr.* **94** (1981) 207161q.

36. A. Fujiwara, T. Hoshimo, T. Sugisawa (Hoffmann-La Roche): Purification and Characterization of L-Sorbose Dehydrogenase from *Gluconobacter oxydans*. *Eur. Pat.* 248 400 (1987); *Chem. Abstr.* **108** (1988) 182808y.

37. D. A. Estell, D. R. Light, W. H. Rasteter, R. A. Lazarus, J. V. Miller (Genentech Inc.): Biosynthetic 2,5-Diketo-gluconic Acid Reductase Recombinant Cells and Expression Vectors for its Production and its Use in Preparing 2-Keto-L-gulonic Acid. *Eur. Pat.* 132,308 (1985); *Chem. Abstr.* **102** (1985) 144140t.

38. S. Anderson, D. R. Light, C. Marks, W. H. Rasteter (Genentech Inc.): Manufacture of Ascorbic Acid Precursor Ketogulonic Acid with Recombinant Microorganisms. *Eur. Pat.* 292,303 (1988); *Chem. Abstr.* **111** (1989) 2173t.

39. J. F. Grindley, M. A. Payton, H. van de Pol, K. G. Hardy: Conversion of Glucose to 2-Keto-L-gulonate, an Intermediate in L-Ascorbate Synthesis, by a Recombinant Strain of *Erwinia citreus*. *Appl. Environm. Microbiol.* **54** (1988) 1770-1775.

14

Studies on Selective Carbohydrate Oxidation

Herman van Bekkum

Laboratory for Organic Chemistry, Delft University of Technology, Julianalaan 136, 2628 BL Delft, The Netherlands

Summary. The oxidation of carbohydrates to carboxylic acids as a means of increasing the biocompatibility and biodegradability is examined. A special aim was the primary alcohol oxidation of carbohydrates on noble metal catalysis with oxygen as the oxidant. Problems to be faced include the activity and selectivity of various noble metals, its dependance on the support, and the deactivation of the catalyst by oxidation of the metal-surface. Another research area was the glycolic oxidation of glucans (e.g. starch) yielding dicarboxy glucans exhibiting excellent calcium complexing properties. Owing to these as well as to the stability at the pH of the washing processs they are potentially attractive cobuilders in phosphate-free detergents.

Introduction

Direct oxidation of carbohydrates to carboxylic acids would seem an obvious transformation; the mass and oxygen content of the systems increase, the character alters substantially whilst some beneficial properties, such as biocompatability and biodegradability, are expected to remain. Intermediate carbonyl compounds may be of value too.

Other arguments in favour of intensified attention for carbohydrate oxidation are the increasing number of oxidation tools and techniques and also the fact that some existing methods are not environmental-friendly and should be replaced by clean procedures. Major carbohydrate oxidations include:

♦ oxidation at an aldehyde carbon,

♦ oxidation of primary alcohol groups,

♦ glycolic oxidation leading – via dialdehydes – to dicarboxylic acids with opening of pyranose or furanose rings.

C$_1$-aldehyde oxidation is relatively well documented, with a choice of several methods. For instance for gluconic acid three types of processes are feasible: bio-catalytic, chemo-catalytic, and electro-chemical. The most economical route is still in debate. There's still progress in selective chemo-catalytic C$_1$-aldehyde oxidation as shown by the recent work on bi- and trimetallic (Pt, Pd, Bi) oxidation catalysts.

Thus, Degussa workers[1] reported recently on glucose oxidation (pH 10, 55 °C, 1 atm O$_2$) with selectivities up to 96 %. Best results were obtained with a 4 % Pd, 1 % Pt, 5 % Bi-on-carbon catalyst. Bismuth, which is in the oxidized state, enhances the selectivity, platinum the activity of the palladium catalyst.

The Eindhoven group[2] obtained highly selective (> 99 %) C$_1$-oxidation in the conversion of lactose towards lactobionic acid, using a special Pd, Bi-on-carbon catalyst.

Dehydrogenations

With respect to C$_1$ oxidation I can't resist mentioning the facile dehydrogenation of aldehyde sugars at high pH (> 12, optimum ~ 13.5) over platinum and rhodium catalysts, a reaction we came across[3] some years ago. Molecular hydrogen is evolved under very mild conditions. The reaction, exemplified by the ensuing formulae,

is generally applicable, e.g. to galactose, lactose, and glucosamine as well, the ease of dehydrogenation of pyranose monosaccharides depending on the sugar structure as follows:

4-ax OH > 4-eq OH > 2-eq OH > 2-H > 2-ax OH > 3-eq OH > 3-ax OH

6-CH$_2$OH > 6-CH$_3$ > H > 6-COO$^-$

These effects can, at least in part, be understood by considering – as proposed by Kieboom – the adsorbed state of the ionized sugar on the Pt-surface, as is shown for the reactive galactose:

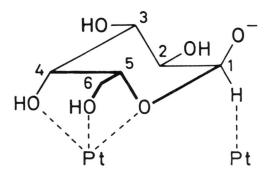

Here, good adsorption is assumed to promote hydride-transfer to the Pt-surface, with formation of the strong carbonyl bond as a driving force. The hydrogen-donor properties of reducing aldehyde sugars can be used in hydogen-transfer reactions as shown by us[4] for the glucose / fructose oxidation / reduction over platinum or rhodium. Essentially all hydrogen is used to hydrogenate fructose, apparently a wonderful balance exists on the noble metal surface.

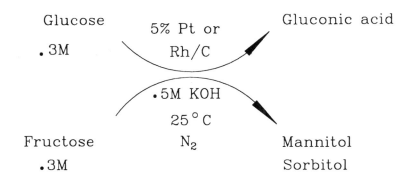

When the driving force is less pronounced as in the case of C=N bond formation, e.g. the conversion of glucose oxime towards the corresponding hydroximolactone, it may be necessary[5] to oxidize the hydrogen in order to make the reaction feasible in a thermodynamic sense.

Oxidations

Turning now to the more difficult primary alcohol to carboxylate oxidation, stoichiometric as well as catalytic methods have been applied as illustrated:

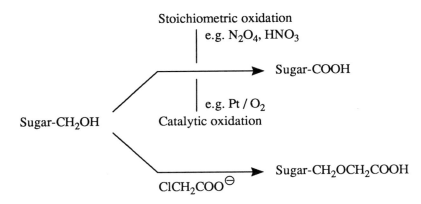

Of course, carboxylates may also be obtained by substituting carbohydrate hydrogens for carboxylate-containing moieties as occurs in the manufacture of carboxymethyl cellulose. Also alkaline oxidative degradation of reducing sugars leading to carboxylates is known. Thus, the high yield (98 %) conversion of glucose into arabonate has been reported.[6] In a recent paper[7] Röger et al. discuss the alkaline oxidative cleavage of reducing disaccharides and report 93 % selectivity for the conversion of isomaltulose into 5-*O*-glucopyranosyl-arabinonic acid.

Direct stoichiometric oxidation reagents include nitric acid (e.g. glucose to glucaric acid) and nitrogen dioxide (manufacture of 6-carboxycellulose from cellulose).[8] In case of protected sugars, many other reagents (e.g. Cr^{VI} and Ce^{IV} compounds) have been applied.

Compared to the methods above, Pt catalyzed oxidation of carbohydrates is a clean method, more up to the standards of today and tomorrow.

First of all I should mention the classical work of Heyns in this area.[9] Combining work on carbohydrate oxidation with studies on the catalytic oxidation of cyclitols the following sequence of oxidation reactivity on Pt evolved:[9]

$$HCO > COCH_2OH > CH_2OH > CHOH_{ax} > CHOH_{eq}$$

Within secondary alcohol groups axial hydroxyls were found to react faster towards ketone functions than equatorial hydroxyls, in other words: equatorial C-H bonds are more prone to dissociation on the catalyst surface than axial C-H bonds, (which is understandable in steric and energetic terms).

Going back in history of catalytic oxidation, other famous names show up: Wieland with his work on Pd-black catalyzed alcohol and aldehyde oxidation (1912 - 1921), and Döbereiner right at the beginning of catalysis in 1845 with the Pt catalyzed deep oxidation of ethanol (pH > 7).

From the viewpoint of clean technology, it is not surprising that in recent years renewed interest in Pt catalyzed oxidation arises. I mention the work of the group of van der Baan and Kuster on the glucose oxidation to glucaric acid[10] and the studies of a Hoechst group, amongst others, on sucrose oxidation.[11] Early patents cover mono- and dicarboxylic acids of sucrose but Fritsche-Lang et al.[11] applied for a patent on the tricarboxylic acid, in which all three primary CH_2OH groups are oxidized. Using a formulation 120 g sucrose and 60 g 5 % Pt / C in 1,5 l water at 80 °C, these authors observed 35 % tricarboxylic acid after 20 h of reaction.

Selective Oxidation of the 6-CH$_2$OH Group

Types of substrates studied at Delft include 1-substituted glucoses, with the target of selective oxidation of the 6-CH_2OH group. A batch liquid phase oxidation reactor is used[12] equipped with thermostat and air-tight stirrer; pH and oxygen pressure were automatically kept constant by suitable devices.

The oxidation of alkyl α- and β-glucopyranosides, with linear alkyl groups, is being studied aiming at a new group of surfactants, combining nonionic and anionic properties in analogy to the ethylene oxide-derived systems $RO(CH_2CH_2O)_nCH_2COO^-$. The following formulae show two examples:

R = CH$_3$ Select.   ~ 70 %
R = n-C$_8$H$_{17}$ "   > 90 %

It was observed[13] that long chain (C_8, C_{10}, C_{12}) glucopyranosides are oxidized with substantially higher selectivity than α-methyl glucopyranoside. In the latter case, the main side product was bicarbonate resulting from a parallel deep oxidation reaction. Böcker and Thiem[14] recently reported high selectivity in the oxidation of dodecyl α-D-glucopyranoside over Pt black. The interesting selectivity enhancement exerted by alkyl chains might be due to

(i) protection against ring oxidation resulting from association of glucosides enhancing the relative accessibility of the 6-CH_2OH at the Pt-surface;

(ii) hydrophobic effects exerted by the alkyl chains in analogy to the effects of 3-, 4- and 3,5-alkyl groups on acidity and reactions of benzoic acids.[15]

Much attention was paid by us to the oxidation of glucose 1-phosphate (α-D-glucopyranose 1-phosphate, Glc-1-P), a compound that can be obtained – together with fructose – from sucrose by enzymatic phosphorylation. Selective C_6-oxidation would lead to glucuronic acid 1-phosphate (GlcA-1-P) which is expected to be a useful intermediate in the production of glucuronic acid, glucaric acid, and ascorbic acid.

Glc - 1 - P   GlcA - 1 - P

First it was observed[12] that for Pt on a given carbon support the particle size and the Pt distribution over the active carbon were of importance for the activity of the catalyst in Glc-1-P oxidation. Fig. 1 shows that small particles (powder Pt / C catalyst) and near-external-surface location of Pt (mantle catalyst, obtained by short impregnation times) show a high initial rate but deactivate rapidly. The Pt-ROX 0.8 catalyst, which worked satisfactory, was based on extrudated gas-activated carbon, cylindrical particles with diameter 0.8 mm and length 3 - 5 mm, BET surface area 700 m^2g^{-1}, impregnated (3 days) with aq. H_2PtCl_6 followed by reduction with formalin.

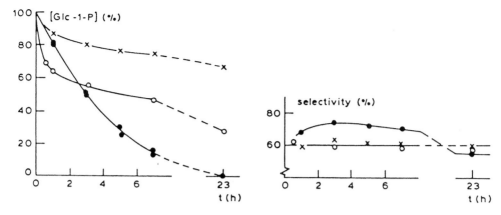

Fig. 1. Oxidation of Glc-1-P. [Glc-1-P]$_O$ 100 mM; P(O$_2$) 0.20 bar; T 60 °C; pH 9.0; catalyst 40 g/l. Catalyst type: powder (5 % Pt/ROX 50-60 μm (O)); scale (2 % Pt/ROX 0.8) (X); 'diffusion stabilized' extrudate (5 % Pt/ROX 0.8; d, 0.8 mm; l, 3-5 mm) (●).

The deactivation of the two former catalyst preparations is due to surface oxidation, which is a general problem when using noble metals in liquid phase oxidation. Deactivation particularly occurs when the substrate (i) reacts (dehydrogenates) slowly and (ii) adsorbs weakly onto the metal. This is the case for Glc-1-P.

Remedies for the formation of an oxide layer on the catalyst surface include the use of

♦ low oxygen partial pressure

♦ low stirring speed

♦ diffusion-stabilized catalyst.

The latter concept of van Dam[12] applies large uniform particles (e.g. extrudates) in which the occurrence of oxygen diffusion limitation leads at a certain depth in the particle, to a proper tuning of reactions and, consequently, to a high steady state reactivity. The principle is illustrated in Fig. 2:

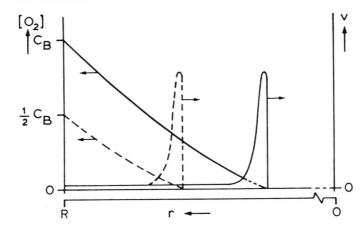

Fig. 2. Schematized oxygen concentration and oxidation rate profiles for a diffusion-stabilized catalyst. C_B: liquid bulk oxygen concentration; $[O_2]$: local oxygen concentration in the pore liquid; v: local reaction rate; r: radius of catalyst particle (from centre to surface).

Regarding the mechanism of the oxidative dehydrogenation of a primary alcohol group on platinum, the following (surface) reaction steps are assumed to play a role (a= adsorbed):

$$RCH_2OH_a \rightleftharpoons RCHO_a + 2\,H_a$$
$$RCHO + H_2O \rightleftharpoons RCH(OH)_2$$
$$RCH(OH)_{2a} \rightleftharpoons RCOOH + 2\,H_a$$
$$RCOOH + OH^- \rightleftharpoons RCOO^- + H_2O$$
$$O_2 \rightleftharpoons 2\,O_a$$
$$2\,O_a + 2\,H_2O \rightleftharpoons 4\,OH_a$$
$$4\,OH_a + 4\,H_a \rightleftharpoons 4\,H_2O$$

$$RCH_2OH + O_2 + OH^- \rightarrow RCOO^- + 2\,H_2O$$

In the dehydrogenation steps – of the alcohol and the intermediate hydrated aldehyde – ionization of the OH group might precede C-H dissociation.

Variables scanned in the liquid phase oxidation of glucose 1-phosphate over Pt catalysts include pH (optimum pH 7-9), temperature, oxygen pressure, and Pt-loading of the active carbon. Substantial, and mostly understandable, differences in reaction rate and catalyst deactivation were observed. Selectivity towards glucuronic acid 1-phosphate was, however, essentially independent of the above variables and amounted to about 70 %. Identified side products[16] are C_1, C_2 and C_3 dibasic acids together with free phosphate. The selectivity is largely independent of the Glc-1-P conversion; the undesired process thus starts with a parallel reaction of the starting material.

Two possibilities can be envisaged for this initial reaction:

(i) (oxidative) hydrolysis of the phosphate bond, and

(ii) oxidation of ring CHOH groups.

For several reasons, the latter reaction, leading to α-ketols which are very sensitive to oxidative C-C bond cleavage, is preferred.

The only variable found to exert an effect on the selectivity in the Pt-catalyzed oxidation of Glc-1-P was the carrier. In Fig. 3 several active carbons are compared as to rate and selectivity.

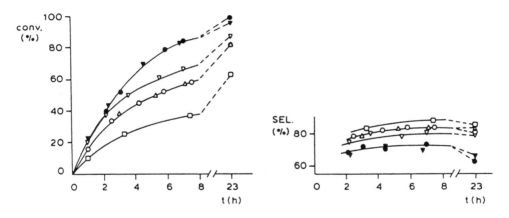

Fig. 3. Effect of carbon support on Glc-1-P oxidation. [Glc-1-P]$_O$ N30 (□); N20 (Δ); N10 (O); N5 (∇); D (▼); A (●).

The carbons are all derived[17] from the same material (Norit ROX 0.8) by nitric acid oxidation, N5 - N30 (the figure denotes the nitric acid content), by air oxidation at 320 °C (carbon A) or by helium treatment at 320 °C (material D). Oxygen content and surface negative charge (at pH 9) are substantially enhanced by the oxidative treatments. Thus the oxygen content which amounts to 3.3 % in the starting material (ROX) goes up to 12.4 % for carbon N30. Selectivity towards GlcA-1-P increases to 87 % when using Pt on N30. The limited insight in metal support interactions does not allow conclusions as to the background of this selectivity enhancement. It is, however, clear that in optimization of Pt / C catalyzed oxidations, the catalyst structure should be regarded as an interesting variable.

In addition it may be noted that a negatively charged support – under working conditions – will be favourable with regard to prevention of consecutive reactions when non-charged sugars are to be converted into mono-carboxylate systems.

The Oxygen Tolerance of the Noble Metals Used for Oxidation

Using methanol as a model substrate for the oxidation of weakly adsorbing but relatively reactive compounds, the oxygen tolerance of the noble metals was compared.[18] The sequence – using 5 % metal on carbon powder catalysts – appeared to be as follows: Pt > Ir > Pd > Rh > Ru.

No methanol oxidation activity was detected for Ru catalysts; apparently this metal is already deactivated at very low oxygen concentrations. Rh and Pd showed an appreciable catalytic activity provided that the oxygen concentration was kept very low, i.e. below 0.1 ppm for Rh and below 0.5 ppm for Pd. Ir and Pt showed a relatively high stability in methanol oxidation at high oxygen concentrations, with Pt being the more active metal. It should be noted, however, that with substrates that are much less reactive than methanol, e.g. gluconic acid or glucose 1-phosphate, Pt catalysts are more sensitive to oxygen.

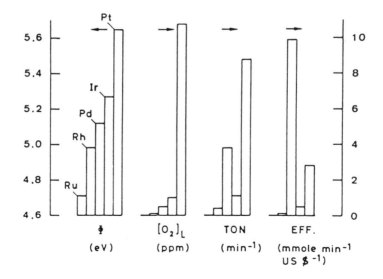

Fig. 4. Comparison of Ru, Rh, Pd, Ir, and Pt as methanol oxidation catalysts. Φ: work function; $[O_2]_L$: optimal oxygen concentration in the liquid phase; TON: maximum turnover numbers (TON) in methanol oxidation (0.5 M CH_3OH, pH 9, 30 °C); EFF: effectivity (max. TON/metal costs assuming metal dispersion = 1).

Fig. 4 compares the behaviour of the five noble metals studied in methanol oxidation. Some correlation seems to exist between the oxygen tolerance and the work function of the metals, i.e. the energy required to remove an electron from the metal to infinity. Pd has the advantage of being relatively inexpensive, the deactivation problems have to be overcome, however, perhaps by varying the support and / or the particle size as discussed above.

Indeed, a 5 % Pd-on-carbon powder catalyst applied in methanol oxidation was found to differ profoundly (see Fig. 5) in its deactivation pattern from a 5 % Pd-on-Al$_2$O$_3$ (powder) catalyst. As the alumina catalyst was found to possess a much lower Pd dispersion than the carbon catalyst, the Pd crystallite size might be another important factor. In harmony with the observations, small Pd crystallites are expected to be more susceptible towards oxidation than larger ones.

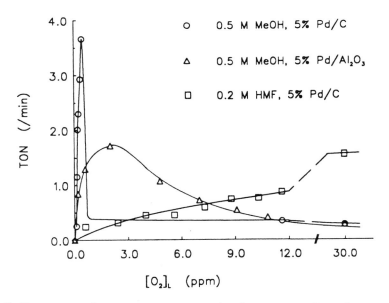

Fig. 5. Turnover number vs. oxygen concentration (kept constant for each measurement) for three oxidation experiments over Pd-catalysts.

Fig. 5 also contains the results of an oxidation experiment on hydroxymethylfurfural (HMF)[19] using the 5 % Pd-on-carbon catalyst. Though HMF is just slowly oxidized, no deactivation of the Pd-on-carbon is observed at higher oxygen concentrations. In fact the reaction is about first order in oxygen. This at first sight surprising result is ascribed[20] to strong adsorption of the HMF system onto the Pd surface.

Oxidation of HMF can afford in principle four products, each of which seems of interest as an intermediate or monomer:

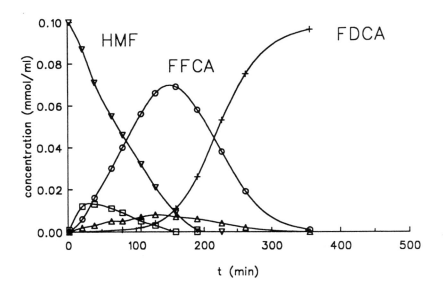

We are studying presently the oxidation of HMF over various noble metal catalysts with the aim of obtaining high selectivity to each particular oxidation product.

Fig. 6. Oxidation of HMF over Pt on alumina. Conditions: 0.1 M HMF; 80 ml H_2O; 1.0 g 5 % Pt/Al_2O_3 (power); 60 °C; pH 9.0; $p(O_2)$ 0.2 atm; HMF (∇); FDC ($\square$); HFCA ($\triangle$); FFCA (O); FDCA (+).

Reasonable to high selectivities towards FFCA, 5-carboxyfurfural, are obtained[20] over Pt as the catalyst. Fig. 6 shows the course of the oxidation over Pt-on-Al_2O_3.

First of all, it may be noted that the final oxidation product is FDCA, furan-2,5-dicarboxylic acid, which is formed in quantitative yield over Pt. The surprisingly high selectivity towards the intermediate FFCA may be due to the stabilizing conjugation between the furan ring and the formyl group in HMF which counteracts hydration of the aldehyde function. The hydrated aldehyde is assumed to be (probably in the ionized form) the direct precursor of the carboxyl group. As hardly any aldehyde hydration takes place, the primary alcohol group is oxidized first, yielding the dialdehyde FDC, in which compound the conjugative effects of the two formyl groups are counteracting, making each group more prone to oxidation than the formyl group in HMF.

This reasoning is not generally applicable to HMF oxidation over noble metal catalysts; Fig. 7 shows the course of the oxidation over Rh-on-Al_2O_3[21] under the conditions of Fig. 6. Major differences between Pt and the oxygen-susceptible Rh are:

    (i)   the low rate of oxidation over Rh and

    (ii) the fact that formyl group oxidation is less pronounced
           than hydroxymethyl group oxidation over Rh.

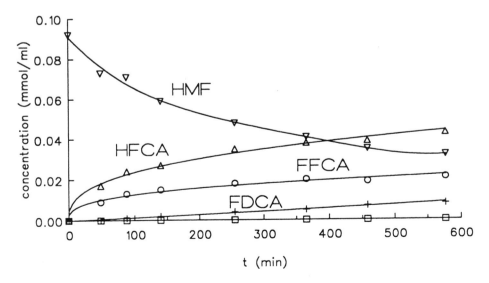

Fig. 7. Oxidation of HMF over 5 % Rh/Al_2O_3 (conditions see Fig. 6).

Additional work on HMF oxidation is in progress to obtain insight into its oxidation behaviour, and in the options for selective conversion.

Chemo-catalytic Oxidation

As a final example of chemo-catalytic oxidation of monosaccharides we mention the selective oxidation of glucose or gluconic acid towards 2-ketogluconic acid[22] using a Pb-catalyst:

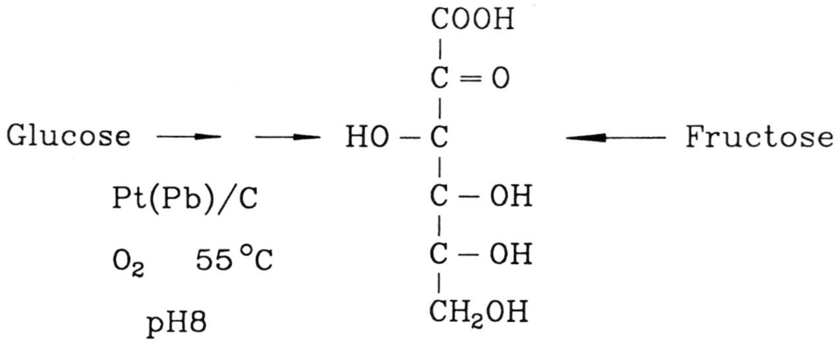

$$
\begin{array}{ccc}
& & \text{COOH} \\
& & | \\
& & \text{C} = \text{O} \\
& & | \\
\text{Glucose} \longrightarrow \longrightarrow & \text{HO} - \text{C} & \longleftarrow \text{Fructose} \\
\text{Pt(Pb)/C} & | & \\
& \text{C} - \text{OH} & \\
\text{O}_2 \quad 55\,^{\circ}\text{C} & | & \\
& \text{C} - \text{OH} & \\
\text{pH8} & | & \\
& \text{CH}_2\text{OH} &
\end{array}
$$

An interesting question is whether fructose could act as starting compound here. For 2-ketogluconic acid also a bio-route is available.

The selectivity enhancement brought about by Pb – and that by Bi mentioned earlier – are far from being fully understood and constitute promising areas for future research. Interaction of carbohydrates with multivalent cation systems seems to have considerable potential for selective activation as well as for selective protection.

Oxidation of Polysaccharides

Turning now to oxidation of polysaccharides both products from CH_2OH group oxidation and from glycolic ring-splitting oxidation are expected to be of value:

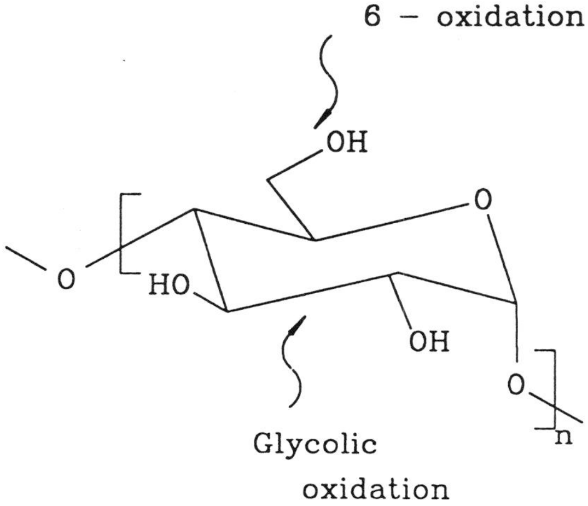

$$6 - \text{oxidation}$$

Glycolic
oxidation

In the case of 6-oxidation of 1,4-glucans alginate- and pectine-type products may arise and the reader may notice that the present price of sodium alginate is about 20 times that of starch.

Obviously, the application of heterogeneous oxidation catalysts will become increasingly difficult with increasing mass and decreasing mobility and solubility of polysaccharides. Aspinall et al.[23] have studied the oxidation of a galactan over Pt black and have achieved partial oxidation of the CH_2OH groups. A (catalytic) homogeneous procedure using a clean oxidant seems preferred but is as yet not available.

We have studied the glycolic oxidation of glucans. Three methods have been compared:

(i)   tungstate-catalyzed oxidation with hydrogen peroxide as the oxidant

(ii) hypochlorite oxidation

(iii) periodate oxidation to the corresponding dialdehyde system followed by chlorite oxidation to the dicarboxy glucans.

The use of the tungstate-H_2O_2 system[24] was studied[25] for starch and maltodextrins as the substrates. The system, which operates at pH 2, was found to effect oxidation of the glucose units via two pathways:

(a) glycol-cleavage of the C_2,C_3-diol moieties in internal glucose units, followed by (undesired) hydrolysis of the ring-opened intermediate and

(b) stepwise decarboxylation at the reducing terminal glucose unit until the glycosidic bond is reached.

Both pathways yield glucose oligomers terminated by erythronic acid at the former reducing end, which is up to 40 % of the oxidation product. Furthermore, decomposition of H_2O_2 is an inevitable side reaction. Altogether the method is not suitable for the dicarboxy glucan preparation.

Hypochlorite glycolic-oxidation of both starch[26] and cellulose[27] has been reported before. Some controversy in the literature concerning the selectivity of the oxidation induced us to re-examine[28] the hypochlorite oxidation of maltodextrins and starch. Applying 3 moles of oxidant per anhydroglucose unit at pH 8-9 dicarboxypolysaccharides were obtained containing up to 45 % ring-opened glucose units. The average degree of polymerization of the products indicated, however, severe chain degradation during the oxidation.

Best results in 1,4-glucan glycolic oxidation were obtained by us using the two-step periodate / chlorite oxidation. The second step, the conversion of dialdehyde glucan into dicarboxy glucan, formerly requiring a molar ratio chlorite : dialdehyde of six according to:

$$RCHO + ClO_2^- \quad \rightarrow \quad RCOO^- + HOCl$$
$$HOCl + 2\,ClO_2^- \quad \rightarrow \quad 2\,ClO_2 + Cl^- + OH^-$$

could be largely improved[29] by adding H_2O_2 as HOCl scavenger:[30]

$$RCHO + ClO_2^- \quad \rightarrow \quad RCOO^- + HOCl$$
$$HOCl + H_2O_2 \quad \rightarrow \quad HCl + H_2O + O_2$$

The first step, the periodate oxidation of glycolic moieties in polysaccharides, proceeds essentially quantitatively. One can economize on this step by chemical or electrochemical regeneration of the iodate formed. The Table gives data on the aldehyde conversion efficiency of the chlorite / H_2O_2 oxidation and the carboxylate content of the products obtained; up to 86 % aldehyde conversion was achieved. The Ca-sequestering ability of the dicarboxy-1,4-glucans obtained in this way was excellent and surpasses that of triphosphate (STP).

Table 1.   Dicarboxy polysaccharides obtained from oxidation of dialdehyde polysaccharides with $NaClO_2 / H_2O_2$ (2 moles of each per mole dialdehyde units)

Dialdehyde substrate	CHO conv. [%]	COONa cont. [mmole/g]	Ca-sequestering capacity [mmole/g]
Starch	80	7.21	2.51
Amylose	75	6.85	2.39
Amylopectin	69	6.53	2.58
Cellulose	83	7.40	2.20
Dextran	69	6.50	1.80
Starch[a]	86	5.43	1.04
STP	--	--	2.00
Sokalan-CP-5[b]	--	11.36	2.75

a) Oxidation with NaOCl (3 moles per mole glucose unit) at pH 8.5 and 293 K

b) Copolymer (70 : 30) of acrylate and maleate (BASF)

Moreover, as shown in Fig. 8, dicarboxy starch is stable at the pH of the washing process but is degraded under the acidic (pH 4-5) waste water conditions, as is to be expected from its polyacetal structure.

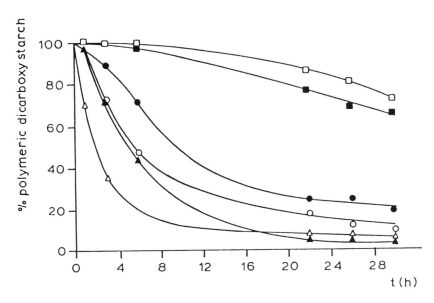

Fig. 8. Hydrolysis of C_2,C_3-dicarboxy starch (10 mg/ml) as a function of pH at 333 K. (% polymeric dicarboxy starch is defined as the fraction with degree of polymerization > 2) pH 2 (●); pH 3 (○); pH 4 (Δ); pH 5 (▲); pH 7 (■); pH 9 (□).

In addition to linear and branched polysaccharides also the cyclic β-cyclodextrin was subjected[31] to periodate / chlorite / hydrogen peroxide oxidation (Fig. 9). As shown by the ^{13}C NMR spectrum (Fig. 10) an essentially pure tetradecacarboxylate is obtained, exhibiting a high affinity for multivalent cations (Ca^{II}, La^{III}).

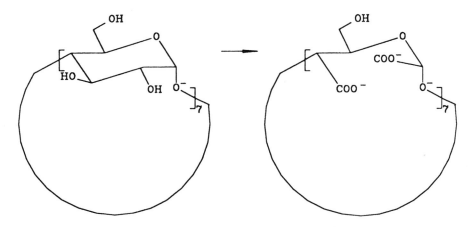

Fig. 9. Glycolic oxidation of β-cyclodextrin.

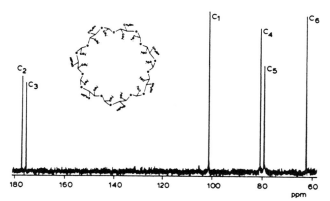

Fig. 10. ^{13}C NMR spectrum of dicarboxy β-cyclodextrin in D$_2$O.

Several techniques were applied[31] to study the dicarboxy glucans and their complexes with Ca^{II}. As an example Fig. 11 shows the differences in optical rotation of some dicarboxy glucans as a function of the amount of Ca^{II} added. This technique is sensitive to conformational changes of solute molecules.

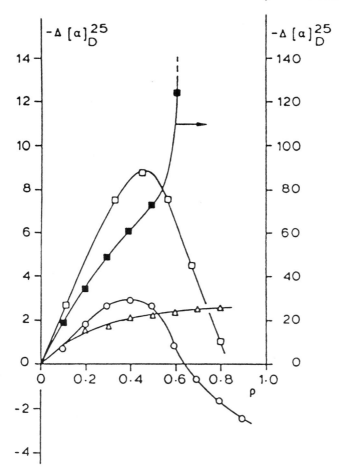

Fig. 11. Variation of the optical rotation of dicarboxy glucans (0.1 M, pH 10.8) upon Ca^{II} addition. $\rho=$ mole Ca^{II} per mole dicarboxy units. Dicarboxy β-cyclodextrin (○); dicarboxy amylose (□); dicarboxy cellulose (■); and dicarboxy amylose upon NaCl instead of $CaCl_2$ addition (△).

The curves for the α-1,4-linked dicarboxy amylose and dicarboxy β-cyclodextrin are indicative for some conformational similarity upon Ca^{II} addition. The β-1,4-linked dicarboxy cellulose, on the other hand, shows a tenfold higher change, indicating an entirely different conformational change upon addition of Ca^{II}. Apparently the energy difference between the conformations of the free and complexed polycarboxylates is sufficiently low to allow still good complexation (cf. Table) in a totally different conformation.

Some Conclusions

♦ Oxidation of carbohydrates leads to interesting classes of compounds for which carbohyrates are the logical starting materials.

♦ Several good methods are available for C_1-aldehyde oxidation.

♦ Catalytic primary alcohol oxidation is still in development. Both the metal and the support are important variables. Deactivation requires precautions.

♦ Glycolic oxidation of polysaccharides can be performed in a satisfactory way. For economic reasons new concepts are required for this type of oxidation.

Acknowledgement. The author gratefully acknowledges cooperation in areas of this article with Drs. A. P. G. KIEBOOM, H.-E. VAN DAM, M. FLOOR, P. VINKE and G. DE WIT.

References

1. B. M. Despeyroux, K. Deller, E. Peldszus: New Developments in Selective Oxidation. *Symp. Proceedings* Rimini, Italy, 1989 (G. Centi, F. Trifiro, Eds.), Elsevier Science Publ., Amsterdam, 1990, p. 159 ff.

2. H. E. J. Hendrikx, B. F. M. Kuster, G. Marin, *Carbohydr. Res.*, in press.

3. G. de Wit, J. J. de Vlieger, A. C. Kock-van Dalen, R. Heus, R. Laroy, A. J. van Hengstum, A. P. G. Kieboom, H. van Bekkum: Catalytic Dehydrogenation of Reducing Sugars in Alkaline Solution. *Carbohydr. Res.* **91** (1981) 125-138.

4. A. J. van Hengstum, A. P. G. Kieboom, H. van Bekkum: Catalytic Transfer Hydrogenation of Glucose – Fructose Syrups in Alkaline Solution. *Starch* **36** (1984) 317-320.

5. D. Beer, A. Vasella: Aldonhydroximino-lactones. Preparation and Determination of Configuration. *Helv. Chim. Acta* **68** (1985) 2254-2274.

6. H. Scholz, G. Gotsmann (BASF AG): Alkali Metal Salt of Arabonic Acid. *Ger. Offen.* 2,628,056 (1976); *Chem. Abstr.* **88** (1978) 170440c.

7. H. Röger, H. Puke, M. Kunz: Untersuchungen zur oxidativen Spaltung von reduzierenden Disacchariden. *Zuckerind.* **115** (1990) 174-181.

8. S. Dimitrijevich, M. Tatarko, R. Gracy, C. B. Linsky, C. Olsen: Biodegradation of Oxidized Regenerated Cellulose. *Carbohydr. Res.* **195** (1990) 247-256 and references cited therein.

9. K. Heyns, H. Paulsen: Selective Catalytic Oxidation of Carbohydrates, employing Platinum Catalysts. *Adv. Carbohydr. Chem.* **17** (1962) 169-221.

10. J. M. H. Dirkx, H. S. van der Baan: The Oxidation of Glucose with Platinum-on-carbon as Catalyst. *J. Catal.* **67** (1981) 1-13; J. M. H. Dirkx, H. S. van der Baan: The Oxidation of Gluconic Acid with Platinum-on-carbon as Catalyst. *J. Catal.* **67** (1981) 14-20; P. J. M. Dijkgraaf, *Doctoral Dissertation*, Eindhoven University of Technology, 1989.

11. W. Fritsche-Lang, I. Leopold, M. Schlingmann (Hoechst AG): Preparation of Sucrosetricarboxylic Acid. *Ger. Offen.* 3,535,720 (1985); *Chem. Abstr.* **107** (1987) 59408v; I. Leopold, M. Wiesner, W. Fritsche-Lang, Eurocarb V Prague, 1989 Abstract D-1.

12. H. E. van Dam, *Doctoral Dissertation*, Delft University of Technology, 1989.

13. A. T. J. W. de Goede, P. Vinke, H. van Bekkum, unpublished results.

14. T. Böcker, J. Thiem, *Tenside, Surf. Det.* **26** (1989) 318.

15. A. J. Hoefnagel, B. M. Wepster, *Collect. Czech. Chem. Commun.* **55** (1990) 119.

16. H. E. van Dam, A. P. G. Kieboom, H. van Bekkum: Glucose-1-Phosphate Oxidation on Platinum-on-carbon Catalysts: Side-Reactions and Effects of Catalysts Structure on Selectivity. *Recl. Trav. Chim. Pays-Bas* **108** (1989) 404-407.

17. H. E. van Dam, H. van Bekkum, *J. Catal.*, in press.

18. H. E. van Dam, L. J. Wisse, H. van Bekkum, *Appl. Cat.*, in press.

19. Hydroxymethylfurfural (HMF) was kindly supplied by Südzucker AG.

20. P. Vinke, H. E. van Dam, H. van Bekkum, *Proc. Symp. New Developments in Selective Oxidation*, Rimini, Italy, 1989 (G. Centi, F. Trifiro, Eds.), Elsevier Science Publ., Amsterdam, 1990, p. 147 ff.

21. P. Vinke, H. van Bekkum, unpublished results.

22. P. C. C. Smits, B. F. M. Kuster, K. van der Wiele, H. S. van der Baan: The Selective Oxidation of Aldoses and Aldonic Acids to 2-Ketoaldonic Acids with Lead-Modified Platinum-on-carbon Catalysts. *Carbohydr. Res.* **153** (1986) 227-235.

23. G. O. Aspinall, A. Nicolson: The Catalytic Oxidation of European Larch ε-Galactan. *J. Chem. Soc.* **1960**, 2503-2508.

24. C. Venturello, M. Ricci: Oxidative Cleavage of 1,2-Diols to Carboxylic Acids by Hydrogen Peroxide. *J. Org. Chem.* **51** (1986) 1599-1602.

25. M. Floor, K. M. Schenk, A. P. G. Kieboom, H. van Bekkum: Oxidation of Maltodextrins and Starch by the System Tungstate-Hydrogen Peroxide. *Starch* **41** (1989) 303-309.

26. S. C. Bright, V. Lamberti, P. J. Powers (Unilever Ltd.): Oxidized Polysaccharide Detergency Builders. *Brit. Pat.* 1,330,122 (1969); *Chem. Abstr.* **80** (1974) 61392c; S. C. Bright, V. Lamberti, P. J. Powers (Unilever Ltd.): Process for Oxidizing Polysaccharides. *Brit. Pat.* 1,330,123 (1970); *Chem. Abstr.* **80** (1974) 61393d.

27. M. Diamantoglou, H. Mägerlein, R. Zielke: Polycarboxylates from Polysaccharides, Wood and Wood-like Composites as novel Sequestering Agents. *Tenside Deterg.* **14** (1977) 250-256.

28. M. Floor, A. P. G. Kieboom, H. van Bekkum: Preparation and Calcium Complexation of Oxidized Polysaccharides. *Starch* **41** (1989) 348-354.

29. M. Floor, L. P. M. Hofsteede, W. P. T. Groenland, L. A. T. Verhaar, A. P. G. Kieboom, H. van Bekkum: Preparation and Calcium Complexation of Oxidized Polysaccharides. Hydrogen Peroxide as Co-Reactant in the Chlorite Oxidation of Dialdehyde Glucans. *Recl. Trav. Chim. Pays-Bas* **108** (1989) 384-392.

30. E. Dalcanale, F. Montanari: Selective Oxidation of Aldehydes to Carboxylic Acids with Sodium Chlorite-Hydrogen Peroxide. *J. Org. Chem.* **51** (1986) 567-569.

31. M. Floor, *Doctoral Dissertation*, Delft University of Technology, 1989.

15

On the Preparation of Polyvinylsugars

Günter Wulff, Jürgen Schmid, and Theodor Venhoff

Institut für Organische Chemie und Makromolekulare Chemie,
Heinrich-Heine-Universität, D-4000 Düsseldorf, Germany

Summary. Starting from easily accessible monosaccharides vinylsugars are prepared, which can be polymerized either by radical or by anionic initiation. The structure of the monomers is such that polymers are generated which have main chains and side chains entirely composed of C-C-chains. After deprotection and reduction, side chains of unbranched sugar alcohol residues are produced. Vinyl ketones and 4-vinylphenyl derivatives of protected monosaccharides obtained from the reaction of aldehydes or amides with vinyl- or 4-vinylphenyl-magnesiumchlorides are especially suitable for polymerization. Depending on the experimental conditions, mole masses between 100 000 and 700 000 can be achieved by emulsion polymerization.

Introduction

For quite some time renewable resources have been used as raw materials in the preparation of polymers. Mono- and oligosaccharides in particular, and their derivatives offer certain advantages. They are produced in large quantities at a relatively low price. They are used mainly as diol compounds in polycondensation or polyaddition polymers.[1] Growing interest is being shown in the use of vinylsugars for polymerization. These polymers are mainly investigated as water-soluble linear polymers. The vinyl group is usually introduced through acrylic acid or methacrylic acid by forming ester or amide bonds.[2] In some cases ether or glycosidic linkages are used to introduce a polymerizable double bond.[2,3]

As far as we know no polyvinylsugars are being used commercially. The reason might be that their use as substitutes for bulk polymers is prohibited by their price. The chance of being used with the present methods of preparation can therefore only be as technical or speciality polymers. Since polyvinylsugars constitute polymers of a new structural type, their possible application in industry cannot fully be foreseen. It is not known what type of properties can be achieved by the various chemical structures possible in the preparation of these polymers. Hence, prerequisite for the introduction of polyvinylsugars as a commercial product will be the systematic investigation of structure-property relationships. As long as products are intended to be used in a conventional way these invetigations might be relatively easy. But because of the very special and new structure of these polymers novel operational areas are to be expected, entailing more difficult systematic investigations.

In our pursuit of varied directions for the preparation of polyvinylsugars we are trying to synthesize readily degradable polymers to be used for short term use articles, as well as stable polymers which are resistant to chemicals, heat, and microorganisms. In the latter case, polymers carrying hydrophobic protecting groups attached to the hydroxyl groups, and alternatively their hydrophilic counterparts with free hydroxyl groups, might be equally valuable. We therefore start from protected sugar derivatives which can be either transformed to polymers with and without protecting groups, i.e. new types of polymers which would fullfill the following prerequisites:

(i) Main chains and side chains should be entirely composed of C-C-connections to avoid any weak bonds for chemical or biochemical decomposition.

(ii) Side chains should not be branched, in order to achieve a comb-like arrangement as in **1** and **2**.

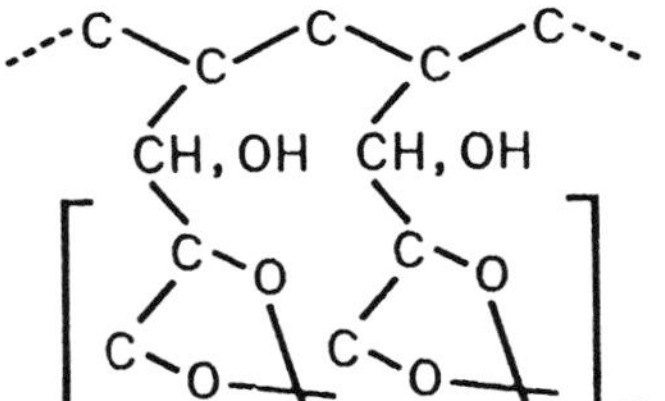

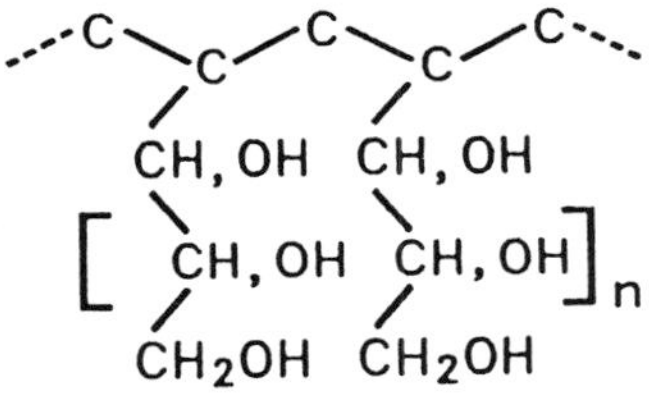

(iii)Monomers should be polymerizable by radical initiation, but also anionically. Anionic polymerization would enable the generation of polymers with defined tacticity (isotactic, syndiotactic) as well as blockcopolymers.

This account deals with the preparation of new vinylsugar monomers meeting the above requirements, and also with their polymerization by radical initiation.

Preparation of New Vinyl-Sugars

Olefinic double bonds directly connected to sugars by C-C bonds usually cannot be polymerized radically or anionically, but have to be activated in a suitable way. We have worked out procedures to prepare vinyl ketones **3** and 4-vinylphenyl derivatives **4** from sugar aldehydes since these two types should be polymerizable by anionic and radical initiation.

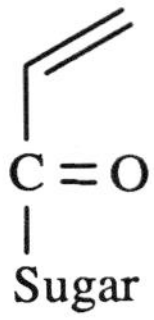

C = O
|
Sugar

Sugar

3

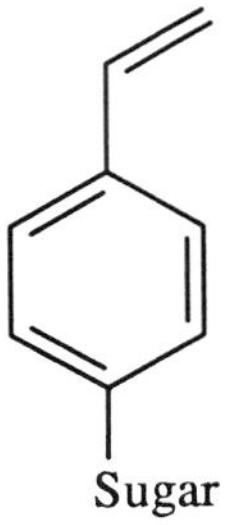

4

The following schemes show the essential steps for the preparation. As an example, reaction of 2,3:4,5-di-*O*-isopropylidene-1-aldehydo-β-D-*arabino*-hexosulose **5**, readily accessible from D-fructose,[4] with the vinyl-magnesium bromide Grignard reagent in tetrahydrofuran yield an 85 : 15 mixture of the two diastereomeric vinyl derivatives **6a** und **6b**. Subsequent oxidation of the diastereomeric mixture with the Swern reagent (DMSO / (COCl)$_2$) affords the vinylketone **7** in good yield. The corresponding styryl derivative **8**, in the form of a 2 : 1 mixture of diastereomers, is obtained in a one step reaction of aldehyde **5** with 4-vinylphenyl-magnesiumchloride. In this case the mixture of the two diastereomers **8** is used for polymerization.

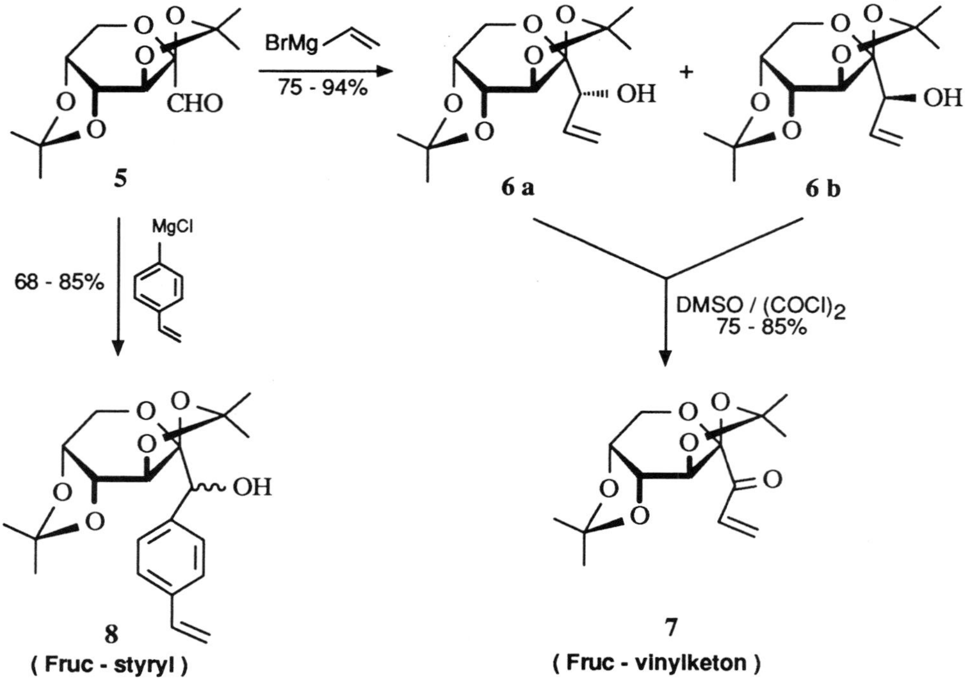
BrMg
75 - 94%
CHO
OH
+
5
6 a
6 b
MgCl
68 - 85%
DMSO / (COCl)2
75 - 85%
8
(Fruc - styryl)
7
(Fruc - vinylketon)

Correspondingly, other aldehydes like 1,2:3,4-di-*O*-isopropylidene-6-aldehydo-D-galactose 9[5], 2,3:4,5-di-*O*-isopropylidene-D-arabinose (**12**)[6], and 2,3-*O*-isopropylidene-D-glyceraldehyde (**16**)[7] are reacted in a similar way and the new monomers **10**,[8] **11**, **13**, **14**, **15**, **17**, and **18** are obtained:

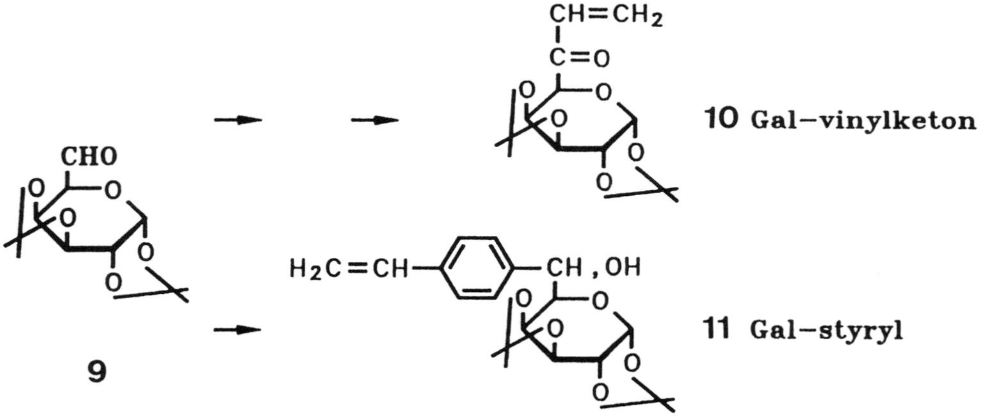
CH=CH2
C=O
10 Gal–vinylketon
CHO
H2C=CH–
CH , OH
11 Gal–styryl
9

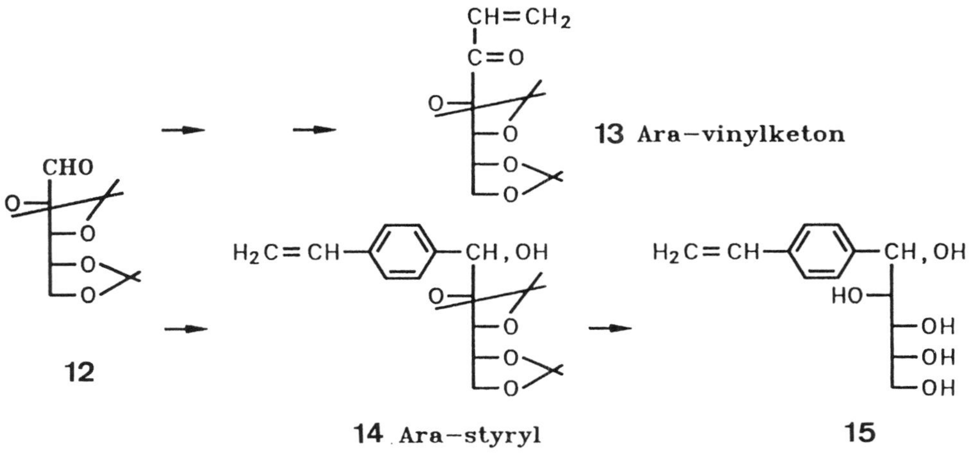

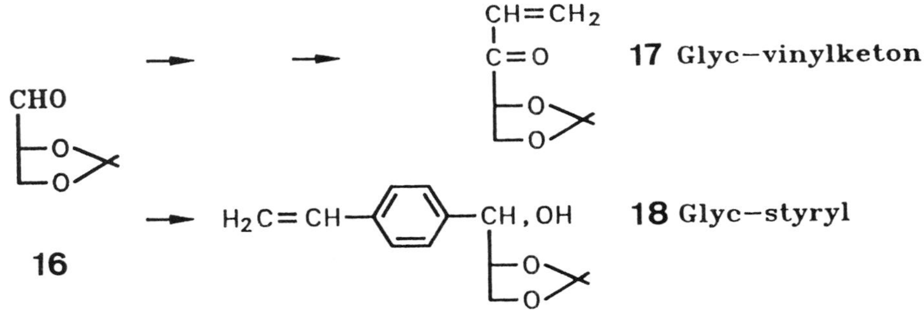

CH=CH₂ / C=O on compound **13 Ara-vinylketon**

12 CHO derivative →→

H₂C=CH—⟨⟩—CH,OH (isopropylidene) **14 Ara-styryl**

H₂C=CH—⟨⟩—CH,OH / HO— —OH —OH —OH **15**

CH=CH₂ / C=O **17 Glyc-vinylketon**

16 CHO derivative →→

H₂C=CH—⟨⟩—CH,OH **18 Glyc-styryl**

To shorten the preparative route for the vinylketones reaction with carboxylic acid derivatives instead of the aldehyde were tried in order to obtain the vinylketone in one step. Accordingly, D-gluconolacton **19**, which is commercially available and produced on a large scale, is converted in a one-step literature procedure[6] into the 3,4:5,6-di-*O*-isopropylidene-D-gluconic acid methylester (**20a**). Our investigations showed that this compound always contained 20 % of the corresponding 2,3:5,6-di-*O*-isopropylidene derivative **20b**. On reaction of **20a** with vinylmagnesium bromide this compound yielded in a ratio of 85 : 15 two substances. The main component proved to be the di-addition product **22** that might find use as a monomer for special application. The minor product **23** is the Michael-addition product of vinylmagnesium bromide to the first step product, the vinylketone. The same products are obtained if tri-*O*-isopropylidene-D-gluconic acid **21** mentioned in the relevant literature[9] is reacted with vinylmagnesium bromide.

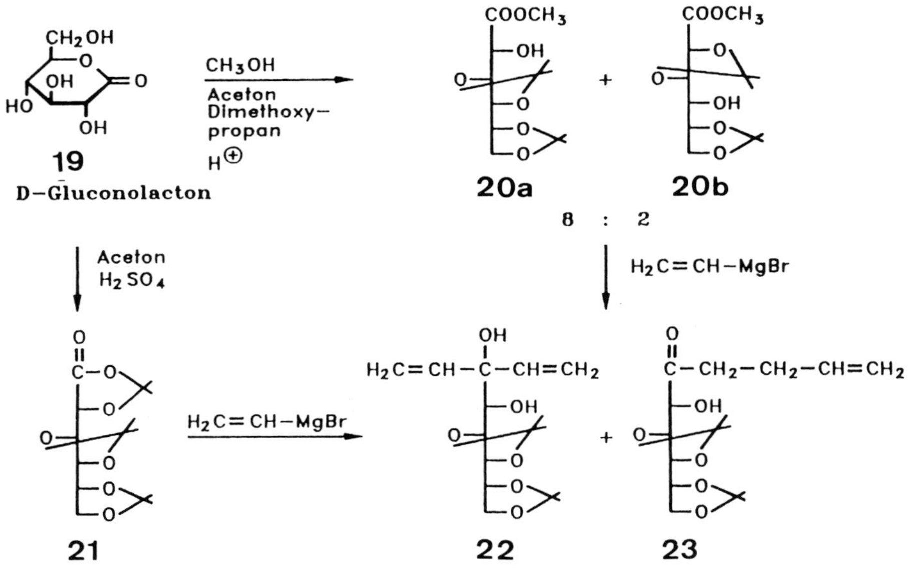

Surprisingly, the acid chloride **25** achieves a particularly high yield of the Michael-addition product **26** of the vinylketone. The potential use of this readily available compound is now being further investigated.

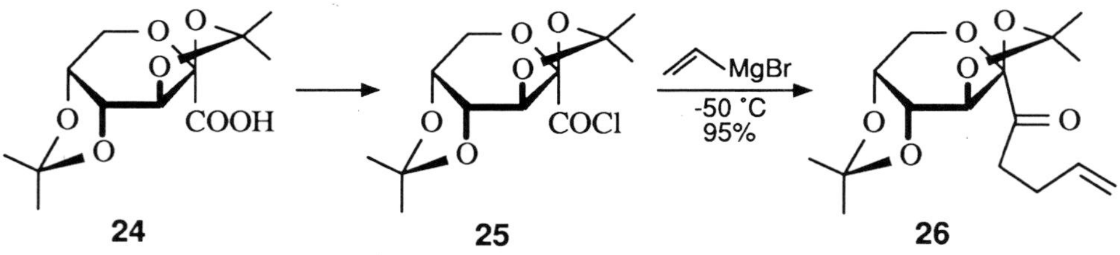

Another possibility of obtaining mono-addition products is to use the corresponding amides, whereby the starting materials can be obtained in a one-pot reaction: treatment of the D-gluconolactone **19** with diethylamine, removal of excess of amine in vacuo and trituration of the residue with acetone / sulfuric acid generates the diisopropylidene amide **27** in 80 % yield. In a second step, **27** is converted with vinylmagnesium bromide in high yield to the desired vinylketone **29**, which thus constitutes a readily accessible highly versatile monomer that regarding to its availability compares favourably with any other monomer prepared so far.

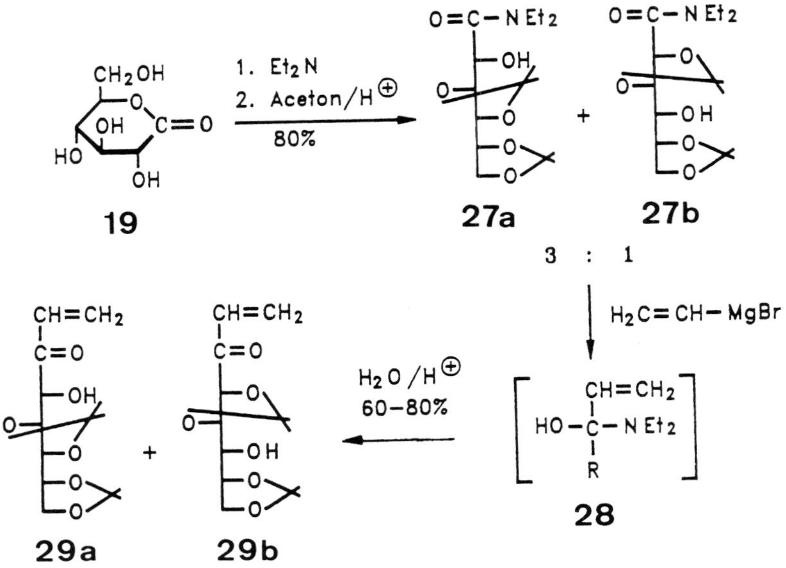

$O=C-NEt_2$ $O=C-NEt_2$

CH$_2$OH

1. Et$_2$N
2. Aceton/H$^\oplus$
   80%

19   **27a**   **27b**

3 : 1

H$_2$C=CH–MgBr

CH=CH$_2$ CH=CH$_2$

C=O   C=O

H$_2$O/H$^\oplus$
60–80%

$$\left[\ HO-\overset{\displaystyle CH=CH_2}{\underset{\displaystyle R}{C}}-NEt_2\ \right]$$

29a   **29b**   **28**

In order to obtain, for comparative purposes, comb-like structures with longer side chains in the polymer, macromers are used which have been prepared in another context.[10,11] By group transfer polymerization of 1,3,2-dioxaborol initiated with 4-vinylbenzaldehyde macromers of type **30** are obtained with varying chain length, whereby the degree of the polymerization can be controlled by the ratio of initiator to monomer. Thus P_n of 3 to 20 can easily be achieved.

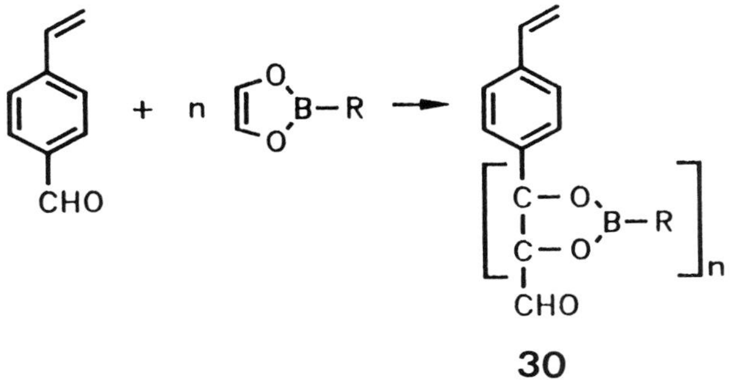

+ n ⟶

CHO

$$\left[\begin{array}{c} C-O \\ |\quad\ \ \diagdown B-R \\ C-O \diagup \\ | \\ CHO \end{array}\right]_n$$

30

Preparation of the Polymers

The monomers **7**, **8**, **10**, **11**, **13** - **15**, **17**, **18**, and **29** were polymerized by radical initiation. The best method was emulsion polymerization with an initiator system $K_2S_2O_8 / K_2SO_3$. Under controlled conditions the molecular weight of the resulting polymers can be regulated. Depending on the condition used, mole masses of 100 000 to 700 000 are obtained. On usual work up procedure yields of 40 - 90 % of the pure polymers are obtained. As an example, in polymer **31**, obtained from monomer **10**, the isopropylidene protecting group can be removed with 80 % formic acid to give **32**. During this procedure complete removal of the isopropylidene groups is possible, but the resulting polymers contained some formyl residues, which are removed though on the subsequent reduction with sodium borohydride. Thus, polymers of type **33** are obtained which contain a polyvinyl backbone with attached sugar alcohol side chains in a comb-like arrangement. Alternately, the sugar-containing polymers **32** can also be oxidized to a polymer with galactonic acid side chains **34** and **35**.

Polymerization of Gal—vinylketon 10

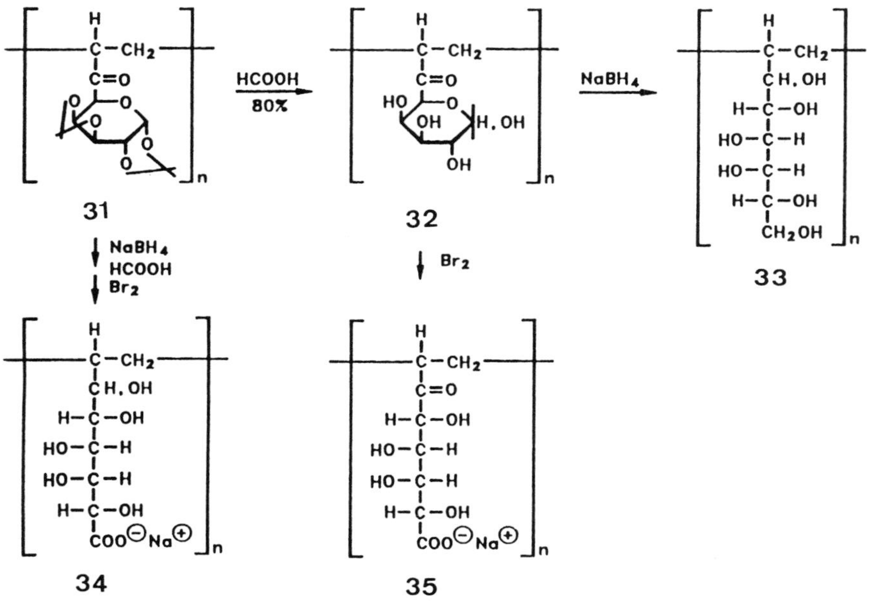

With the fructose-containing polymer **36**, obtained on polymerization of the fructosyl-vinylketone **7**, similar transformations can be effected, e.g., **36** → **37** → **38**:

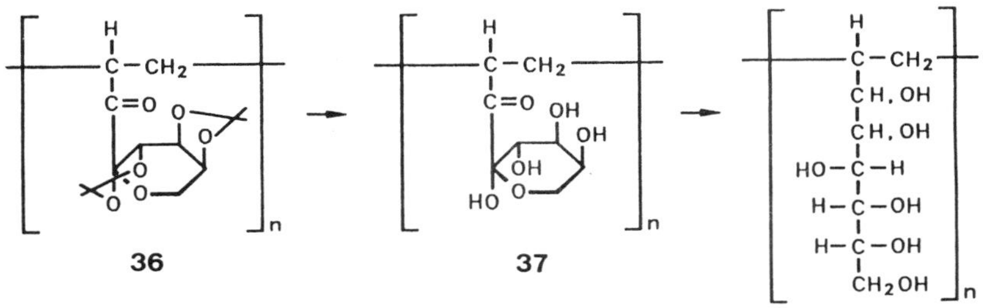

36 **37** **38**

The last scheme is a summary of the principle different structural types of polymers which have been obtained in our group so far. From open chain vinylketones the polymers **39** with m= 1 and 2 with an additional CHOH group are obtained. Deprotected water soluble polymers with varying chain length are represented by **40**, in which the sugar alcohol can consist of 3, 5, and 6 carbon atoms. Alternatively, in some cases, the end groups can be oxidized to a carboxylic acid. The corresponding polymers with a styrene group are represented by formulae **41** and **42**.

New Types of Polymers Prepared:

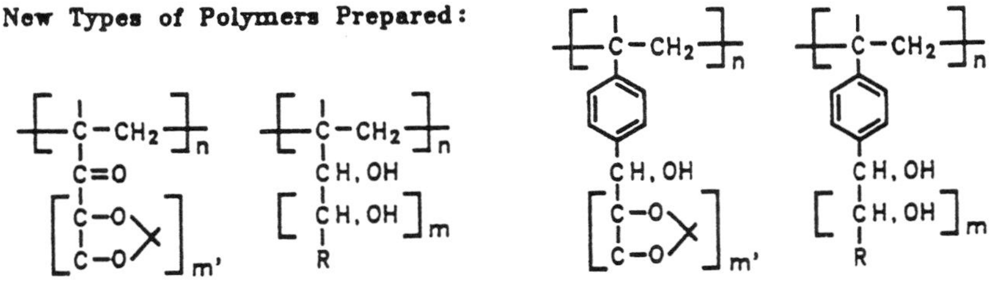

39 m'=1,2,2(+1C)

40 m=1,3,4 R=CH_2OH, R=COOH

41 m'=1,2,2(+1C)

42 R=CHO, m=1,3,4 R=CH_2OH, R=COOH

In the case of the deprotected linear side chain polymer the influence of the side chains carrying 3, 5, 6, and 21 carbon atoms can be compared. The solubility in water is highest for a side chain of 5 carbon atoms (ca. 90 g/l), whereas the solubility drops for polymers with longer and shorter side chains. With long side chains (C-21) only a very low solubility is present. Formulae **43** and **44** represent polymers with protected and deprotected cyclic sugar side chains.

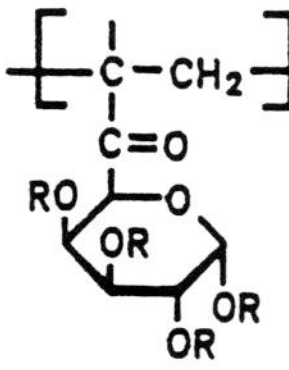

43 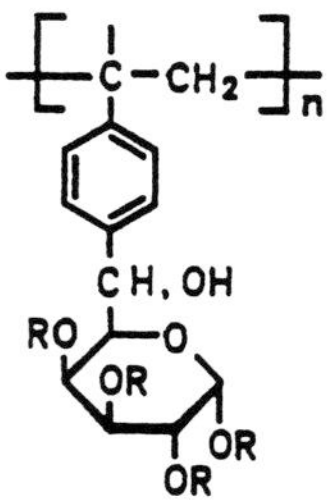   44

R=H
R=isopropylidene$_{/2}$

Some data on the polymers are given in Table 1 and 2. Molecular weight is determined by G.P.C. and the numbers are correlated to polystyrene standards. In several cases the determination was also performed by membrane osmometry. In addition, values of optical rotation and of the intrinsic viscosity are given. The viscosity of the compound carrying galactonic acid side chains, e.g. **35**, is especially high. Most of the polymers show interesting circular dichroism spectra. Further investigations on the properties of these polymers are under way. We expect that the use of light scattering methods, in particular, will produce important results.

Table 1. Polymers prepared by emulsion polymerization of vinyl-saccharides

Monomer	Yield (%)	M GPC	M_n Osm	$[\alpha]_D$ (°)	$[\eta]_{lim}$ $c \to 0$
Gal-vinylketone (**10**)	66	160 000	200 000	-60	17
"	69	420 000	480 000	-60.5	40
"	77	–	620 000	–	42
Gal-styryl (**11**)	77	110 000	–	-37	30
"	71	210 000	–	–	–
Fruc-vinylketone (**7**)	70	–	650 000	–	48
Fruc-styryl (**8**)	72	–	160 000	–	12
"	61	190 000	200 000	–	15
Ara-vinylketone (**13**)	36	316 000	300 000	–	67
"	71	600 000	–	–	–
Ara-styryl (**14**)	91	270 000	–	+22	103
Glyc-vinylketone (**17**)	50	250 000	–	+130	–

Table 2. Polymers obtained by polymer-analogous transformation

Starting Polymer Prepared from 4	Molecular Weight	Reaction Condition	Polymer	$[\alpha]_D$ (°)	$[\eta]_{lim\ c\to 0}$ in H_2O	0.1m Na_2SO
Gal-vinylketone (**10**)	100 000	80% HCOOH NaBH$_4$	**33**	-65	67	45
Gal-vinylketone (**10**)	420 000	80% HCOOH Br$_2$, NaOH	**35**	-474	500	100 (pH11.6)
Fruc-styryl (**8**)	200 000	80% HCOOH	Type **44** R=H	-25.6	16	–
Fruc-vinylketone (**7**)	650 000	80% HCOOH NaOH	**37**	–	–	23
Fruc-vinylketone (**7**)	650 000	80% HCOOH NaBH$_4$	**38**	–	–	26
Ara-styryl (**14**)	270 000	80% HCOOH	Type **42** R=CH$_2$OH	+14.5	–	20
Glyc-vinylketone (**17**)	250 000	80% HCOOH	Type **42** R=CH$_2$OH	+11.4	–	12

Acknowledgement. This work was supported by *Bundesminister für Forschung und Technologie* under the registry number 0319057A. The responsibility for this paper is completely due to the author.

References

1. J. Feldmann, H. Koch: Stärke und deren Folgeprodukte als Bausteine. *Schriftenreihe des Fonds der chemischen Industrie*, Heft **25** (1986) 49-59.

2. Compilation of literature see: J. Klein, D. Herzog: Poly(vinylsaccharides). 2. Synthesis of some Poly(vinylsaccharides) of the Amide-Type and Investigation of their Solution Properties. *Makromol. Chem.* **188** (1987) 1217-1237.

3. Y. Koyama, K. Kurita: *Current Topics in Polymer Science* (R. M. Ottenbrite, L. A. Utraki, S. Inoue, Eds.) Vol. I, Karl Hanser Verlag, Munich 1987, p.130ff.

4. R. W. Lowe, W. A. Szarek, J. K. N. Jones: Conversion of 2-Hexuloses into 3-Heptuloses: Synthesis of D-*manno*-3-Heptulose. *Carbohydr. Res.* **28** (1973) 281-293.

5. R. E. Arrick, D. C. Baker, D. Horton: Chromium Trioxide-Dipyridine Complex as an Oxidant for Partially Protected Sugars; Preparation of aldehydo and certain keto Sugar Derivatives. *Carbohydr. Res.* **26** (1973) 441-447.

6. H. Regeling, E. de Rouville, G. J. F. Chittenden: The Chemistry of D-Gluconic Acid Derivatives. Synthesis of 3,4:5,6-Di-*O*-isopropylidene-D-glucitol and 2,3:4,5-Di-*O*-isopropylidene-*aldehydo*-D-arabinose from D-Glucono-1,5-lactone. *Recl. Trav. Chim. Pays-Bas* **106** (1987) 461-464.

7. D. Y. Jackson: An Improved Preparation of (+)-2,3-*O*-Isopropylidene-D-glyceraldehyde. *Synth. Commun.* **18** (1988) 337-341.

8. This compound has been prepared in another context and by another synthetic route: D. Horton, J. H. Tsai: Routes to Higher-carbon Sugar Derivatives having Keto-acetylenic and -alkenic, and α,β-Unsatured Aldehyde Functionality. *Carbohydr. Res.* **75** (1979) 151-174.

9. J. W. W. Morgan, M. L. Wolfrom: Lithium Aluminum Hydride Reduction of 3,4:5,6-Di-*O*-isopropylidene-D-gluconamide and Di-*O*-isopropylidene-galactaramide. *J. Am. Chem. Soc.* **78** (1956) 2496-2497.

10. G. Wulff, P. Birnbrich, A. Hansen: Synthese von oligomeren und polymeren Monosacchariden durch Aldol-Gruppentransfer-Polymerisation. *Angew. Chem.* **100** (1988) 1197-1198; *Angew. Chem., Int. Ed. Engl.* **27** (1988) 1158-1159.

11. G. Wulff, P. Birnbrich, unpublished results.

16

From Carbohydrates to Pigments : An Exercise in Molecular Material Science and Material Transformation[1-3]

Jörg Daub, Knut M. Rapp, Josef Salbeck, and Ulrich Schöberl

Institut für Organische Chemie, Universität Regensburg,
D-8400 Regensburg, Germany

Summary. The utilization of saccharides as basic materials for the synthesis of high-grade chemicals, optional in electronic and opto-electronic applications, is outlined. A general scheme for the overall procedure is given. The conversion of 5-hydroxymethyl-2-furaldehyde (HMF) – a key chemical of industrial sugar chemistry – into molecular components with electron-transfer and light-sensitive properties is presented. The syntheses of electrochemichromic and photochromic molecular components are described and their application in chemical and physical switching devices is given. Results are presented concerning the chemistry of HMF-based poly(arylene vinylenes) and carbohydrate-modified conducting polymers.

1. Introduction

The quest for renewable resources as substitutes for the fossil raw materials coal, crude oil or natural gas has its ups and downs depending on economic and political actualities.[4] Nevertheless, there is need and interest to outline strategies providing the chemical conversion of renewable resources into compounds and materials which might be of significance for trendsetting technologies. In the following account possibilities are examined to utilize sucrose, glucose, and fructose as the source for molecular materials suitable for electronic and opto-electronic devices.[5,6] Compounds that meet these requirements are pigments with either one of the following properties: (1) electron transfer acitivity, (2) electrochemichromism, (3) photochromism, (4) luminescence, (5) non-linear optical response, and (6) magnetic properties.

Correspondingly, a general description is given on the conversion of "natural" resources into "molecular" materials as exemplified by the use of *sucrose* and *sucrose-derived saccharides* for the preparation of molecular materials of electronic and opto-electronic interest. In addition, the elucidation of the properties of those molecular components by physicochemical methods is described.[7]

2. Sucrose : Depot-compound of Solar Energy – Source-compound for Chemical Use

To evaluate the use of sucrose as basic source-material, some general data are given in Table 1 highlighting the energetics of solar irradiation and the availability of sucrose as a renewable energy resource which is industrially exploited by high-standard technology.[8]

Table 1. Sucrose as a biochemically renewable deposit of solar energy on earth

approximate total radiation energy of sun; in W	$3.8 \ 10^{26}$	Ref.[a]
solar radiation energy received by earth; in W	$1.7 \ 10^{17}$	Ref.[b]
production rate of biomass on earth (total land regions); to/sec	3 170	Ref.[c]
approximate total amount of carbohydrates produced by photosynthesis	(90 % of biomass)	
estimate of total amount of sucrose industrially produced		
worldwide (1989/90); to/sec	3.4	Ref.[d]
total amount of sucrose used for industrial purpose worldwide; to/sec	~0.025	Ref.[4]

a) Enzyklopädie Naturwissenschaft und Technik (K.-H. Schriever, F. Schuh, Eds.), Zweiburgen Verlag, Weinheim, 1981.
b) A. Goetzberger, V. Wittwer, Sonnenenergie – Physikalische Grundlagen und thermische Anwendungen, Teubner, Stuttgart, 1986.
c) W. H. Bloss: Solar, Wind, Waves, Biomass, Heat pumps, Hydraulic energy. Renewable Energy Resources, World Energy Conference, Survey of Energy Resources (F. Bender, K. E. Koch, U. Ranke, W. Süss, Eds.), Munich, London, 1980.
d) Wirtschaftsübersicht: *Zuckerind.* 115 (1990) 215.

Solar energy is an inexhaustible source of energy as is demonstrated by the total amount of irradiation energy generated by nuclear fusion on sun ($> 10^{26}$ W) and the solar irradiation energy received by earth ($> 10^{17}$ W). The biomass produced by photosynthesis, which largely consists of carbohydrates, exceeds 3 000 to/sec. Sucrose industrially produced amounts to more than 3 to/sec worldwide. So far only a minor share is used industrially.

The utilization of sucrose in food and non-food industry is outlined in the following scheme.[5,9] Most of the sucrose is consumed in the food section, other major fields of application are biotechnology and fermentation processes. Possible applications of sucrose as tensides,[10], surfactants, or liquid crystals[11] are under investigation at different places. Fine chemicals prepared from sucrose are of interest as chiral synthetic units.[12]

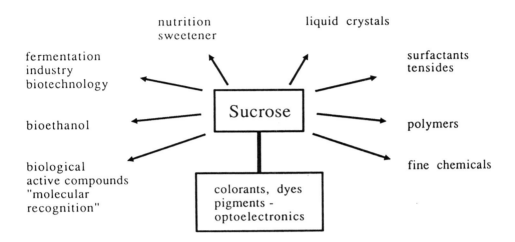

Sucrose as raw material for food and non-food utilization

As of now, little emphasis was given to saccharides as source-chemicals of functionalized non-carbohydrate polymers or functionalized dyes and pigments with applications in electronics and opto-electronics. In dyes and colourants it is the electronic structure of the chromophors that makes them electron-transfer active as well as suitable for energy transfer interaction. Both of those properties are essential prerequisites for electronic and opto-electronic materials.[13,14]

In the future, new organic opto-electronic materials are expected to gain increasing market shares and pay-offs in technologies for the storage, transformation, and processing of information and energy.[15] Because of its molecular structure and of the high standard of industrial processing, sucrose must be a rich starting material for those chemical components. The same applies to glucose, fructose, cellulose, and starch.

A general strategy for the chemical conversion of carbohydrates into pigments may be outlined as follows:

poly-, oligo-, or monosaccharides $\rightarrow \rightarrow$ π-heterocycles $\rightarrow \rightarrow$
molecular units (molecular materials) for electronic and opto-
electronic applications $\rightarrow \rightarrow$ materials

In the first step, agriculturally harvested materials like poly-, oligo-, or monosaccharides are processed and converted into π-heterocycles which afterwards are transformed into molecular units with dye-behaved functionalization in order to absorb visible light or near infrared irradiation. In the next assembling step, the molecular units attain cooperative behaviour (molecular materials). A final finishing process makes materials ready for practical use.[16] In other words, the conversion of natural materials into those for technical use entails a multistep procedure, of which in this account only the synthesis and characterization of molecular units with electronic and opto-electronic characteristics will be discussed.

3. Five and Six-membered π-Heterocycles from Saccharides

It is well established that a manifold of five and six-membered π-heterocycles can be prepared from sucrose, heptulose, glucose, and fructose, some of which are shown in the following scheme:

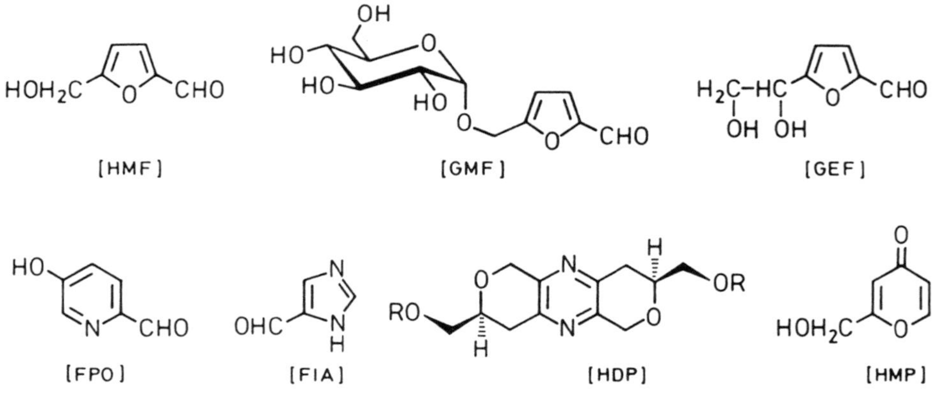

π-Heterocycles with potential use as precursor compounds for electronic and opto-electronic applications

All of these compounds constitute molecular components which, by proper functionalization, can be converted into pigments. The furan and imidazole derivatives 5-hydroxymethyl-furaldehyde (HMF),[17,18] 5-(α-D-glucosyloxymethyl)-furaldehyde (GMF),[19] 5-(dihydroxyethyl)-2-furaldehyde (GEF),[20] and 4-formylimidazole (FIA)[21] are obtained in good yields from glucose (→HMF, FIA), isomaltulose (→GMF), or from sedoheptulose (→GEF). 6-Formyl-3-pyridinol (FPO) is available from (HMF) by electrochemical or chemical conversion.[22]

1,4-Diazines as, for example, palythazin (HDP), again are molecular units with good prospects for reversible electron-transfer behaviour.[23,24] The same applies to the yet unknown 2-hydroxymethyl-4H-pyran-4-one (HMP) which is expected to be convertible into the interesting class of pyrylium salts.[25]

4. 5-Hydroxymethyl-2-furaldehyde (HMF) : Basic Compound for High-grade Chemicals – Molecular Materials for Electronic and Opto-electronic Devices

5-Hydroxymethyl-2-furaldehyde (HMF) turns out to be a versatile π-heterocyclic compound for the conversion into electron-transfer-active (chapter 5) or light-sensitive (chapter 6) molecular components as will be shown by HMF-derived compounds in the ensuing scheme:[26]

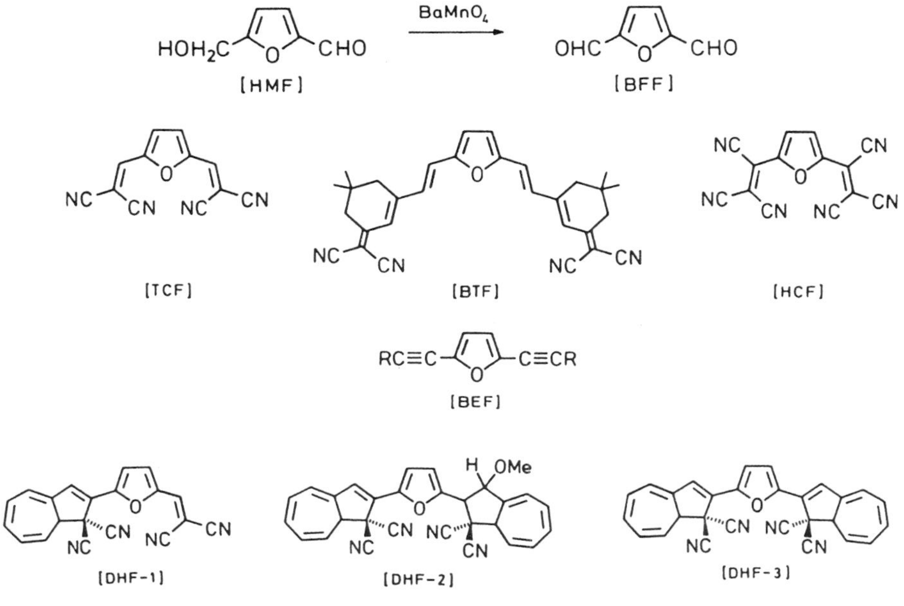

Molecular units derived from 5-hydroxymethyl-2-furaldehyde (HMF)

All of them are prepared from furan-2,5-dicarboxaldehyde (BFF)[27,28] which is synthesized from HMF by oxidation.[29] Tetracyano compound TCF,[30] a key-substance, results by condensation with malononitrile in the presence of base. In the solid state TCF has a nearly planar structure with an approximate C_{2v}-symmetry and *s-trans* stereochemistry around C-1–C-10[31] (cf. Fig. 1).

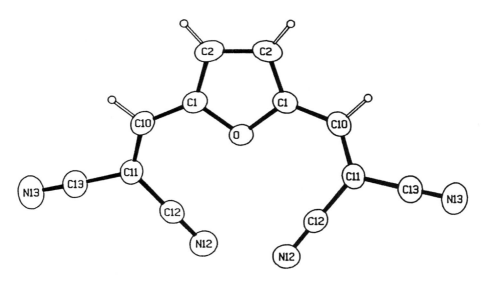

Fig. 1. Crystal structure of TCF. Significant bond lengths: C(1)-C(2) 1.372 Å; C(2)-C(2) 1.401 Å; C(1)-C(10) 1.423 Å; C(10)-C(11) 1.353 Å; C(11)-C(12) 1.429 Å; C(12)-N(12) 1.148 Å

The hexaene furan BTF is formed from the dialdehyde BFF in a two-step sequence: base catalyzed condensation of BFF with isophorone in the first step and subsequent reaction with malononitrile. It turned out, by electrochemical investigations, that TCF and BTF exhibit quite contrasting electron-transfer properties (chapter 5). The hexacyano derivative HCF was synthesized from the tetracyano derivative TCF by twofold addition of cyanide anion (potassium cyanide)[32] and subsequent oxidation of the dianion.[33] The bisacetylene BEF, accessible from BFF in a multistep sequence, is anticipated to be convertible into polyacetylenes and poly(phenylenes), which are of interest as conducting and electrochemichromic polymers.[34,35] Dihydroazulenes DHF-1, DHF-2, and DHF-3 again were synthesized from the tetracyanovinyl derivative TCF in a multistep sequence[34] and differ significantly in their photochromic behaviour as will be shown in chapter 6.[36]

5. Electron-transfer Compounds → Electrochemichromism

The difference of furanoid and benzenoid chemistry, the former representing renewable-resources-chemistry, the latter a typical building block originating from fossil materials, is strikingly substantiated by the electron-transfer behaviour of 2,5-bis(dicyanovinyl)-furan (TCF) and 1,4-bis(dicyanovinyl)-benzene (TCB), which is obtained from terephthal-dialdehyde. Both, cyclic voltammetry[37] and spectroelectrochemistry,[38] disclose significant differences on electrochemical reduction. The results are given in Fig. 2. Unlike the benzenoid TCB, the furan derivative TCF exhibits clear reversible formation of the radical anion (TCF)$^{\bullet-}$ and the dianion (TCF)$^{2-}$. The curvature of the cyclic voltammogram of TCB[39] is best explained by an EC-type mechanism, representing a two-step-reaction consisting of a one-electron reduction to the radical anion (TCB)$^{\bullet-}$ in the first step and a subsequent chemical step leading to dimerization under C-C-bond formation.[40]

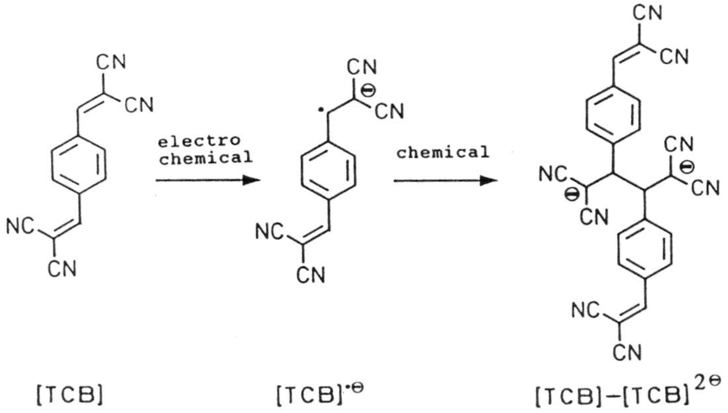

Electron-transfer chemistry (EC-behaviour) of 1,4-bis(dicyanovinyl)-benzene (TCB)

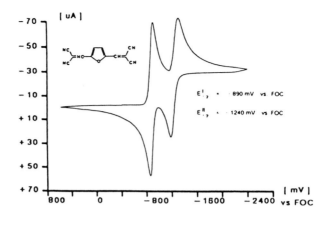

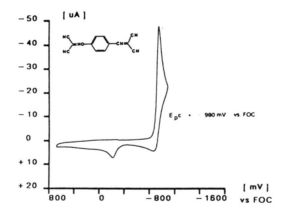

Fig. 2. Cyclic voltammograms of TCF (above) depicting the reversible formation of radical anion and dianion, and of TCB (below) showing the formation of the radical anion and subsequent fast C-C-dimerization[27]

The reversibility of the reduction sequence

$$[TCF] \rightleftharpoons [TCF]^{\bullet-} \rightleftharpoons [TCF]^{2-}$$

is analytically proven by UV/VIS-spectroelectrochemistry (Fig. 3) and by the reversible change of colour (electrochemichromism). In solution, the neutral compound TCF appears to be yellow and turns blue on reduction to the radical anion. Further reduction to the dianion leads to an orange solution (Table 2).

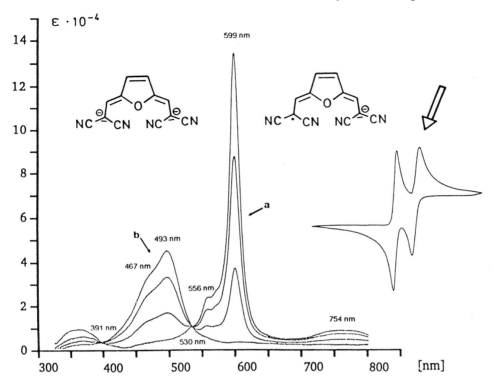

Fig. 3. Electrochemichromism of radical anion / dianion conversion of [TCF]•⁻ and [TCF]²⁻ as shown by UV/VIS-spectroelectrochemistry (Taken from ref.[27]). Insert: Cyclic voltammogram with an arrow indicating the potential of generating the electronic spectra.

Table 2. Electrochemical potential [vs. ferrocene (FOC)] and absorption spectra on reduction of the electron-transfer compounds TCF, HCF, BTF, and TCNQ

Compound	neutral compound λ_{max}(nm)	$E_{1/2}$(mV)	radical anion λ_{max}(nm)	$E_{1/2}$(mV)	dianion λ_{max}(nm)
TCF	393(4.5)	-890	599(5.0)	-1240	493(4.6)
HCF	423(4.5)	-205	663(5.1)	-570	603(4.5)
BTF	514(4.3)	-1190	mixed potential	-1010	617(4.5)
TCNQ	392(4.8)	-190	842(4.7)	-740	330(4.8)

The molecular description of this electrochemichromic phenomenon is given in the following scheme. Electrochromism (electrochemichromism)[41] is gaining application in optical displays, light valves, electrooptical switching, and as photoconductives.[42]

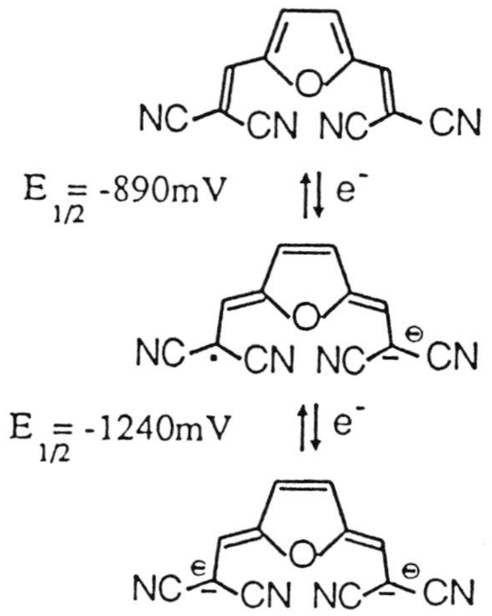

NC CN NC CN

393nm (4.5)   bright yellow
409nm (4.5)

$E_{1/2}$ = -890mV   ↑↓ e⁻

NC CN NC CN

358nm (4.1)
556nm (4.3)
599nm (5.0)   deep blue
754nm (4.1)
875nm (4.1)

$E_{1/2}$ = -1240mV   ↑↓ e⁻

NC CN NC CN

467nm (4.5)
493nm (4.6)   orange

Electron-transfer behaviour, contrasting to [TCF], is exhibited by the polyene BTF (Table 2).[43] The reduction gives a single wave, analyzed as a two-electron reduction, leading in one step to the dianion [BTF]²⁻ with no indication of the paramagnetic radical anion [BTF]•⁻ as an intermediate. Therefore, on the basis of these results, the tetracyano compound TCF is found to be a one-electron mediator, whereas the hexaenic derivative BTF appears to be a two-electron transfer reagent.[44] Applications of the different behaviour in the realm of electron-transfer catalysis[45] are under investigation.

The bis(triscyanovinyl) derivative HCF was synthesized in order to explore access of carbohydrate-derived heterocycles into the chemistry of stable radical ion salts. It is well established that in the solid state radical ions may show electrical conductivity.[46] Due to the six cyano substituents, HCF is reduced at less negative potential compared to the tetracyano compound TCF (Table 2). Furthermore, in comparison with the reduction potential of tetracyanoquinodimethane (TCNQ) it turns out that the radical anions of HCF and TCNQ have rather the same reduction potential. Tetracyanoquinodimethane (TCNQ) is one of the prototypes of "organic metals" with conducting or semiconducting properties.[47] The findings that [HCF]•⁻ and [HCF]²⁻ as well as [TCNQ]•⁻ and [TCNQ]²⁻ are reversibly formed by electrochemical reduction of the neutral compounds is again verified by UV/VIS/NIR-spectroelectrochemistry as exemplified by the one-electron reduction to the radical anions in Fig. 4 and Fig. 5.

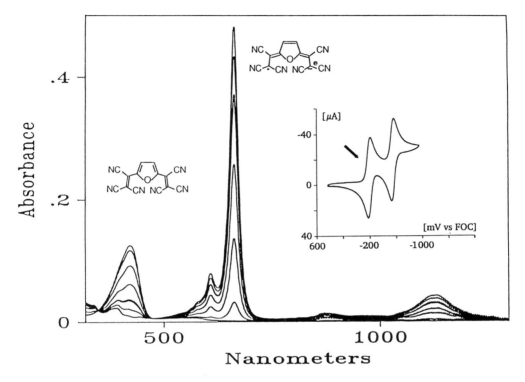

Fig. 4. UV/VIS-spectroelectrochemistry of the formation of the radical anion [HCF]•‾ from HCF by electrochemical reduction. Insert: Cyclic voltammogram of radical anion and dianion formation of HCF, the arrow indicating the scan of potential under which the absorption spectra were monitored

TCNQ

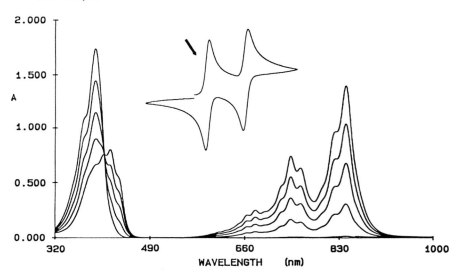

Fig. 5. UV/VIS-spectroelectrochemistry of the formation of the radical anion [TCNQ]$^{\bullet-}$ from TCNQ by electrochemical reduction. Insert: Cyclic voltammogram of radical anion and dianion formation of TCNQ, the arrow indicating the scan of potential under which the absorption spectra were monitored

6. Light-sensitive Compounds – Photochromism

The phenomenon which describes change of colour caused by irradiation and thermal or photochemical bleaching is termed photochromism.[48] Photochromic materials offer interesting applications including the storage and processing of information and the conversion of irradiation energy into heat.[49] Photochromism is abundant in nature, as for example in the processes of vision or photomorphogenesis.[50] Photochromes are expected to be capable of the storage of information as high as the theoretical storage density of 6.5×10^{12} bits/cm^3 at $\lambda_{max} = 532$ nm in a 3D memory.[51]

In his doctoral dissertation, Knöchel[52] made the observation, that a solution of dihydroazulene (DHA), when exposed to sun light, turns deeply red and reverts to yellow in the dark. It turned out that a photorearrangement of dihydroazulene (DHA) to vinylheptafulvene (VHF) occurs, i.e. an electrocyclic 10π-electron arrangement:[53]

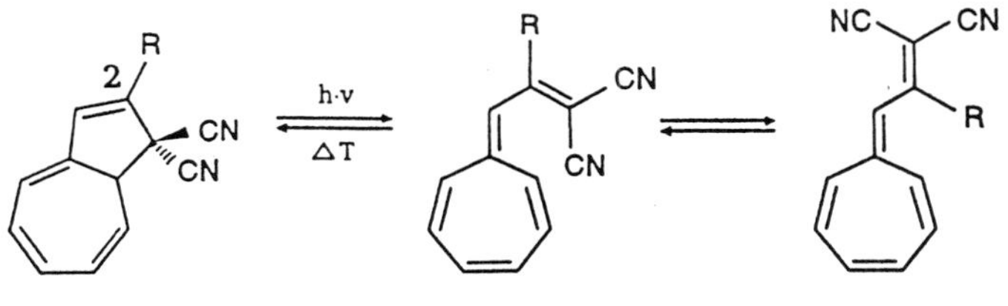

By increasing temperature fading of the colour is achieved by the thermal electrocyclization VHF → DHA. The process is reversible and, since on irradiation a long-wave length absorption appears, the process is termed a "normal" or "positive" photochromism.

A furan substituent at C-2 in DHA has a significant effect on the kinetics of the photochemical and thermal reactions as shown by the photochromes DHF-1, DHF-2, and DHF-3. To observe colouration on irradiation, DHF-1 has to be cooled down to -50 °C. This is explained by a fast thermal back-reaction leading to immediate bleaching at room-temperature. The same is true for DHF-3. In contrast, DHF-2 rearranges at room-temperature photochemically to the red-coloured VHF-2 and, thus, exhibits room-temperature photochromism.

Table 3. Long-wavelength absorption maxima of the photochromic system DHF / VHF

compounds	solvent / temp.	λ_{max}(nm)		$\Delta\lambda_{max}$(nm)
		DHF	VHF	
DHF-1 / VHF-1	EtOH / 200 K	440	548	108
DHF-2 / VHF-2	MeCN / r.t.	369	494	125
DHF-3 / VHF-3	EtOH / 200 K	454	500	46

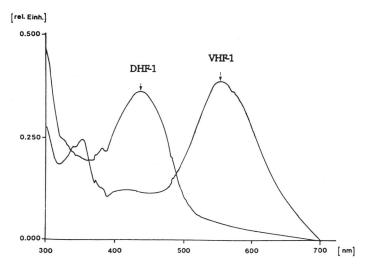

Fig. 6. Photochromism at -50 °C in ethanol, as exemplified by the HMF-derived system DHF-1 ⇌ VHF-1. Light induced process DHF-1 → VHF-1, thermal backreaction VHF-1 → DHF-1 (taken from ref.[27])

7. Light-sensitive Electron-transfer Compounds – Chemical and Physical Switching

The furan derivatized photochromic molecular components led us into an exiting field of new chemistry and physics dealing with functionalized pigments for the chemical and physical switching.[27,54] The underlying concept is to synthesize multifunctional dye compounds contained in light-sensitive and electron-transfer active subunits and to subject those molecular components to physico-chemical investigation by combination-techniques.[55] The HMF-derived photochromic compound DHF-1, composed of the light-sensitive dihydroazulene substructure and the electron-transfer active dicyanovinylfuryl group, were selected for testing. Since, on conversion of DHF into the deeply coloured vinylheptafulvene VHF, the electron acceptor strength of the dicyanovinylfuran is increased, the electrochemical reduction of the light-adapted vinylheptafulvene form must occur at lower negative reduction potential. This presumption could be proven by cyclic voltammetry of a structurally related compound.[54] A new technique (photomodulation amperometry, cf. Fig. 7), the instrumentation comprising of an electrochemical cell equipped with optically transparent electrodes as working electrodes, a platinum counter electrode, and a reference electrode, was employed. Subjecting a solution of the electron-transfer active photochromic compound to a light-pulse gives, under appropriate instrumental conditions, the current / time (i/t)-diagram as shown on the right in Fig. 7.

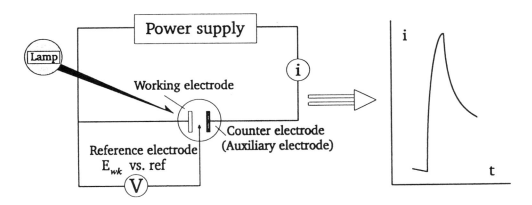

Fig. 7. Schematic representation of the instrumental set-up for photomodulation amperometry

Fig. 8 gives a schematic representation of the molecular processes involved in the described photomodulation-amperometry technique. It is important to note that the electrode potential has to be adjusted so that under "dark"-conditions no response is given. In the first step dihydroazulene DHF-1 is photochemically rearranged into vinylheptafulvene VHF-1, generating a stronger electron-acceptor and leading to an onset of the electric current as long as electroactive species are formed. In the dark the current flow gradually decreases. A sequence of photomodulation obtained by DHF-1 is given in Fig. 9.

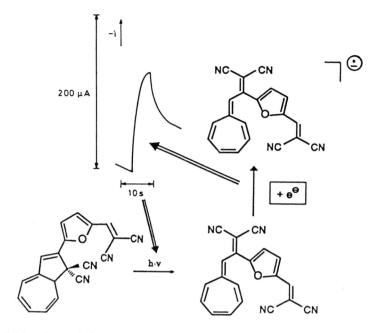

Fig. 8. Light-triggered electron-transfer, monitored by photomodulation amperometry. Instrumental set-up as displayed in Fig. 7.

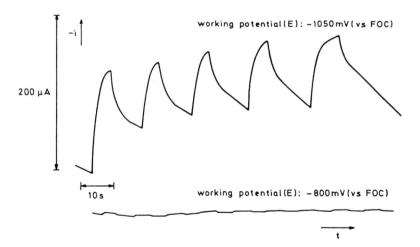

Fig. 9. Photostimulated electron-transfer activation of DHF-1 $\rightleftharpoons$ VHF-1 monitored by photomodulation amperometry (figure taken from ref.[27])

These observations can be explained in a qualitative way by simple molecular-orbital considerations as described in Fig. 10: the occupied energy level, representing the cathodic electrode potential, is maintained constant under this approximation, whereas the energy of the lowest unoccupied orbital (LUMO), which is assigned to the π-system of the pigment, depends on the structure as indicated in Fig. 10. By light-induced rearrangement, DHF-1 → VHF-1, the energy of the LUMO decreases and the transfer of electrons becomes favourable thermodynamically.

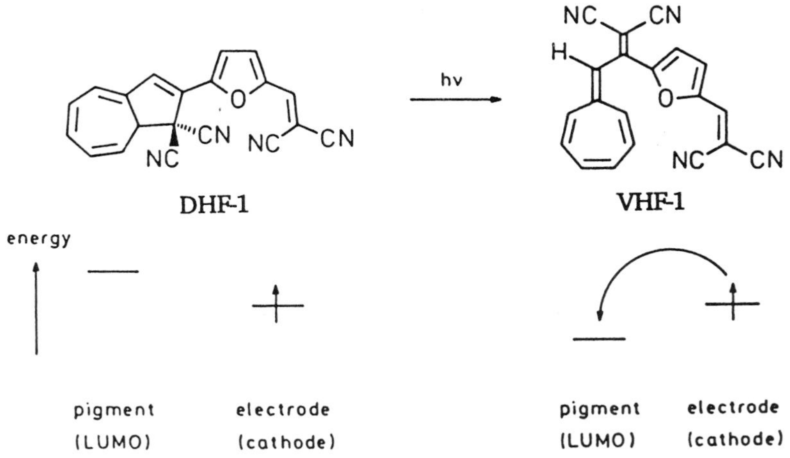

Fig. 10. Schematic description of the electronic changes by photostimulated electron-transfer

This process of switching, at least on a molecular level, allows to transcribe light-puls-information into electrical signals.

8. Conjugated Polyenes from Furan-2,5-dicarbaldehyde – Poly(Arylene Vinylenes)

Poly(arylene vinylenes), being composed of alternating and conjugated arylene and vinylene groups, are of interest with respect to macromolecular semiconductors, photoconductors, and basis material for conducting polymers.[56]

Conjugated polyenes of a structurally related type were obtained by condensation of *p*-xylylene dicyanide (XDC) with 2,5-furan-dicarbaldehyde (BFF) under basic conditions (sodium ethoxide in ethanol).

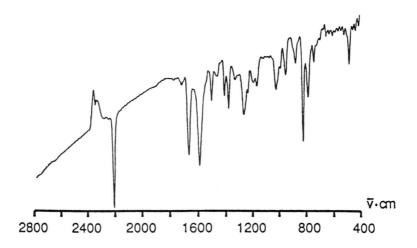

The resulting, high-melting (mp > 300 °C) and insoluble material, separated from low-molecular weight compounds by extraction with toluene, awaits scrutinized structural characterization. Characteristic bands obtained from the FT-IR-spectrum (Fig. 11) give some information on the structure of the polymer. A high-intensity band at 2208 cm⁻¹ is typically for the stretching mode of a nitrile group conjugated with a carbon-carbon double bond. The stretching mode for the C=C-double bond is found at 1597 cm⁻¹. The intense band at 1667 cm⁻¹, which must be assigned to an aldehyde group, indicates a terminal furan carboxaldehyde substructure.[57]

Fig. 11. FT-IR-spectrum of the oligomer or polymer formed by condensation of XDC and BFF

Investigation of the chemical and physical properties of poly(FVP) are in progress.

9. Carbohydrate-modified Conducting Polymers – Functionalised Polyazulenes

In this final section recent results will be presented on carbohydrate-modified conducting polymers. This topic unites several fields of interest:

(i) the chemistry of conducting polymers,

(ii) the relevance and importance of carbohydrates in intermolecular interaction and recognition, and

(iii)the use of carbohydrates as auxiliaries to increase solubility in water.

Conducting organic polymers have potential technical prospects in the following fields: rechargeable batteries, chemical sensors, conducting films, electronic and memory devices,[58] or chemically modified electrodes.[59]

In previous studies we found that C-2 substituted azulenes are versatile monomers for the generation of chemically modified polyazulenes by electropolymerization:[60]

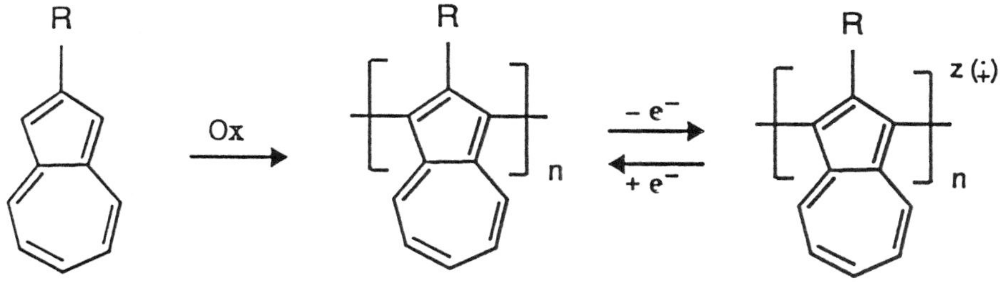

monomer	polymer	electrochemical oxidation
		(anion doping)
GMA	poly- GMA	conducting poly- GMA

Azulenes are assumed to electropolymerize under bond formation at C-1 and C-3 whereas radical cations are formed in the first step which subsequently dimerize or react with a neutral azulene. From azulene chemistry it is well established that oxidative dimerization occurs regiospecifically at C-1 or C-3. Therefore, substituents at C-2 do not interfere with polymerization. Charging of the polymeric film can be obtained by oxidative or reductive doping furnishing cations or anions, respectively. Doping, in general, leads to an increase in conductivity.

An efficient synthetic route for the monomeric species was developed recently.[61] An extension of this reaction sequence to carbohydrate chemistry gives the azulene-carbohydrate-conjugate GMA with glycosidically linked glucose:[62]

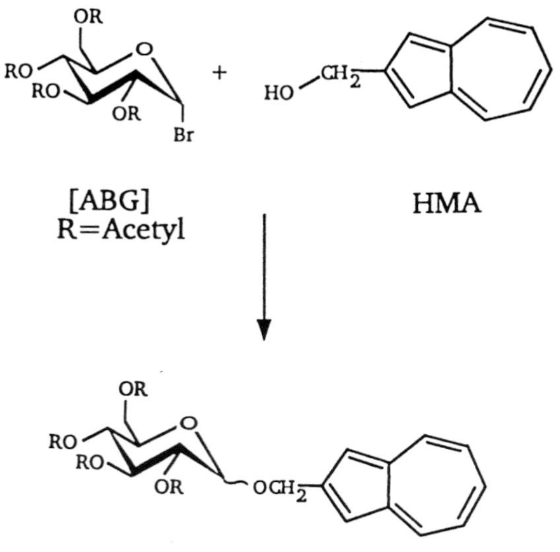

[ABG]
R=Acetyl

HMA

R=H [GMA]

The glucosidation was carried out under Koenigs-Knorr conditions, i.e. by reaction of acetobromoglucose (ABG) with 2-hydroxymethylazulene (HMA) in the presence of silver carbonate. The anomers were formed in a ratio $\alpha:\beta = 1:9$ as shown by ^{1}H-NMR spectroscopy. The removal of the protecting groups was achieved with 10 % methanolic KOH. Unlike other azulenes, GMA, melting at 189 °C, is soluble in water.

The electropolymerization of the monomeric glucosyloxymethyl-azulene (GMA) is shown in Fig. 12, analytically monitored by cyclic voltammetry. It is interesting to note that GMA can be electropolymerized in organic solvents (acetonitrile) and in water as well. The irreversible electrooxidation is demonstrated by the initiation curves in Fig. 12 which clearly show different peak currents for the oxidative and the reductive steps. The overall curvature of the diagram in Fig. 9 clearly indicates the formation of an electrically conducting film with conductive activity at about 600 mV (vs. Ag/AgCl):

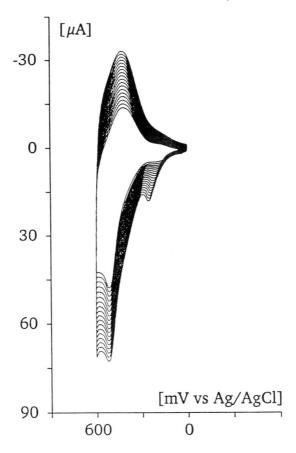

Fig. 12. Electropolymerization of GMA under formation of poly(GMA) and electrochemical activation under formation of an electrically conducting polymer. The polymerization was undertaken in water / 0.1 mol NaClO$_4$ on a Pt-electrode

Investigations on the structure of poly(GMA) and on the physical and chemical properties of carbohydrate modified polymers are in progress.

10. Summary and Conclusions

In the introduction, the chemical conversion of carbohydrates into high-grade chemicals (molecular units) with option and opportunities in the chemistry of information- and energy-transfer, is outlined. Emphasis is given to conceptional strategies. π-Heterocycles play the role of key-intermediates. First encouraging results are given, demonstrating preparation and characterization of electrochemichromic, photochromic, and electrically conducting molecular components. Further intensive cooperative inquiries at university and industry have to be undertaken in the future. Especially in regard of transforming molecular units into molecular materials. Chirality as an outstanding property of carbohydrates was given only modest consideration at this stage of the study, but will receive stronger accentuation in future work.

Acknowledgements. This research is supported by Bundesminister für Forschung und Technologie, Volkswagen Stiftung, and Fonds der Chemischen Industrie. Cooperation with Zentrallaboratorium of Südzucker A.G., Mannheim/Ochsenfurt (Director Dr. H. SCHIWECK) is gratefully acknowledged. The skilled assistance of Petra SEITZ, Dr. Thomas KNÖCHEL, Christian FISCHER, Michaela FEUERER, and Peter A. BROSS have been crucial to the success of this work. Artwork was performed in part by Mrs. G. SUREK.

References

1. J. Daub: Molekulare Bausteine aus niedermolekularen Kohlenhydraten. *Nachr. Chem. Tech. Lab.* **36** (1988) 896-899.

2. J. Daub: Abstract. *Zuckerind.* **115** (1990) 204.

3. F. W. Lichtenthaler: Niedermolekulare Kohlenhydrate als Rohstoffe für die Chemische Industrie. *Nachr. Chem. Techn. Lab.* **38** (1990) 866.

4. For Reviews on Renewable Resources, see:

    a) H. Baumann, M. Bühler, H. Fochem, F. Hirsinger, H. Zoebelein, J. Falbe: Natürliche Fette und Öle – nachwachsende Rohstoffe für die chemische Industrie. *Angew. Chem.* **100** (1988) 41-62; *Angew. Chem., Int. Ed. Engl.* **27** (1988) 41-62.

    b) D. Osteroth: Von der Kohle zur Biomasse – Chemierohstoffe und Energieträger im Wandel der Zeit. Springer Verlag, Berlin, 1989.

    c) Der Bundesminister für Forschung und Technologie: Nachwachsende Rohstoffe. *Bundesministerium für Forschung und Technologie*, Bonn, 1986.

    d) B. A. Tokay: Biomass Chemicals. *Ullmann's Encyclopedia of Industrial Chemistry* (W. Grehartz, Ed.), 5th Ed., Vol. A4, p. 99 ff., VCH Verlagsgesellschaft, Weinheim, 1985.

5. For Reviews on the Utilization of Saccharides, see:

    a) H. Schiweck, K. Rapp, M. Vogel: Utilization of Sucrose as an Industrial Bulk Chemical – State of the Art and Future Implications. *Chem. Ind (London)* **1988**, 228-234.

b) J. L. Hickson (Ed.): Sucrochemistry. *ACS Symposium Series* **41**, Am. Chem. Soc., Washington D.C., 1977.

6. M. G. Clark: Materials for Optical Storage. *Chem. Ind. (London)* **1985**, 258-263.

7. a) J. Daub, K. M. Rapp, J. Salbeck, H. Schiweck, P. Seitz, R. Wild: Organic Materials Based on Fructose Chemistry: Photochromics and ET-reagents Using 5-Hydroxymethyl furfuraldehyde as Starting Compound. 4th European Carbohydrate Symposium, Darmstadt, 1987, Abstract D-18.

    b) J. Daub, T. Knöchel, K. M. Rapp, J. Salbeck, R. Wild: Materials Science Opportunities in Renewable Resources Chemistry: Molecular Materials Based on Carbohydrate Chemistry. 14th International Cabohydrate Symposium, Stockholm, 1988, Abstract p. 401.

    c) J. Daub, J. Salbeck, K. M. Rapp, H. Schiweck: Furans from Fructose: HMF Derivatives with Special Electrochemical and Electrochromic Properties. 15th International Carbohydrate Symposium, Yokohama, 1990, Abstract D-028.

8. G. G. Birch, K. J. Parker (Eds.): Sugar: Science and Technology. Applied Science Publ., London, 1979.

9. a) A. Gaset, M. Delmas: Products Derived from Sugars. *Comm. Eur. Communities* **1986**, 46; *Chem. Abstr.* **106** (1986) 10603.

    b) R. Khan: Chemistry and New Uses of Sucrose: How Important?. *Pure Appl. Chem.* **56** (1984) 833.

    c) C. E. James, L. Hough, R. Khan: Sucrose and Its Derivatives. *Progr. Chem. Org. Nat. Prod.* **55** (1989) 117-184.

10. T. Böcker, J. Thiem: Synthese und Eigenschaften von Kohlenhydrat-Tensiden. *Tenside* **26** (1989) 318.

11. G. A. Jeffrey: Carbohydrate Liquid Crystals. *Acc. Chem. Res.* **19** (1986) 168-173.

12. N. R. Williams (Ed.): Carbohydrate Chemistry – A Specialist Periodical Report. The Royal Society of Chemistry, London, annually.

13. a) H. Zollinger: Colour Chemistry. VCH Verlagsgesellschaft, Weinheim, 1987.

    b) J. Rochlitz: Farbstoffe für die Energie-Umwandlung. *Chimia* **34** (1980) 131-144.

14. a) F. L. Carter, (Ed.): Molecular Electronic Devices. Marcel Dekker, Inc., New York, 1982.

    b) R. W. Munn: Molecular Electronics. *Chem. Britain* **1984**, 518.

    c) A. Aviram: Molecules for Memory, Logic, and Amplification. *J. Am. Chem. Soc.* **110** (1988) 5687-5692.

346   *J. Daub, et al.*

15. H. Steppan, G. Buhr, H. Vollmann: Resisttechnik – ein Beitrag der Chemie zur Elektronik. *Angew. Chem.* **94** (1982) 471-485; *Angew. Chem., Int. Ed. Engl.* **21** (1982) 455.

16. For a Similar Approach, see: J. Simon, J.-J. André, A. Skoulios: Molecular Materials. I. Generalities. *Nouv. J. Chim.* **10** (1986) 295.

17. T. El Hajj, A. Masroua, J.-C. Martin, G. Descotes: Synthèse de l'Hydroxyméthyl-5 Furane Carboxaldehyde-2 et de ses Dérivés par Traitement Acide de Sucres sur Résines Échangeuses d'Ions. *Bull. Soc. Chim. Fr.* **1987**, 855-860.

18. O. Theander, D. A. Nelson: Aqueous, High-temperature Transformation of Carbohydrates Relative to Utilization of Biomass. *Adv. Carbohydr. Chem. Biochem.* **46** (1988) 273.

19. F. W. Lichtenthaler: Large-scale Adaptable Routes from Monosaccharides to Versatile Six-Carbon Building Blocks. *Zuckerind.* **115** (1990) 198 and 790-798. See also: *These Proceedings*, p. ... ff.

20. C. Fayet, J. Gelas: Access to a Chirally Substituted Furan. Synthesis of 5-(D-*glycero*-1,2-Dihydroxyethyl)-2-furaldehyde from a Natural Heptulose. *Heterocycles* **20** (1983) 1563-1566.

21. J. R. Totter, W. J. Darby: 4-Hydroxymethyl-imidazole Hydrochloride. *Organ. Synth. Coll.* Vol. III, **1955**, 460-464.

22. a) N. Elming, N. Clauson-Kaas: Transformation of 2-(Hydroxymethyl)-5-(aminomethyl)furan into 6-Methyl-3-pyridinol. *Acta Chem. Scand.* **10** (1956) 1603-16.. .

    b) U. Schöberl: Unpublished work, University of Regensburg, 1989.

23. P. Jarglis, F. W. Lichtenthaler: Eine stereospezifische Synthese von (S,S)-Palythazin aus D-Glucose. *Angew. Chem.* **94** (1982) 140-141; *Angew. Chem., Int. Ed. Engl.* **21** (1982) 141-142.

24. W. Kaim: Die vielseitige Chemie der 1,4-Diazine – organische, anorganische und biochemische Aspekte. *Angew. Chem.* **95** (1983) 201-221; *Angew. Chem., Int. Ed. Engl.* **22** (1983) 171.

25. E. Fischer, F. W. Lichtenthaler: Ergiebige Synthese von Zucker-3,2-enolonen aus 2-Hydroxyglykalen und ihre Umwandlung in γ-Pyrone. *Angew. Chem.* **86** (1974) 590-592; *Angew. Chem., Int. Ed. Engl.* **13** (1974) 548-549.

26. J. Daub, K. M. Rapp, P. Seitz, R. Wild (Süddeutsche Zucker AG): Dicyanvinylsubstituierte Furanderivate, Verfahren zu ihrer Herstellung und deren Verwendung. D.B.P. 3 718 917 (1988); *Chem. Abstr.* **110** (1989) P 212593k.

27.   J. Daub, J. Salbeck, T. Knöchel, C. Fischer, H. Kunkely, K. M. Rapp: Lichtsensitive und elektronentransferaktive molekulare Bausteine: Synthese und Eigenschaften eines photochemisch schaltbaren, dicyanvinylsubstituierten Furans. *Angew. Chem.* **101** (1989) 1541-1542; *Angew. Chem., Int. Ed. Engl.* **28** (1989) 1494-1495.

28.   C. Dominguez, A. G. Csaky, J. Magano, J. Plumet: Reactions of 2,5-Furandicarbaldehyde with Stabilized Phosphonium Ylides. Applications to the Synthesis of 5-Vinyl-2-furaldehyde and 2,5-Divinylfuran Derivatives. *Synthesis* **1989**, 172-175.

29.   H. Firouzabadi, E. Ghaderi: Barium Manganate. An Efficient Oxidizing Reagent for Oxidation of Primary and Secondary Alcohols to Carbonyl Compounds. *Tetrahedron Lett.* **1978**, 839-842.

30.   K. Yu. Novitskii, V. P. Volkov, Yu. K. Yurev: Furan-2,5-Dialdehyde and its Reactions with Malonic Acid Derivatives. *J. Gen. Chem. USSR (Engl. Transl.)* **31** (1961) 494-497.

31.   U. Schöberl: Organische Materialien aus niedermolekularen Kohlenhydraten – Wege zur Herstellung lichtsensitiver Verbindungen. *Diploma Thesis*, University of Regensburg, 1989; Structure Determination by Dr. Klement, University of Regensburg and Prof. G. Maas, University of Kaiserslautern.

32.   B. B. Corson, R. W. Stoughton: Reactions of Alpha, Beta-Unsaturated Dinitriles. *J. Amer. Chem. Chem. Soc.* **50** (1928) 2825-2837.

33.   a) U. Schöberl: *Doctoral Dissertation*, University of Regensburg, in preparation.

      b) On the synthesis of functionally related benzenoid compound, see: P. M. Allemand, P. Delhaes, Z. G. Soos, M. Nowak, K. Hinkelmann, F. Wudl: *Synth. Meth.* **27** (1988) B-243.

34.   U. Schöberl: Organische Materialien aus niedermolekularen Kohlenhydraten – Wege zur Herstellung lichtsensitiver Verbindungen. *Diploma Thesis*, University of Regensburg, 1989.

35.   a) C. K. Chiang, A. J. Heeger, and A. G. MacDiarmid: Synthesis, Structure, and Electrical Properties of Doped Polyacetylene. *Ber. Bunsenges. Phys. Chem.* **83** (1979) 407-417.

      b) M. Mastragostino, A. M. Marinangeli, A. Corradini, S. Giacobbe: Conducting Polymers as Electrochromic Materials. *Synth. Meth.* **28** (1989) C501.

36.   U. Schöberl, J. Daub: Manuscript in preparation.

37. Cyclic Voltammetry: In an electrochemical cell containing working electrode, counter electrode and reference electrode, the potential of a stationary electrode in an unstirred solution of the electroactive compound and an inert electrolyte is varied linearly with time by a defined sweep rate. After reaching a certain potential, the potential scan is reversed. The resulting current/potential (i/e)-diagram is monitored. J. Heinze: Cyclovoltammetrie – die "Spektroskopie" des Elektrochemikers. *Angew. Chem.* **96** (1984) 823-840; *Angew. Chem., Int. Ed. Engl.* **23** (1984) 831.

38. Spectroelectrochemistry combines electrochemical and spectroscopic techniques. For measurements an electrochemical cell containing preferentially optically transparent electrodes (OTes) is applied. Electronic spectra depending on the electrochemical potential are obtained:

   a) T. Kuwana, W. R. Heineman: Study of Electrogenerated Reactants Using Optically Transparent Electrodes. *Acc. Chem. Res.* **9** (1976) 241-248.

   b) J. Salbeck, I. Aurbach, J. Daub: CD-Spektroelektrochemie – Eine neue Methode zur Charakterisierung chiraler Elektronentransferreagenzien und chiraler Zwischenstufen – Vorstellung einer spektroelektrochemischen Zelle. *Dechema-Monographien* **112** (1988) 177.

39. a) R. O. Loutfy, C. K. Hsiao, B. S. Ong, B. Keoshkerian: Electrochemical Evaluation of Electron Acceptor Materials. *Can. J. Chem.* **62** (1984) 1877.

   b) M. Sertel, A. Yildiz, H. Baumgärtel: Mechanism of the Electroreduction of 1,1-Dicyanoethylene Derivatives. *Electrochimica Acta* **31** (1986) 1625.

40. On EC-type Mechanisms, see: J. Salbeck, J. Daub: Methoxy-substituierte Acenazulene: Chemische und elektrochemische Reduktion in aprotischem Lösungsmittel – Charakterisierung von Radikalanionen, s-verknüpften Dianionen und cyclisch konjugierten Dianionen. *Chem. Ber.* **122** (1989) 1681-1690, and literature cited there.

41. Short definition of these two terms: Electrochromism denotes the change of a compound's colour if subjected to an electrical field. Electrochemichromism means the change of colour on reductive or oxidative electron transfer.

42. F. G. K. Baucke, J. A. Duffy: Darkening Glass by Electricity. *Chem. Britain* **21** (1985) 643-646.

43. J. Salbeck, U. Schöberl, K. M. Rapp, J. Daub, manuscript submitted.

44. More detailed information on one-electron and two-electron transfer is contained in:

   a) L. Eberson: Electron Transfer Reactions in Organic Chemistry. *Reactivity and Structure Concepts in Organic Chemistry* **25** (1987) 1;

   b) J. K. Kochi: Elektronen- und Ladungsübertragung: Zur Vereinheitlichung der Mechanismen organischer und metallorganischer Reaktionen. *Angew. Chem.* **100** (1988) 1331-1372; *Angew. Chem., Int. Ed. Engl.* **27** (1988) 1227-1268.

c) D. Astruc, M. Lacoste, L. Toupet: How to Design a Fast Two-electron Transfer: Structural Rearrangement in the Second Electron Transfer provides Stabilization. *J. Chem. Soc., Chem. Commun.* **1990**, 558;

d) K. Hinkelmann, J. Heinze: Analysis of "Two-electron" Transfer Processes by Cyclic Voltammetry. *Ber. Bunsenges. Phys. Chem.* **91** (1987) 243-249.

45. H. Lund: Catalysis by Electron Transfer Reagents in Organic Electrochemistry. *J. Mol. Cat.* **38** (1986) 203.

46. For reviews on organic metals, see:

    J. H. Perlstein: "Organische Metalle" – Die intermolekulare Wanderung der Aromatizität. *Angew. Chem.* **89** (1977) 534-549; *Angew. Chem., Int. Ed. Engl.* **16** (1977) 519-... .

    A. J. Epstein, J. S. Miller: Eindimensionale Leiter. *Spektrum der Wissenschaft* **1979**, 63.

47. For a pertinent review on salts of tetracyanoquinodimethane, see: J. B. Torrance: The Difference between Metallic and Insulating Salts of Tetracyanoquinodimethane (TCNQ): How to Design an Organic Metal. *Acc. Chem. Res.* **12** (1979) 79-86.

48. a) Y. Hirshberg, E. Fischer: Low-temperature Photochromism and its Relation to Thermochromism. *J. Chem. Soc.* **1953**, 629-636.

    b) Y. Hirshberg: Reversible Formation and Eradication of Colours by irradiation at Low Temperatures. A Photochemical Memory Model. *J. Am. Chem. Soc.* **78** (1956) 2304-2312.

49. a) R. Dessauer, J. P. Paris: Photochromism. *Adv. Photochem.* **1** (1963) 275.

    b) H. Dürr: Perspektiven auf dem Gebiet der Photochromie: 1,5-Electrocyclisierung von heteroanalogen Pentadienyl-Anionen als Basis eines neuartigen Systems. *Angew. Chem.* **101** (1989) 427-445; *Angew. Chem., Int. Ed. Engl.* **28** (1989) 413.

50. K. Schaffner, S. E. Braslavsky, A. R. Holzwarth: Photophysics and Photochemistry of Phytochrome. *Adv. Photochem.* **15** (1990) 229.

51. D. A. Parthenopoulos, P. M. Rentzepis: Three-Dimensional Optical Storage Memory. *Science* **245** 843-845.

52. T. Knöchel: Hydroazulene – Synthese, Funktionalisierung, Photochromie. *Doctoral Dissertation*, University of Regensburg, 1990.

53. J. Daub, S. Gierisch, U. Klement, T. Knöchel, G. Maas, U. Seitz: Lichtinduzierte reversible Reaktionen: Synthese und Eigenschaften photochromer 1,1-Dicyan-1,8a-dihydroazulene und thermochromer 8-(2,2-Dicyanvinyl)heptafulvene. *Chem. Ber.* **119** (1986) 2631-2646.

350   *J. Daub, et al.*

54.   J. Daub, C. Fischer, J. Salbeck, K. Ulrich: Molecular Materials for Chemical and Physical Switching IV: Photomodulation of Electric Current by Light-Sensitive Electron-Transfer-Compounds – Aryl/Dihydroazulene-Conjugates. *Adv. Mater.* **2** (1990) 366.

55.   C. Fischer, *Doctoral Dissertation*, University of Regensburg, in preparation.

56.   H.-H. Hörhold, M. Helbig, D. Raabe, J. Opfermann, U. Scherf, R. Stockmann, D. Weiß: Poly(phenylenvinylen); Entwicklung eines elektroaktiven Polymermaterials vom unschmelzbaren Pulver zum transparenten Film. *Z. Chem.* **27** (1987) 126.

57.   For the related poly(p-phenylene vinylene), see: H.-H. Hörhold, D. Gräf, J. Opfermann: Poly(arylenecyanoethylenes)-Structural Effects and Optical and Electrophysical Properties. *Plaste und Kautschuk* **17** (1970) 84.

58.   a) R. S. Potember, R. C. Hoffman, H. S. Hu, J. E. Cocciaro, C. A. Viands, R. A. Murphy, T. O. Poehler: Conducting Organics and polymers for Electronic and optical Devices. *Polymers* **28** (1987) 574.

       b) D. C. Bott: Structural Basis for Semiconducting and Metallic Polymers. Conducting Polymers (T. A. Skotheim, Ed.), M. Dekker, New York, 1986, p. 1191.

       c) G. P. Evans: The Electrochemistry of Conducting Polymers. Advances in Electrochemical Science and Engineering (H. Gerischer, C. W. Tobias, Eds.), VCH Verlagsgesellschaft, Weinheim, 1990, p. 1.

59.   J. Schreurs, E. Barendrecht: Surface-modified electrodes (SME). *Recl. Trav. Chim. Pays-Bas* **103** (1984) 205-219; A. Merz: Polymer-modifizierte Elektroden. *Nachr. Chem. Tech. Lab.* **30** (1982) 16-23.

60.   a) A. Mirlach, J. Salbeck, J. Daub: Cyclovoltammetrie und elektrochemische Polymerisation substituierter Azulene. *Dechema-Monographien* **117** (1989) 367.

       b) P. A. Bross, A. Mirlach, J. Salbeck, J. Daub: Azulenkonjugate mit photochromen Teilstrukturen: Vom Monomer zum Polymer – Eine elektrochemische und spektroelektrochemische Untersuchung. *Dechema-Monographien* **121** (1990) 375.

       c) A. F. Diaz, J. Bargon: Electrochemical Synthesis of Conducting Polymers. *Handbook of Conducting Polymers* (T. A. Skotheim, Ed.), M. Dekker, New York, 1986, p. 81.

       d) J. Daub, M. Feuerer, A. Mirlach, J. Salbeck: Functionalized Conducting Polymers with Polyazulene Backbone. *Synthetic Meth.*, submitted.

61.   P. A. Bross: Präparative Nutzung der [8+2]-Cycloaddition – Synthese und Eigenschaften mono- und disubstituierter Azulene. *Diploma Thesis*, University of Regensburg, 1990.

62.   P. A. Bross, U. Schöberl, J. Daub, submitted for publication.

Subject Index

Enzymes (*continued*)
dextran sucrase
in oligosaccharide synthesis 161-165
of dextran 165, 184
of gentiobiose 162
of leucrose 162, 185
of panose 162-164
kinetics 163, 164
reaction mechanism and pathways
163, 185, 186
transformation rates 187
endo-inulinase 75, 76
fructansucrase 172, 176
pH optimum 172
fructosyl transferase 73
galactose oxidase 271
glucose oxidase 274
Co-immobilization with Catalase 274
isolation from *Aspergillus niger* 274
glycoside-3-dehydrogenase 156, 157
kinetics 157
gulonic acid dehydrogenase 276
inulin-fructotransferase 77, 78
lipase 68-70
pyranose-2-oxidase 269, 270
isolation 269
glucose oxidation 269, 270
pH-optimum 269
pyranoson-dehydratase 270, 271
sucrose→isomaltulose mutase 60

Epoxides:
propylene oxide 270

F

Fatty acid derivatives:
amides 113
as detergents 41, 113
lipase hydrolysis of esters 68-70
sucrose monoesters 40-42
sucrose polyesters (Olestra) 42, 68-72
preparation 42, 71, 72
properties 42, 70-72

Fermentation (*see* Bacteria)

Flow sheets:
isomalt production 65
isomaltulose production 59
leucrose production 188
Olestra preparation 72
sucrose sulfation 120

Food additives:
coatings 41, 42
emulsifiers 40
for lowering colesterol level 42, 71, 72
low calorie fats and oils 42
sweeteners 45, 46, 62, 66-68

4-Formyl-imidazole 326

6-Formyl-3-pyridinol 326

Fructans / Fructo-oligosaccharides
applications 77, 78, 179
biosynthesis 170
degree of polymerization 73-76
fermentation process 171-173
formation by bacteria 73, 170
hydrolysis 175-178
by acid 175, 177, 178
by enzymes 175, 176
low caloric food-ingredient 77
molecular weight 175, 176
NMR spectra 174, 175
occurance in nature 73
production
by fructosyltransferase 73
by hydrolysis of inulin 75, 76
HPLC profile 74, 76, 172
reaction conditions 172, 173
properties 76, 77, 175, 177, 178
effect on intestinal flora 77
non-digestability 76
sweetness 77
viscosity of solutions 178
purification 173
structure 174, 175

D-Fructofuranosyl carbocation 198, 34, 35

D-Fructose
annual production rate 209
bulk price 208, 209
preparation from 2-ketoglucose 270, 271

D-Fructose dianhydrides
as low-calorie food-additives 255
dehydration to HMF 214
derivatization of tautomers 214
di-β-D-fructofuranose 1,2':2,1'-
dianhydride 77, 78, 254, 255
applications 78, 255
by enzymatic treatment of inulin
77, 78
from fructose in HF-solution 254, 255
stability and conformation 254, 255

K

Bl